Hunger, Deprivation and Social Exclusion

Hunger, Deprivation and Social Exclusion

A Political Economy Perspective

R.K. Kumbhar

RAWAT PUBLICATIONS

Jaipur • New Delhi • Bangalore • Guwahati • Kolkata

ISBN 978-81-0316-0816-6

The publication of this book was financially supported by the ICSSR. The responsibility for the facts stated, opinions expressed, or conclusions reached, is entirely that of the author and that the ICSSR accepts no responsibility for them.

ICSSR Consultant: Professor M.K. Agarwal, Department of Economics, Lucknow University, Lucknow.

Published by
Prem Rawat for **Rawat Publications**
Satyam Apts, Sector 3, Jawahar Nagar, Jaipur 302 004 (India)
Phone: 0141 265 1748 / 7006 Fax: 0141 265 1748
E-mail: info@rawatbooks.com
Website: www.rawatbooks.com

New Delhi Office
4858/24, Ansari Road, Daryaganj, New Delhi 110 002
Phone: 011 2326 3290

Also at *Bangalore, Guwahati* and *Kolkata*

Typeset by Rawat Computers, Jaipur
Printed at Chaman Enterprises, New Delhi

To
the people
suffering with
hunger and *deprivation*

Contents

Preface

Hunger and deprivation are undesirable in any society from the viewpont of economic growth, social justice and global peace. However, large number of people are suffering because of being below the poverty line in India. The highest proportion of population below poverty line amongst all Indian states is found in Odisha (around 47%) and that has not reduced from 1993–94 to 2004–05 (as per the estimation of Planning Commission of India). The incidences of basic deprivation like health, education and housing also remain high in this state. This study, therefore, has been taken up to examine the problems of hunger and deprivation (nature, extent, and causes) in the context of Odisha.

While hunger implies a 'discomfort or painful sensation caused by want of food' (Dreze and Sen, 1989), deprivation implies a state of observable and demonstrable disadvantage relative to the reference group or local community (Townsend, 1987). Both hunger and deprivation are all-encompassing and crosscutting themes, and they require a multidisciplinary approach to understand the phenomena. Therefore, a multidisciplinary approach has been considered here to study the problem. The political economy approach has been emphasised and is a dominant approach in economics, often adopted to study hunger and deprivation, where it examined the role of the structural and institutional process that generated hunger and deprivation in the state of Odisha in India. Collecting facts and figures both from primary and secondary sources and using simple arithmetic, statistic (wherever required, econometrics) and visual technique, the main findings of the study are given in succeeding paras.

(i) Chapter 2 found a higher variation of hunger and deprivation across social groups, geographical regions, occupations, education status, landholding and dependency ratio. The reasons are (a) failure of agriculture, (b) the institutional and structural rigidities, and (c) the colonial rule. However, the single most important variable that can explain the higher incidence of hunger and deprivation in Odisha to a great extent is the social structure and social hierarchy prevailing in the state, where the SCs/STs, women and children faced most disadvantages. As the SC/ST communities concentrate in specific regions (southern and part of northern Odisha) regional disparities in hunger and deprivation have been observed there. (ii) Chapter 3 found a slow and lopsided structural transformation, which failed to bring trickle-down effect. Further, the rate of change in output has not had much effect on the rate of change in hunger and deprivation as there is less employment growth in the primary sector, on which most of the hungry and deprived people depend. The regional picture shows that in all respects, southern region and SC/ST social groups are the most vulnerable and prone to hunger and deprivation. (iii) Chapter 4 found that absence of access to land and related resources is the reason behind hunger and deprivation. (iv) Chapter 5 inferred from the micro level case studies that there is prevalence of exclusion (in explicit or implicit form) of Dalits and SCs/STs from various socio-economic spheres, such as exclusion in the socio-cultural institutions like educational institutions, religious institutions like temples and economic institutions like labour market, etc. And that might be the reason why Dalits and SCs/STs are more prone to hunger and deprivation compared to any other community of other castes. (v) Chapter 6 the role of public provisioning in reducing hunger and deprivation is examined and it is found that at macro level some of the factors responsible for the ineffectiveness of public provisioning are inadequate entitlement and inadequate coverage. The micro surveys indicate existence of various public provisioning institutions, less in number, their functioning to achieve the desired goals are very poor and highly biased towards the upper castes. From the household perspective, some of the reasons for the ineffective functioning of the schemes are inadequate coverage, irregular and insufficient entitlement, wrong targeting of household, wrong targeting of commodities and meagre assistance (subsidy). Therefore, when the support-led development has been considered an alternative to achieve development quicker, there must be adequate coverage with adequate amount. (vi) Chapter 7 found that drought and flood are a regular phenomenon and are also responsible for hunger and deprivation. Cyclones, though not a regular phenomenon, also harm the society and economy a lot. Though the impacts of natural calamities are hazardous for all, it hits the poor most due to the political economic reasons. (vii) Chapter 8 'Displacement in the Name of Development' by the neo-liberal prescriptions of special economic zone and export promotion zone is found to be a reason producing hunger and deprivation and subsequently lumpenisation in the state. The analysis suggests that the impact of displacement is felt disproportionately among the SCs/STs, Dalits or untouchables as the displacement regions are the inhabitation of these population groups.

The study inferred that failure of both growth and distribution (necessary and sufficient conditions respectively) contributed to higher incidence of hunger and deprivation in the state. However, given the subsistence level of growth, the impact of hunger and deprivation could have been less had there been an egalitarian distribution. The distribution remains highly biased towards upper social groups. This biasness is also a reason for regional disparities. For instance, southern Odisha, where there is a higher percentage of socially disadvantaged groups, faces the maximum level of hunger and deprivation in the state. Therefore, it is required to reduce the disparities to reduce the higher incidence of hunger and deprivation in the state. The disparities can be reduced through land redistribution followed by other agrarian reforms (as landlessness is increasing and as there is low productivity and many dependents in agriculture), through effective implementation of public provisioning and through positive discriminations.

This book is a revised and updated version of my Ph.D. thesis submitted to Jawaharlal Nehru University (JNU), New Delhi. The research work was carried out at Centre for Development Study, Thiruvananthapuram, under the joint supervision of Professor K.P. Kannan and Dr P.K. Panda. Dr Kannan, who always advised that carrying out research, has to be your own passion and it is like 'fighting your own battle', has given me the strength to go to the depths of the subject matter without any fear. When I returned (from field research) with almost 500 pages of draft, I understood that mere possession of valuable facts and figures though necessary is not sufficient to produce a thesis, I was aware of that fact; but not in the lucid manner as Dr Kannan's classical tune for the thesis made it explicit; without which the thesis would have remained a report rather than a Ph.D. thesis in the political economy framework. Dr Panda's close involvement in my work and his possessiveness over my academic and personal success always provided me extra strength and confidence. I feel proud to have Dr Kannan and Dr Panda as my supervisors. I am grateful to them.

I am thankful to Mrs Sobha Kannan whose sweet voice, inspiration, encouragement and care are a motherly support to me. I am also thankful to Dr Madhavi Panda for her sisterly care, valuable support and encouragement at various stages of the thesis.

At CDS, I had received invaluable support and encouragement from various quarters. I wish to place on record my acknowledgement and gratitude to faculty, visiting scholars and guest faculty of CDS. I have benefitted immensely from various academic and policy related discussions by intellectuals and professors visiting CDS from time to time. At an early stage of my Ph.D. work, Dr Kannan arranged a valuable discussion with Prof. Gerry Rodger. At that time itself Prof. Rodger suggested to see if the problem can be explained through social exclusion framework and later on I found social exclusion is one of the major factors that contribute to hunger and deprivations in Odisha.

There are many individuals and institutions, who cordially helped in the preparation of this study. I am also grateful to the library staff, the administration and the computer centre at CDS (Thiruvananthapuram); at ISID, JNU, CSO and ISI, New Delhi; and NCDS, Bhubaneshwar. I am grateful to the households of the villagers for their cooperation in providing me primary data collection.

I also worked at Institute for Studies in Industrial Development and got the opportunity to interact with Dr T.S. Papola, Dr Jesim Pais and Dr Partha Pratim Sahu. I have received lots of intellectual inputs and inspiration from them. My present work place is Sambalpur University. I am thankful to all my colleagues and students.

I am thankful to Prof. D. Narasimha Reddy and Prof. Kailas Sarap, the external evaluator of the thesis for their valuable suggestions and advice to publish the thesis in book form.

My thanks are due to Professor R.K. Sharma, meeting with whom renewed the idea of publishing the thesis in the form of a book – an idea which had been dormant for some time.

Last but not the least, without the love, affection, support and constant encouragement of my parents, relatives, wife Seema and her parents, and daughter Vidushi, it might have been difficult for me to complete the thesis. Hailing from a rural backward, and moving out to pursue higher studies, there has definitely been the backing and support of many persons, but I would like to make a special mention of the blessings, unflinching support and high aspirations and expectations of my mother, Mrs Bhama Kumbhar, who dared to dream beyond limits. Finally, I offer my humble obeisance to the divine almighty for having provided me with good health and spirit, which enabled me to complete this work.

Rathi Kanta Kumbhar

Abbreviations

BMI	Body Mass Index
BPL	Below Poverty Line
CED	Chronic Energy Deficiency
CPI	Communist Party of India
CPR	Common Property Resources
CSO	Central Statistical Organisation
CV	Coefficient of Variation
CWR	Child Women Ratio
EGS	Employment Guarantee Scheme
EPZ	Export Promotion Zone
FAD	Food Availability Decline
FDI	Foreign Direct Investment
GDP	Gross Domestic Product
GNP	Gross National Product
HCR	Head Count Ratio
HDR	Human Development Report
ICDS	Integrated Child Development Schemes
IMR	Infant Mortality Rate
JRY	Jawahar Rozgar Yojana
Kcal	Kilo Calorie
LAA	Land Acquisition Act
ME	Middle English
MDM	Mid-day meal
MoM	Mission of Mercy
MPCE	Monthly Per Capita Consumer Expenditure
MTA	Mother Teacher Association
NDP	National Domestic Product

NFHS	National Family on Health Survey
NREGA	National Rural Employment Guarantee Act
NSDP	Net State Domestic Product
NSSO	National Sample Survey Organisation
OBC	Other Backward Caste
OREGS	Orissa Rural Employment Guarantee Scheme
PCU	Per Consumer Unit
PHC	Primary Health Care
PTA	Parent Teacher Association
RLE	Rural Labour Enquiry
RR	Rehabilitation and Resettlement
SC	Scheduled Caste
SD	Standard Deviation
SDP	State Domestic Product
SEZ	Special Economic Zones
SHGs	Self-Help Groups
ST	Scheduled Tribes
UN	United Nation
UN-DMT	United Nations Disaster Management Team
UNDP	United Nation Development Programme
UNICEF	The United Nations Children's Fund
UP	Upper Primary
UPSS	Usual Principal and Subsidiary Status
VEC	Village Educational Committee
WDR	World Development Report
WFP	World Food Programme
WFPR	Work Force Participation Rate
WHO	World Health Organisation
WPR	Worker Population Ratio

1

Introduction

Statement of the Problem

Hunger and deprivation are undesirable in any society from both economic growth and social justice point of view (Rawls, 1971; Dasgupta, 1993; Dreze and Sen, 1989). Hunger and deprivation are all-encompassing and crosscutting themes, and hence require a multidisciplinary approach to understand the phenomena. The political economy approach is a dominant approach in economics and often adopted to study hunger and deprivation (Dreze and Sen, 1989, 1991; Griffin, 1976; El-Ghonemy, 1990; Franke, 1996).

Hunger and deprivation have remained stagnant or increased among some social groups, especially since 1990s in some regions of Odisha. The longstanding incidence and stagnation of hunger and deprivation in Odisha forms the basic premise of the study. The nature, extent and causes of hunger and deprivation in Odisha are examined in relation to the country as a whole through an interdisciplinary political economy framework. The focus of the study is the political economy approach, where it examines the role of the structural and institutional process that generated hunger and deprivation in Odisha.

Concepts and Measurement of Hunger and Deprivation

What is Hunger?[1]

According to the dictionary meaning, 'hunger is a kind of discomfort or painful sensation occurred due to want of food'. While hunger implies a 'discomfort or painful sensation caused by want of food' (Dreze and Sen, 1989), deprivation implies a state of observable and demonstrable disadvantage relative to the reference group or local community (Townsend, 1987). Dreze and Sen (1989: 7) have elaborated the definition further, which stated 'to end hunger, the concern is not only merely avoidance of that discomfort or pain, but also to conquer its manifold consequences, such as under nourishment, debilitation, fatigue, morbidity and possibly mortality with obvious effects on human well-being and productivity'. Newman (1990: 3) has put it in a more formal way that 'hunger could be defined as an inadequacy in individual's dietary intake relative to the kind and amount of food required for growth, for activity and for maintenance of good health'.

Identification and Measurement of Hunger

There are four ways by which the hunger-affected persons can be identified. These measures are based on (a) nutritional intake, (b) perception, (c) anthropometry and (d) income.

In nutritional intake approach, a specified amount of calorie requirement for a person is determined. Generally, the nutritionists determine this specific calorie requirement. Sen (1981) stated that the direct method, (i.e. based on calorie intake) is the best one to identify hunger. However, as he pointed out, it is difficult to construct a universal demarcation line between hunger and non-hunger based on normative specification of the nutritional intake for the entire population due to differences in tastes, preferences, cultures and food intakes across society.

For perception-based measurement of hunger, the household is asked, 'whether all members of the household get two square meals a day in the specified period'. If the response is no, then it implies the presence of hunger in the household. However, there is a methodological difficulty in capturing the extent of hunger, as a person will be identified as hungry regardless of the quantity of meals consumed (e.g. a full meal or one-tenth of a full meal).

The above two procedures of identifying hunger can specify the current situations of the person with regards to hunger but it does not indicate if a person is chronically affected with hunger.[2] The anthropometric approach tries to capture this aspect of hunger through Body Mass Index.[3] If a person has a BMI below 18.5, then he/she is considered as undernourished.[4]

In income-based approach, the required level of calories would be expressed in income based on the food prices (locally available food) and if the per capita expenditure of a person is below the specified level of income then he/she would be identified as hungry.

The following section examines the concept of deprivation.

What is Deprivation?

'Deprivation may be defined as a state of observable and demonstrable disadvantage relative to the local community or the wider society or nation to which an individual, family or group belongs' (Townsend, 1987: 125). In the applied sense, 'people are said to be deprived if they lack the types of diet, clothing, housing, household facilities and fuel, and environmental, educational, working and social conditions, activities and facilities which are customary, or at least widely encouraged and approved, in the societies to which they belong' (Townsend, 1979: 413). Therefore, deprived people are those who fall below the standards of living, which are either socially accepted or institutionalised or can be demonstrated to be attaining by a majority of the national population. Townsend discussed the relativist approach of deprivation as an individual's or household's deprivation determined by its position relative to others in society. However, two points are important to consider such as 'commodities' and quantities of commodities available to persons under deprivation.

Identification and Measurement of Deprivation

Broadly, there are two ways by which one can identify the number of people who are deprived. These are (a) absolute deprivation and (b) relative deprivation.

As per the concept of 'absolute deprivation' the standard would be fixed as a norm by an expert group. Individuals who cannot achieve that specified amount are considered as deprived persons. For relative deprivation, the standard is set according to the society or class to which it refers or belongs and persons who cannot achieve that standard are called relatively deprived.

Approaches to the Study of Hunger and Deprivation

A number of studies have been conducted across the disciplines to study deprivation, such as Runciman, 1966; Townsend, 1979; Sen, 1981; 1983; Nolan and Whelan, 1996; Srinivas, 1996; Saha, 1991. In the context of poverty literature, Sen (1981) defines absolute deprivation as 'an irreducible core of absolute deprivation in our idea of poverty, which translates reports of starvation, malnutrition and visible hardship into a diagnosis of poverty, without having ascertained first the relative picture'. Sen provides this statement in the context of famine, but this irreducible core need not be time or space invariant, i.e. absolute needs can vary over time and from society to society.

Sen (1983; 1985) also explains 'the absolute satisfaction of certain needs might depend on a person's relative positions vis-à-vis others'. Also, Sen argues 'absolute deprivation in terms of a person's capabilities relates to relative deprivation in terms of commodities, incomes and resources'.

Runciman talks about the social deprivation from economic, social and political perspectives; Townsend approached deprivation from standard of living perspectives. Similar approach is followed by Nolan and Whelan, 1996. Sen discusses the deprivation of entitlement that causes hunger, starvation and followed by famines and the capability deprivation of individuals and lack of freedom for development. In India, literature on deprivation also deals with caste or social deprivation (Saha, 1991; Srinivas, 1996).

In India, studies examine deprivation in three different ways.

(a) Putting different socio-economic indicators as a prefix to deprivation individually (i.e. with separate indicator or commodities; especially deprivation due to education, health etc.).

(b) Taking a host of socio-economic indicators together. It is more similar to Townsend's work but in India these kinds of studies have been termed as multiple dimensions of poverty or of well-being (WDR, 2001). The Human Development Report, 2001 discusses three important kinds of deprivation, such as income, health and education often known as capability deprivation.

(c) Social deprivation such as caste and gender (de Haan and Nayak, 1995; Praxis, 2001; Panda and Agarwal, 2005).

Dreze and Sen, (1991) also adopted the interdisciplinary approach in their famous work of *Political Economy of Hunger*. They stated, 'it does not, of course, really matter whether political, social and cultural influences on economic matters are counted inside or outside the discipline of economics, but it can be tremendously important not to lose sight of these influences in analysing many profoundly important economic problems' (1991: 3–30).[5] Although specifically the work does not provide the connotation of interdisciplinary approach, it can be well understood that, it includes a variety of aspects, such as economics, politics, socio-cultural activities, international relations, role of civil societies, etc.

Lal and Myint (1996), in their book *The Political Economy of Poverty, Equity and Growth,* have used the term 'political economy' to direct attention to the political, institutional and historical factors affecting economic growth and income distribution.

Kannan (1990) narrated the interdisciplinary approach in more detail, while he was exploring the relevance of the approach in the context of understanding the dynamics of rural labour markets. As he stated, interdisciplinary approach incorporates the perspectives of the different social science disciplines such as economic history, political economy and economics, sociology, social anthropology and political science while analysing a research issue. The idea behind examining an issue from economic history perspective is that it would help appreciate the role of 'initial conditions'. This would provide the potentialities of the economy to tackle hunger and deprivation. Along with this, political economy would look into the structural constraints in an economy and within which it looks into the growth process (or the absence of it) and it is the distribution among different groups of households depending on their economic interests, surplus extraction and resource endowments, etc. These structural factors of growth and distribution could be the reason of supply-side problem or demand side problem of hunger and deprivation in further analysis. In addition, the movement of wages, employment and earnings becomes more meaningful.

From the sociological perspective, it helps in understanding the social structure and institutions (such as caste/gender based occupation, different system and surplus extractions and different survival strategies of the people) and its impact on hunger and deprivation. This implies hunger and deprivation are not caused by economic factors alone, rather they are reared by the social institutions as well. The perspective of social anthropology would lead us a step forward. This would help to understand the culture of the society and its impacts on group of households and with that across gender and age.

Last, but not the least, is the political science perspective to the problem of hunger and deprivation, where the role of the state and that of political organisations and civil societies are assessed. What role do political organisations play to reduce hunger and deprivation in the economy? The position, nature and character of the state or these organisations and their direct and indirect interventions over the time to reduce hunger and deprivation are highly important. Political economists look at the matter from the class relation's point of view. The concentration of resources among a few persons would cause poverty (Marx, 1956). Moreover, the proletariat struggle is an important way to get rid of poverty (Kannan, 1988). The other way to eradicate poverty is, by improving purchasing power, according to Keynes. Some others presumed poverty can be eradicated by the active roles of the state and civil societies (Smith, 1976; Dreze & Sen, 1989; Currie, 2000), if necessary by offering rights to human beings (UN, 1948).

Further, while the social anthropologists considered the social and cultural diversified practices as the reasons for hunger and deprivation, the political economists hinted at the inefficient role of the state as the reason for poverty.

Rationale of the Study

Many of the studies conducted in the field of political economy emphasised on economic growth, social disparities and human rights. The economic growth dimension, which accepts, a better-fed person can yield a higher productivity than the underfed person (Leibenstien, H. 1957; Dasgupta, 1993; Fogel, 1994).[6] Therefore, hunger must be addressed so that every individual participates and benefits from economic growth.

The prevalence of hunger and deprivation in a society may also lead to social problems like delinquency and violence, Hence, in an inter-temporal situation to reduce the risk of social loss, the problem of hunger and deprivation must be eradicated. The issues considered in these studies are relevant to a country like India with its multiple deprivations at different levels, despite its higher economic growth and consequent characterisation as an emerging economic power. Within India, the regional differences are quite significant. Odisha provides an example of one of the poorest states and presents a unique setting for the study from various perspectives. Historically, instances of famine and drought have been occurring and added to this is the regional, topographical and economic disparities. Reports of hunger and starvation are frequent in the state, especially in southern Odisha. The data on hunger and deprivation in the state of Odisha indicates a bleak scenario with vast majority of population living below poverty line and consuming less than 2,200 kcal in a day. Hence, the need for the study arises out of the fact that despite major attempts to bring to light causal factors in hunger and deprivation, there are few studies in Odisha conducted from multiple disciplinary perspectives. The aim of the study therefore, is to examine the issue of hunger and deprivation in Odisha through a multidisciplinary lens encompassing social factors, historical and topographical features, and economic factors. National and state-level policy analyses are conducted to explore the gaps in implementation of programmes addressing the issues.

Major Problems of Odisha: Then and Now

Odisha emerged as a separate state in 1936. At the time of India's independence in 1947, the major problem in Odisha was the poor living conditions of the people (Rath, 1977), which might be the cumulative effect of the economic backwardness, social-economic and political inequality in the state, and of natural calamities. While the inequality or the distributional problem and natural calamities directly affected the life of the common people leading to hunger and deprivation, the insufficient growth pattern also contributed to hunger and deprivation.

Post-independence scenario:

(i) *Low per capita state domestic product:* The Odisha economy was backward in terms of her base (per capita SDP) and pace (the growth rate).[7] The long run annual growth rate of the state domestic product remained at around 2 to 3 per cent (between 1950 to 2006). Moreover, since 1980, the net per capita state domestic product as a share of national per capita domestic product has been deteriorating. Therefore, one of the major problems the Odisha economy has been facing is the low base and pace of the economy, which also is responsible for the tardy economic growth with low productivity (Padhi, 2000).

(ii) *Predominant agriculture:* Odisha's economy has a feature of predominant agricultural economy. However, over a period (i.e. from 1950–51 to 1999–2000), though the structural transformation has happened in output composition (due to rapid growth in service sector), agricultural sector remained the major source of employment. During 1950–51, the share of primary sector in the Net State Domestic Product (NSDP) was almost 65 per cent, whereas, the share of primary sector in the Net Domestic Product of the country was around 60 per cent. In the following decades, the shares have declined for both Odisha and all-India, but the rate of decline in all-India is much faster than the rate of decline in Odisha.

By 2004–05, the share of primary sector in the NSDP in Odisha has come down to 39 per cent and for all-India, it has declined to 25 per cent. However, as employment is concerned, more than 63 per cent of the total labour force in Odisha depends on primary sector, whereas such share in the national context is around 57 per cent. Therefore, the state is confronted with a paradoxical situation of absence of agricultural development and large dependence on agriculture.

(iii) *Low industrial base and pace:* The Odisha economy was mostly depending on cottage and village industries with the traditional technology. Those outputs are mostly produced for self-consumption rather than market. Mostly mines (iron, coal and aluminium) lead the modern industrialisation in the state. Moreover, as these industries are mostly producing raw materials, it could not expand sufficiently in terms of output and employment. Therefore, the industrial base of the economy could not make a significant difference in the overall economic growth and the sector is stagnant in its share (at around 17%) to NSDP for more than last forty years.

(iv) *Socio-economic and political (regional) inequality:* Inequalities in society and economy exist in several parts of the country, but the intensity of inequality in Odisha is deep-rooted. The wealth and knowledge have been concentrated with only one group of the society, resulting in huge differences in social and economic well-being of social sub-groups. The SC, ST and females are specifically vulnerable as compared to other groups. For instance, literacy rates among the ST and SC in Odisha are 37 per cent and 55 per cent respectively, compared to 74 per cent among 'other' social groups, during 2001. Similarly, the level of literacy among the females in Odisha is 50.51 per cent compared to 75.96 per cent among males.

Similarly, the regional differences in Odisha are observed. As compared to the Northern and Coastal parts of Odisha, the Southern part continues to remain most backward in terms of poverty, education, health and housing.[8] The inequalities have become much more persistent with passage of time and have not improved drastically despite the fact that several programmes

(Navodaya schools, Sarva Shiksha Abhiyan, JRY, NREGA, Indira Awas Yojana, other targeted programs) have been implemented.

(v) *Natural calamities:* Natural calamities such as drought in the south-western part of Odisha and floods/cyclones in the coastal part of Odisha have been occurring frequently, e.g. droughts in 2001 and super-cyclone in 1997. The natural calamities not only affect the life and living conditions of the current generation, they also destabilise the pace of the economy and affect the living conditions of the future generation (Bhatta, 1997).

(vi) *Education and health:* Education and health are crucial for a healthy economic development. In primary education, though the state performs at par with the country as a whole (with around 63 per cent of literacy rate each), in terms of health achievement, Odisha is far behind all-India level.

In the line of above discussion, it is found that the incidences of hunger and deprivation have remained at high level in Odisha than in any other Indian states. The following sections provide some figures on hunger and deprivation in Odisha.

Hunger: The Facts and Figures

As per the above discussion, at least five indicators can assess hunger in India, such as (a) the protein calories undernutrition, (b) the income poverty, (c) household's perception on calorie adequacy (chronic poverty), (d) the anthropometrical indicators, such as Body Mass Index (BMI), wasting, stunting and underweight; and (e) the health indicators.[9] Similarly, given the multiple dimensions of deprivation, the indicators chosen here are housing, clothing, health, education and other social deprivation.[10] These indicators have been presented in Table 1.1.

Table 1.1 *Indicators considered for hunger and deprivation*

Hunger	*Deprivation*
Undernourished persons by nutritional inadequacy	Housing
Chronic hunger by household's perception	Health and clothing
Below poverty line population	Education
Anthropometrical measures	Social deprivation

Sources: Sen, 1983; Townsend 1984

(i) *Hunger in terms of Protein Calories Undernutrition[11]:* Table 1.2 and Appendix A1.1 assess the situation of hunger from 1972–73 to 2004–05, across the Indian states based on the calories intakes of individual, which reveals the following:

- In rural sector, not only the *level* of the calories intake per consumer unit (PCU) remained lower in Odisha during 1972–73, compared to many other Indian states and all-India level, but the calories intake of Odisha has also remained much lower than the required level of 2,700 kcal.[12] However, this situation had improved during 1993–94, but declined in 2004–05. The rank of the state in terms of calories intake remained almost stagnant: 13th or 12th out of 17 states or 11th out of 15 states.

- In urban Odisha, the case is different where the *level* of PCU intake of calories was higher in Odisha during 1972–73, compared to many other Indian states. However, Odisha's relative position has declined in 2004–05 (from Rank 6 to 4 during 1972–73 and 2004–05).

Overall, as compared to rural India, rural Odisha's position has improved in terms of calorie intake during 1972–73 to 2004–05. In 2004–05 it is catching up with rural India. On the other hand, as compared to urban India, urban Odisha did better throughout the period (1972–73 to 2004–05). Since only 13 per cent of the population in Odisha is urban, the better performance in urban areas will not improve the overall situation of Odisha.

Table 1.3 and Appendix A1.2, reveal the situation of hunger based on the protein intake of individuals for the urban and rural areas of Odisha, all-India and other major states. The results show the following:

In rural sector, Odisha's PCU intake of protein has remained much lower than all-India and many other states. Also, Odisha's relative position has remained almost the same through 1972–73 and 2004–05. The trend is similar in the urban sector of Odisha as well. The only difference is that protein intake is relatively better in the urban sector than the rural sector in Odisha.

Table 1.2 *Calorie intake by the people of Odisha and her position compared to major Indian states in rural and urban sector, 1973–74 to 2004–05*

	Rural				*Urban*			
State	*1972–73*	*1983*	*1993–94*	*2004–05*	*1972–73*	*1983*	*1993–94*	*2004–05*
ORI	2494	2624	2740	2512	2845	2735	2754	2596
Rank	13	12	7	11	6	4	3	4
IND	87.3	94.4	102.1	98.90	108.0	106.3	108.3	104.89

Notes: Odisha's figures have been expressed in K.Calories. India's figures express what percentage of India that Odisha is consuming. For 2004–05, rank is calculated from 15 states. For other periods, rank is calculated from 17 states.

Source: Various NSSO reports.

Table 1.3 *Protein intake among all the Indian states in rural and urban sector, 1972–73 to 2004–2005*

	Rural				*Urban*			
State	*1972–73*	*1983*	*1993–94*	*2004–05*	*1972–73*	*1983*	*1993–94*	*2004–05*
ORI	62.38	65	65.6	59.9	69.87	68	69.7	67
Rank	15	13	13	14	12	11	9	11
IND	79.6	84.4	87.5	84.60	98.6	97.1	99.3	95.85

Notes: Odisha's figures have been expressed in grams. India's figures express what percentages of India that Odisha is consuming. 1972–73 estimated by using conversion factor of (per capita = 0.8 of PCU). For 2004–05, rank is calculated from 15 states. For other periods, rank is calculated from 17 states.

Source: Various NSSO reports.

Table 1.4 *Fat intakes among all the Indian states in rural and urban sector, 1972–73 to 2004–2005*

	Rural				*Urban*			
State	*1972–73*	*1983*	*1993–94*	*2004–05*	*1972–73*	*1983*	*1993–94*	*2004–05*
ORI	11.13	18	18.5	22.1	29.625	28	34.2	34.4
Rank	17	17	17	15	16	17	17	14
IND	36.3	52.9	47.3	50.23	64.6	65.1	66.3	59.11

Notes: Odisha's figures have been expressed in grams. India's figures express what percentage of India that Odisha is consuming. 1972–73 estimated by using conversion factor of (per capita = 0.8 of PCU). For 2004–05, rank is calculated from 15 states. For other periods, rank is calculated from 17 states.

Source: Various NSSO reports.

Table 1.4 and Appendix A1.3 show the situation of hunger based on the fat intakes of individuals. The level and trend in the rural and urban areas show the following:

- In rural sector, as the level of the PCU intake of fat is concerned, in Odisha, this has remained much lower than all-India and many other Indian states. In fact, Odisha has never been able to consume even 50 per cent of the all-India PCU consumption of fat. In urban sector too, Odisha's PCU intake of fat remained much lower than other Indian states and from all-India.
- As the trend is concerned, in rural sector, there is a steady improvement in fat consumption of both Odisha and all-India over the whole time period. But in urban sector, while all-India has shown a marginal improvement over the period, Odisha has shown an overall decline.

From the above analysis of hunger in terms of calorie, protein and fat intake, Odisha has remained in a disadvantaged position than all-India and many other states. Though the incidence of hunger has been a little lesser based on calorie intakes, in case of protein and fat intakes, the gaps between Odisha and all-India have been much more.

Further, the above analysis is based on only the average (or aggregate) calorie, protein and fat intake. These figures do not ensure an equitable amount for all the individuals of a household especially, where there would be gender disparities in food distribution. Though it is difficult to capture such dynamics and complex situations from a macro-level data, to an extent, this can be examined from the anthropometrics indicators, which are discussed in the later part of this chapter. The following section looks into the money-metric measurement of hunger, i.e. income poverty.

(ii) *Hunger in terms of Income Poverty:* Income poverty otherwise known as the money-metric measurement of poverty is used to identify the poor. The poverty line represents the level of income required for meeting the minimum level of nutrition as suggested by the Planning Commission of India (Dandekar and Rath, 1971). The poverty line can be considered as hunger specifying line due to its narrow specifications and overemphasis on nutritional intake rather than standards of living or deprivation of people.

Table 1.5 and Appendix A1.4 shows the relative position of Odisha vis-à-vis other Indian states based on the proportion of population living below the poverty line.

- Table 1.5 indicates that as the level of overall picture of poverty is concerned, during 1973–74, it was highest in Odisha (66%) followed by West Bengal (63%) and Kerala (60%) and all three of them remained much higher than the level of poverty in all-India (55%). Whereas, as the trend is concerned, while Kerala, West Bengal and all-India have shown a sharp decline, Odisha remained way behind. In 2004–05, the level of poverty has remained highest in Odisha (46.4%) compared to all-India (27.5%), West Bengal (24.7%) and Kerala (15.0%).
- As the level of rural poverty is concerned, this has remained higher in Odisha (67%) when compared to all-India (56.4%) during 1973–74. Although there has been continuous decline in poverty for both, there has been marginal increase of poverty in rural Odisha from 1973–74 to 1977–78. However, in the later part of 1990s (till, 2004–05), the poverty level between Odisha and India took a divergent path, (due to stagnation in rural poverty reduction of Odisha and faster decline of the same in all-India). During 1973–74, West Bengal was the only state, which was having more percentage of rural poor (73.2%), as compared to Odisha (67%). West Bengal has made a significant achievement in reducing rural poverty and has brought it down to 28.6 per cent compared to rural Odisha's 46.8 per cent in 2004–05. The other states, which are having a higher percentage of poverty and are closer towards Odisha, but have made significant improvements are Bihar, Kerala and Madhya Pradesh.
- The level of urban poverty also remained higher in Odisha compared to all-India. Although there has been almost a similar rate of decline in poverty rate for Odisha and India from 1973–74 to 1987–88, since then, while the rate of decline has been faster for all-India, it has remained slow for Odisha, leading to widening gap between these two poverty rates. Comparing urban poverty of Odisha (56.3%) with other Indian states, in 1973–74, there were other three states closer to the poverty level of Odisha, namely Kerala (62.2%), Madhya Pradesh (58.3%) and Uttar Pradesh (59.5%). These states were at similar rate for poverty reduction till 1987–88, but after that the rate of decline is higher for Kerala and UP as compared to Odisha and Madhya Pradesh and finally the urban poverty also remained at highest level in Odisha (42.8%). Here too, though the overall trend of poverty has declined for all the states, it has declined at a relatively slower rate and has almost remained stagnant in Odisha since early 1990s. Thus, Odisha's relative rank has remained consistently lowest throughout the period (1973–74 to 2004–05).

Table 1.5 *Percentage of population in income poverty by states 1973–74 to 2004–05*

		1973–74	*1983*	*1993–94*	*2004–05*
ORI	R	67.3	67.5	49.7	46.8
	U	56.3	50.6	41.6	44.3
	C	66.18	65.29	48.6	46.4
Rank of Odisha	R	24	25	24	25
	U	22	24	24	25
	C	25	25	25	25
IND	R	56.4	45.6	37.3	28.3
	U	49.2	42.2	32.4	25.7
	C	54.88	44.48	35.9	27.5

Note: The rank order is calculated from 25 Indian states.

Source: Complied from Planning Commission of India and www.indiastat.com.

As it has been observed, not only poverty in general (both at intertemporal and spatial context) has remained at a much higher level in Odisha than in all-India, but also the same has been increasing (in absolute number) in the state during 1993–94 to 2004–05.[13] During 1993–94 to 2004–05, while the all-India Below Poverty Line (BPL) population reduced by 10 per cent point, the same remained almost stagnant for Odisha. Among the major Indian states, Kerala has been effective in bringing down overall poverty from 59.79 per cent in 1973–74 to 15 per cent in 2004–05. The worst performer in poverty reduction is Odisha, which showed a reduction in overall poverty from 66.2 per cent to 46.4 per cent in the same time. Further, when hunger has been assessed based on the perception of the households, it shows a different picture, which has been discussed below.

(iii) Hunger in terms of Households' Perception on Calorie Adequacy: Apart from the conventional prescriptions of calorie requirements to find out hunger, another method is by asking the household's perceptions. Table 1.6 and Appendix A1.5 present the magnitude of hunger based on the households' responses to the question, 'whether all members of the household get two square meals a day', for the period 1983 to 2004–05. The table reveals the following:

Table 1.6 *Percentage distribution of households by the type of response on the question whether all members of the household get two square meals a day, by states 1983 to 2004–05*

	Rural						*Urban*					
	Some Months			*No Month*			*Some Months*			*No Month*		
States	*1983*	*1993–1994*	*2004–2005*	*1983*	*1993–1994*	*2004–2005*	*1983*	*1993–1994*	*2004–2005*	*1983*	*1993–1994*	*2004–2005*
ORI	31.8	14.9	4.8	5.02	0.5	1.3	11.33	3.3	0.4	0.84	0.7	0.2
Rank	15	15	14	13	9	15	14	14	10	10	10	12
IND	16.2	4.2	2.0	2.4	0.9	0.4	5.6	1.1	0.4	0.77	0.5	0.1

Note: Rank is from 15 Indian states. Sources: NSSO, various Calorie adequacy reports.

- In rural sector, during 1983, the proportion of households getting two square meals a day only for some months has remained the highest in Odisha (31.8%) and Bihar (31.8%) followed by West Bengal (31%) compared to all-India (16.2%). In 2004–05, this proportion remained at the second highest (4.8%) in Odisha after West Bengal (10.6%) and the same for all-India remained at 2 per cent. Also, there has been a declining trend for all the states in terms of getting two square meals a day. Further, as the Table reveals even during 2004–05, there are households which do not get two square meals throughout the year.
- In urban sector, though the trend (1983 to 2004–05) has improved a lot for all the states, during 1983, the level of household's response for getting two square meals a day only for some months has remained highest in Odisha (11.33%) and Kerala (11.7%) compared to all-India (5.6%). However, some households did not get two square meals throughout the year even in 2004–05. This indicates that chronic poverty has increased in the state.

Since the NSSO data on income poverty and the concept of 'getting two square meals a day' are based on household data, they cannot precisely measure an individual's actual state of hunger. To fill such a gap, we have used the NFHS data, which provides individual's health status. As considered in some of the research work (Vaidyanathan, 2001; Deolalikar, 1988), poor anthropometric indicators are the result of poverty in general and sustained inadequacy in nutritional intake in particular.

(iv) *Hunger in terms Anthropometrical Indicators:* The high magnitude of hunger in the state reveals different forms of malnutrition.[14] The data collected by National Family Health Survey (NFHS) provides the nutritional status of women and children in terms of different anthropometric measures, such as Body Mass Index (BMI), underweight (weight for age), stunted (height for age) and wasted (weight for height). Table 1.7(a) and (b) and Appendix A1.6 (a) and (b) have been compiled from NFHS 1992–93, 1998–99 and 2005–06 data. Table 1.7(a) shows the mean BMI of the eligible women across the Indian states.[15]

Table 1.7(a) *Body mass index of women across the indian states, 1998–99 and 2005–06*

Name of the States	*Mean BMI and percentages of less than 18.5 of women*			
	1998–99		*2005–06*	
	Mean BMI	*< 18.5*	*Mean BMI*	*< 18.5*
ORI	19.2	48	19.7	41.4
Rank	24	24	24	22
IND	20.3	35.8	20.5	35.6

Notes: The Body Mass Index is the ratio of the weight in kilograms to the square of the height in metres (kg/m2).

Source: NFHS, 1998–99, 2005–06.

During 1998–99, the mean BMI for Odisha was 19.2 compared to 20.3 for India. Odisha's BMI was lowest among the Indian states and closer towards 18.5. The same during 2005–06 is 19.7 compared to 20.5 for all-India level. In terms of average BMI, the relative position of the state has remained the same (rank, 24).

As per the norm, if the BMI is less than 18.5, then it is a sign of chronic energy deficiency (CED). The proportion with CED has remained very high in Odisha, although it has declined from NFHS-2 to NFHS-3 (48% to 41%). In contrast, 36 per cent had reported CED in case of all-India level for both the periods.

Table 1.7(b) *Nutritional status of children across Indian states, 1992–93 to 2005–06*

(All are in percentages)

Name of the States	*Weight for age*						*Height for age*						*Weight for height*					
	% <3SD			*% <2SD*			*% <3SD*			*% <2SD*			*% <3SD*			*% <2SD*		
	1	*2*	*3*	*1*	*2*	*3*	*1*	*2*	*3*	*1*	*2*	*3*	*1*	*2*	*3*	*1*	*2*	*3*
ORI	22.7	20.7	13.4	53.3	54.4	40.7	25.2	17.6	19.6	48.2	44	45.0	3.6	3.9	5.2	21.3	24	19.5
Rank	22	20	17	21	23	19	12*	13	14	14*	19	17	14*	21	13	18*	24	16
IND	20.6	18	15.8	53.4	47	42.5	28.9	23	23.7	52	45.5	48	3.2	2.8	6.4	17.5	16	19.8

Notes: Each index of the children nutritional status is expressed in standard deviation units (SD) from the median of the international references population. * Ranking is taken from 19 states, all others are from 24 states.

Source: NFHS, 1992–93, 1998–99, 2005–06.

Table 1.7(b) shows the nutritional status of children, which depicts, irrespective of the type of measures (wasted/stunted/underweight) taken, the incidence of nutritional deficiency remains high in Odisha compared to many other Indian states.

The above analyses (Tables 1.2 to 1.7) have revealed that irrespective of the indicators taken the intensity of hunger has remained high in Odisha. However, as it has shown below, Odisha is also way behind the other states in fulfilling other socially necessary consumptions like housing, education and health, though the situation is worse in case of hunger. Consumption of non-food items may be a reason of food deprivation. Such dynamic situations are described below with the help of data. Therefore, not only food deprivation but also fulfilments of basic non-food deprivation are also necessary to avoid hunger.

Deprivation: The Facts and Figures

The four kinds of non-food deprivation considered are shelter (housing) and clothing, health, education and social deprivation. Tables 1.8 to 1.12 state the magnitudes of these variables except social deprivation.

(i) *Housing deprivation:* Table 1.8 and Appendix A1.7 present the housing conditions in Odisha, which indicates that in 1973–74 and in 2002, the housing conditions in rural Odisha remained worse as compared to other Indian states, except Assam in 2002. During 1973–74 in rural Odisha, while there were only around 6 per cent of pucca houses, 79 per cent of the houses were kutcha. The corresponding figures for all-India were around 18 per cent and 49 per cent respectively. After 28 years, in 2002, the percentage of pucca houses in rural Odisha reached to 18 per cent, but 53 per cent of the houses were kutcha. During the same time, in rural India, the numbers of pucca houses were 36 per cent, which is twice than that of Odisha and number of kutcha houses in rural India is 21 per cent.

However, in case of urban areas, Odisha is not only better from her rural counterparts but is also better than many other Indian states. But as stated earlier, these conditions in urban areas would not compensate much to the overall housing conditions in the state, as the share of urban households in Odisha is around 15 per cent. In terms of Odisha's rank, the relative position on housing conditions in Odisha has remained more or less the same during 1973–74 to 2002.

Table 1.8 *Housing conditions across the Indian states*

		1973–74		*1988*			*1993*			*2002*	
State	*Region*	*Pucca*	*Katcha*	*Pucca*	*Katcha*	*% of KHHB*	*Pucca*	*Katcha*	*% of KHHB*	*Pucca*	*Katcha*
ORI	R	5.97	78.78	9.36	71.43	51.76	12.06	59.59	43.07	18	53
	U	49.53	34.33	53.15	29.46	63.92	55.42	16.78	58.26	68	11
Rank	R	15	14	14	14	13	14	14	11	14	14
	U	14	14	14	14	12	14	14	9	9	14
IND	R	18.42	49.17	27.15	39.35	41.88	32.28	31.72	34.91	36	21
	U	64.52	15.84	71.08	10.92	55.71	73.84	8.27	52.63	77	3

Notes: KHHB – Kutcha households in bad conditions; ORI–Odisha, IND–India, R–Rural, U–Urban.

Source: NSSO various rounds.

Table 1.9 *Percentage distribution of the conditions of houses across the major Indian states, 2001*

States		*Good*	*Liveable*	*Dilapidated*
ORI	R	22.26	67.45	10.29
	U	46.31	45.85	7.84
	C	25.64	64.41	9.95
Rank	R	15	1	13
	U	15	1	14
	C	15	1	13
IND	R	44.89	48.88	6.23
	U	64.03	32.34	3.63
	C	50.21	44.28	5.51

Note: Rank is from 15 major Indian states

Source: Census of India, 2001

The 2001 census data is useful to examine the housing condition in terms of whether it is good, liveable or dilapidated. Table 1.9 and Appendix A1.8

present the conditions of houses across the major Indian states from the census data. While only one-fourth (26%) of the houses in Odisha can be considered as 'good' condition in Odisha, this proportion is twice higher in India as a whole (50%). As one would expect, there are huge rural-urban differences in housing conditions as well. While urban houses in Odisha are twice as likely as rural houses to be in 'good' condition, such difference is less sharp for all-India. In comparison to other major states in India, Odisha ranks last in terms of 'good' housing condition.

The above data leads to the question, 'whether these poor housing conditions led to expenditure on housing and thereby brought about hunger'.

(ii) *Health deprivation:* Life expectancy and infant mortality rate are the two indicators considered to understand the health deprivation. Table 1.10 and Appendix A1.9 show the expectation of life at birth among the major Indian states from 1970–75 to 1999–2003. It shows, during 1970–75 in rural areas, the life expectancy for Odisha was 45.1 years, which was better than only Uttar Pradesh (42 years) among the Indian states. For India, the rural life expectancy was 48 years during 1970–75. By 1999–2003, the life expectancy for rural Odisha increased to 58 years, which was similar to the life expectancy of Assam, MP, UP, Bihar and Rajasthan and was lower than all-India (61.4 years). In the urban areas, the life expectancy of Odisha was 65.8 years, which was 3rd from below after Uttar Pradesh (63.3 years) and Madhya Pradesh (64.4 years) and around 2 years less than India's life expectancy (68.1 years). Irrespective of rural or urban areas, the life expectancy of Odisha remained around 2 to 3 years behind than that of India.

Table 1.11 and Appendix A1.10 present the infant mortality rate (IMR) across the major Indian states by their place of residence from 1971 to 2005 in each five years interval. In rural Odisha, the IMR was 131 during 1971, which was better than that of all-India (138) and some other Indian states, such as UP (173), MP (144) and Gujarat (155). Over a period, the rest of the Indian states have improved a lot. During 2005, the highest IMR has been observed in Odisha (78) and MP (80). The same for rural India remained at 64.

Table 1.10 *Expectation of life at birth among the major Indian states, 1970–75 to 1997–2001*

		1970–75	*1981–85*	*1991–95*	*1997–2001*	*1999–2003*
ORI	R	45.1	52.4	55.7	57.5	58
	U	55.4	59.6	64.4	65.4	65.8
	C	45.7	53	56.5	58.1	58.7
Rank	R	11	10	13	13	13
	U	11	13	9	12	13
	C	11	11	13	13	13
IND	R	48	53.7	58.9	61	61.4
	U	58.9	62.8	65.9	67.6	68.1
	C	49.7	55.5	60.3	62.2	62.7

Note: R – Rural, U – Urban, C – Combined Rank is calculated from 15 states.

Source: Registrar General of India, (1999 and 2006).

As the urban areas are concerned, though the absolute magnitude of IMR remained at lower level compared to rural counterparts, the relative position of Odisha has worsened. During 1971, the IMR of urban Odisha was 84, which was better than Gujarat (110), Maharashtra (88) and UP (119). The same for urban India was 82. Whereas during 2005, in urban area, the highest IMR has been seen in Odisha (55), which is much higher than all-India (40).

In comparison to other states in India, Odisha's rank in IMR has deteriorated over time, whereas there has been a slight improvement in Odisha's rank for life expectancy. The higher health deprivation might increase a household's health-related expenditure and this in turn might reduce households' disposable income, leading finally to hunger.

Table 1.11 *Infant mortality rate for selected years among the major Indian states from 1973 to 2005*

		1971	*1981*	*1991*	*2001*	*2005*
ORI	R	131	140	129	90	78
	U	84	68	71	56	55
	C	127	135	124	87	75
Rank	R	10	13	15	15	14
	U	10	11	13	14	15
	C	10	13	15	15	14
IND	R	138	119	87	69	64
	U	82	62	53	40	40
	C	129	110	80	63	58

Note: R – Rural, U – Urban, C – Combined, for 1971, rank is calculated from 12 states, all other are from 15 states.

Source: Registrar General of India, (1999 and 2006).

(iii) *Literacy rate:* After examining the housing and health deprivation, Table 1.12 provides the literacy rates across the major Indian states from 1951 to 2001.

- During 1951, the level of literacy rate in Odisha (15.8%) was lower than that of all-India (18.3%). During 2001, the literacy rate for Odisha was 63.08 per cent and for all-India, it was 65.38, which is due to faster rate of growth in literacy during 1980s and almost equal rate of growth (13%) in 1990s in Odisha. Further, as the rank order shows among the 15 major Indian states, Odisha's position in literacy remained in between 8 to 11, unlike income poverty during 1973–74 to 1999–00 where it was always in the last rank.
- If Odisha can do well in non-food consumption like education, then why does it fail to achieve the similar success in reducing income poverty? Moreover, there are states (Bihar and UP) whose literacy rates are even lower than Odisha but have done well in reducing income poverty.[16] It is interesting to note that, as shown in Table 1.13, BPL households also spend significantly on education, health, clothing and other miscellaneous goods and services. Hence, it seems, not only food, but rather a few non-food items are essential to live in a society.

Table 1.12 *Literacy rates across the major Indian states, 1951 to 2001*

States	*1951*	*1961*	*1971*	*1981*	*1991*	*2001*
ORI	15.8 (8)	21.66 (11)	26.2 (9)	33.62 (11)	49.1 (10)	63.08 (11)
AP	13.2 (11)	21.19 (13)	24.57 (12)	35.66 (10)	44.1 (12)	60.47 (13)
ASM	18.3 (7)	32.95 (5)	33.94 (8)	NA	52.9 (9)	63.25 (10)
BIH	13.49 (10)	21.95 (10)	23.17 (14)	32.32 (13)	37.49 (15)	47 (15)
GUJ	21.82 (4)	31.47 (7)	36.95 (5)	44.92 (6)	61.3 (4)	69.14 (6)
HAR	NA	24.1 (9)	25.71 (11)	37.13 (9)	55.8 (8)	67.91 (7)
KAR	19.3 (6)	29.8 (8)	36.83 (6)	46.21 (5)	56 (7)	66.64 (8)
KER	47.18 (1)	55.08 (1)	69.75 (1)	78.85 (1)	89.8 (1)	90.92 (1)
MP	13.16 (12)	21.41 (12)	27.27 (8)	38.63 (8)	44.67 (11)	63.74 (9)
MAH	27.91 (2)	35.08 (3)	45.77 (2)	57.24 (2)	64.9 (2)	77.27 (2)
PUN	15.2 (9)	31.5 (6)	34.12 (7)	43.37 (7)	58.5 (5)	69.95 (4)
RAJ	8.5 (14)	18.1 (15)	22.57 (15)	30.11 (14)	38.6 (14)	61.03 (12)
TN	20.8 (5)	36.4 (2)	45.4 (3)	54.39 (3)	62.7 (3)	73.47 (3)
UP	12.02 (13)	20.87 (14)	23.99 (13)	32.65 (12)	40.71 (13)	56.27 (14)
WB	24.61 (3)	34.46 (4)	38.86 (4)	48.65 (4)	57.7 (6)	69.22 (5)
All IND	18.3	28.3	34.45	43.57	52.2	65.38

Source: Office of the Registrar General of India.

Notes: Literacy for 1951, 1961 and 1971 Census relate to population aged five years and above, the rates for the 1981, 1991 and 2001 Census relate to the population aged seven years and above, the literacy rates for West Bengal 1951 relates to total population including 0–4 age group. Figures in bracket are the ranks of the states.

Table 1.13 indicates that households below poverty line spend around 11–14 per cent of their expenditure on a few selected non-food items (like education, health and clothing) in rural areas. The same for the urban areas is around 13 per cent for both Odisha and India. Therefore, if food is the first priority then these households should have spent all of their expenditure on food, but the data does not show the same. The households below poverty line do consume other non-food items as well, despite their food requirement remaining unfulfilled. Hence, in such a case, possibly the relative deprivation of other essential commodities may also become the cause of hunger. Therefore, to study hunger, we need to examine the basic non-food deprivation as well.

(iv) *Social Exclusion:* Social exclusion is another aspect of hunger and deprivation. 'Social exclusion is a process, which restricts the access of certain social groups to valued resources and entitlements, relating to them the status of social outsiders (Rodgers and Figueirede, 1995: 237)'. In its widest connotation, social exclusion encompasses theories of poverty, inequality, racial/ethnic cleavages, long-term unemployment and citizenship. But all differences that occur (in the socio-economic indicators) in a society does not imply social exclusion, rather social exclusion 'presupposes an active process of social distancing both covert and overt. It attributes agency to groups closely guarding their domain of privilege and profit from exclusion'. Moreover, the socially excluded are

Table 1.13 *Share of total expenditure spent by the household on education, health and clothing*

Monthly Per Capita Consumer Expenditure (MPCE) classes	Rural						MPCE classes	Urban					
	% Expenditure		Calorie intake		% of households			% Expenditure		Calorie intake		% of households	
	India	Odisha	India	Odisha	India	Odisha		India	Odisha	India	Odisha	India	Odisha
0–225	10.54	8.87	1,755	1,967	4.0	14.2	0–300	11.31	9.58	1,745	2,100	3.8	9.3
225–255	11.31	9.98	2,023	2,220	3.4	9.5	300–350	11.54	9.56	2,047	2,452	3.7	8.0
255–300	11.81	11.07	2,173	2,384	8.9	13.3	350–425	12.57	10.79	2,143	2,510	7.5	11.5
300–340 PL	12.59	11.66	2,330	2,627	10.0	13.8	425–500 PL	12.90	13.14	2,359	3,028	8.6	11.8
340–380	12.55	12.86	2,433	2,696	11.1	9.6	500–575	14.01	15.70	2,417	2,762	8.6	10.9
380–420	13.05	13.11	2,548	2,879	11.5	9.0	575–665	13.97	12.58	2,557	2,896	9.1	9.1
420–470	13.39	14.58	2,698	2,898	10.6	8.1	665–775	14.89	14.83	2,665	3,017	9.6	9.1
470–525	13.99	16.13	2,834	3,105	10.5	6.0	775–915	15.56	14.19	2,788	3,014	10.3	8.8
525–615	14.66	17.33	2,967	3,200	11.2	5.7	915–1120	15.88	16.66	2,990	3,074	11.3	8.9
615–775	16.07	17.41	3,177	3,374	9.0	5.6	1120–1500	16.37	14.15	3,063	3,390	13.0	8.2
775–950	16.92	23.5	3,355	3,560	4.3	2.3	1500–1925	16.71	17.58	3,309	3,412	6.7	1.9
950-more	20.04	22.27	3,893	3,771	5.6	2.8	1925-more	17.23	17.11	3,578	4,592	7.7	2.5
All	14.86	14.26	2,668	2,635	100	100	All	15.45	13.99	2,637	2,802	100	100

Notes: For rural area the below poverty line is ₹ 323.92 for Odisha and ₹ 327 for India, whereas for urban area it is ₹ 473.12 for Odisha and 454.11 for India.

Source: NSSO, 1999–2000.

not excluded because of their cumulative individual failing, rather certain structural factors cause the exclusion and reiterated losses, whose sum is exclusion itself. In the Indian context, such differences exist on account of the institution of social hierarchy based on caste. It is also recognised that certain castes, especially, 'untouchables' and indigenous people (referred to in India as Scheduled Castes and Scheduled Tribes), suffer from the cumulative effects of the systematic exclusion from land ownership, education, access to common resources and work.

As literature shows the prevalence of these social exclusions often seen in Odisha (Praxis, 2001) prevent some communities to have access even to labour market, to educational institutes, to temple, to public tube well and to many other public places (Sainath, 1996; Bailey, 1963). Hence, such exclusion based on one's community may also become a cause of hunger and deprivation in many cases.

The above data analysis and literature review suggest that –

(1) Based on all almost all indicators of hunger and deprivation, Odisha is the worst performer state.
(2) There is no substantial improvement either in terms of level or in relative position (rank) of the state.
(3) There exist higher variations in the rural and urban sector of the state.

In spite of all the aforementioned issues, there are few attempts to understand the causal factors of hunger and deprivation. Also there are fewer remedial measures taken against it by the State. Hence, our attempt here is to understand the incidence of hunger and deprivation and to examine the extent, patterns and causes of hunger and deprivation in Odisha.

Objectives and Hypothesis of the Study

Objectives

- Examine the mass (inadequate growth) and class (unequal distributions) character of hunger and deprivation in Odisha.
- Examine social exclusion dimensions of hunger and deprivation in Odisha.
- Examine the role of state (democratic civil society), institution (public action) to reduce hunger and deprivation and formulate policy for the same for the state?
- Examine the external sources of hunger and deprivation in Odisha.
- Examine the displacement induced hunger and deprivation.

Hypothesis

- Economic backwardness (growth and distribution) leads to hunger and deprivation.
- Caste-based social exclusion leads to hunger and deprivation.
- Institutional failure leads to hunger and deprivation.
- External shocks such as natural calamities and displacement causes hunger and deprivation.

Theoretical Framework

Growth and distribution have been considered as the principal determinants of hunger and deprivation. But the growth and distribution itself is the outcome of certain long standing institutional and structural process or the political economy of a society. Therefore, to assess the existing hunger and deprivation, understanding of the study of internal social system and other structural features of an economy is required. Below, we have explained the framework in a lucid way.

H and D =	Primary factors	Secondary factors
H and D =	Internal Social System	External shocks
H and D =	(Necessary condition, Sufficient condition)	(Natural calamities, Displacement)
H and D =	[(Economic growth), (Distribution)]	[(Flood, Drought/Famine, Cyclone/Tsunami, Earthquake), (Displacement)]
H and D =	[(growth in primary, secondary and tertiary sectors) (Distributive justice, second order distribution)]	[(Flood, Drought/Famine, Cyclone/Tsunami, Earthquake), (Displacement)]
H and D =	(growth and productivity, in primary, secondary and tertiary sectors) (land reforms, public provisioning)	[(Flood, Drought/Famine, Cyclone/Tsunami, Earthquake), (Displacement)]
H and D =	[(growth and productivity in primary sector) (land reforms and public provisioning across class, caste, region and gender)]	[(Flood, Drought/Famine, Cyclone/Tsunami, Earthquake), (Displacement)]
H and D =	[(growth and productivity) (capital formations) (Investment) (distribution through assets and employment) (distribution through PDS, ICDS, mid-day meal)]	[(Flood, Drought/Famine, Cyclone/Tsunami, Earthquake), (Displacement)]
H and D =	[(growth and productivity) (capital formations) (Investment) (distribution of assets, higher employment with low elasticity of demand) (distribution through welfare schemes)]	[(Flood, Drought/Famine, Cyclone/Tsunami, Earthquake), (Displacement)]

Note: H–Hunger; D–Deprivation.

As shown in the table above, hunger and deprivation may arise broadly from two sources:[17]

(a) The major and frequent contributors of the hunger and deprivation arise from the Internal Social System (ISS) of an economy, which is the primary factor behind the outcome.

(b) The external shocks, i.e. external to the state or to the local economy including drought/famine, cyclone, flood and displacement also contributes to hunger and deprivation in a society, which can be the secondary factors for the outcome.

Economic growth is the principal and necessary factor of ISS that may cause hunger and deprivation (Lal and Myint, 1996). Growth in any form (growth and productivity either in primary, secondary or tertiary sector) is not sensitive to reduce poverty in general or hunger and deprivation in particular

(Ravallion and Datt, 1996). Rather, growth should appear in that sector on which most of the poor depend. As Clark (1932) and Kuznets (1971, 1973) observed, poverty in underdeveloped countries is a result of two factors: (i) if the structure of the economy is such that a large share of the labour force is in agriculture and (ii) very low product per worker in the agricultural sector both absolutely and relative to the total per worker productivity. Hence, to reduce poverty, agricultural growth and labour productivity in agricultural sector should increase.[18]

Many rural-urban interactions or dualistic models are also prescribed to reduce hunger and deprivation by achieving economic development through structural transformations, e.g. Lewis model of economic development (1954). Lewis pointed out – while the agricultural sector can supply labour and food to the industrial sector, the industrial sector may fail to supply machinery if the capital available in this sector is scarce. The capital generation in the industrial sector depends on the increasing rates of saving and investment in an economy. Hence, the tardy movements of labour from agriculture to non-agriculture as well as low capital formation and investment in any regional economy could be a cause of hunger and deprivation. A similar logic is provided also in Fie and Reins model.

The structural transformation and reduction of hunger and deprivation hypothesis can be tested in the regional economy like Odisha. Nayak and Sarangi (1993) have carried out the economic development and structural transformation exercise in Odisha. Meher (2003) attempted to understand the dynamics of economic development in Odisha.

However, with regards to food non availability as a reason of hunger in Odisha, Kumbhar (2001) has shown that during 1960–98, the net per capita availability of food grain remained higher in Odisha for most of the years than in all India.[19] But as that study has shown the food availability remained at the subsistence level and any reduction from that level could be a threat and cause of hunger and deprivation in the state. However, if the distribution could have been proper; even at that amount of food availability, the extent of hunger could have remained at a low level. Therefore, it seems, hunger and deprivation in Odisha could be question of distribution as well. Moreover, growth in an economy also depends on the distributions of other factors of production among the people; or on the prevailing mode of production in the economy. Therefore, fulfilment of this first and necessary condition alone may not be adequate to reduce hunger and deprivation and hence the distributional dimension is important.[20]

Distribution: The question in distribution is with respect to resources and/or output.

Distribution of land is important as this is the only factor from which human being can directly or indirectly find out their survival.[21] Further, land distributions have also played a significant role in achieving human capability (Bagchi, 2004). Hence, a land entitlement is necessary for poverty reduction (Sen, 1981). In Odisha, where social hierarchy is so deeply rooted, some communities are significantly landless and hence lacks the primary source of livelihood, especially in rural economy. Therefore, land redistribution may provide them livelihood to get rid of hunger and deprivation. The skewed distributions of resources may also lead to a mode of production, which may create class disparities, and hence hunger and deprivation may prevail among the citizens of underclass sections. Therefore, land distribution may also be a necessary to remove hunger and deprivation. But land distribution alone is not a panacea to get rid of hunger and deprivation, rather agrarian reform (other agrarian infrastructure) has to follow it, where the distribution can also be facilitated through employment. However, here too, it is not only quantum of employment, but an adequate amount of wage rate also is important to remove hunger and deprivation (Mitra, 1990). But in case of lack/inefficient land reforms, the socio-economically disadvantaged will be still in a backward stage. Therefore, the second order redistribution that is the capability enhancing provisioning through public support is necessary, which may be provided through public action model. The public action model to reduce hunger and deprivation is presented below.

Public Action Model

The other way through which distribution can reduce hunger and deprivation is through social democracy. Here the democratic organisations attempt to distribute minimum necessary resources to the poor and vulnerable group of population through various social welfare schemes. Therefore, hunger and deprivation also can be reduced through state provisioning in the form of social security schemes.

In India, the states could ameliorate the incidence of hunger and deprivation substantially through different public provisioning (Kannan, 1995; Seeta Prabhu and Sudarshana, 2002; Dev et al., 2003). In India, the role of public distribution system (e.g. the Antyodaya scheme, the Annapurna scheme etc.) have been implemented for the poorer sections of the communities to ameliorate their conditions of hunger and other basic deprivation, like health, housing and education. But in all these schemes, two main demerits are – negligible assistance and the targeting errors (Kumbhar, 2001).[22] In the state like Odisha, where there is regular occurrence of starvation deaths (Lok Sabha Debates, 1986: 409–10, cited in Currie, 2000), efficient functioning of these schemes are necessary to reduce both hunger and deprivation.

However, all these macro-level indicators can ensure supply of food or other facilities at reasonable prices but the real problem arises when a households fails to have effective demand.[23] In the southern part of Odisha, the prevalence of hunger is high; there is also evidence of the frequent occurrence of starvation death (in Kashipur block and Kalahandi district) in spite of high net per capita food availability. If inadequate growth alone is the cause of hunger and deprivation then the incidence would have been less in Odisha as well as in southern Odisha or in Kalahandi. Hence, the distribution aspects remain important to ensure household's effective demand. However, in the absence of ensuring land and employment with wages to all distributions of output (food, health and education), different schemes (Public Distribution System–PDS, mid-day meals, Integrated Child Development Services–ICDS, etc.) are necessary to protect people from hunger and deprivation.

Social Exclusion: Myrdal (1968) has pointed out – the 'inefficiency, rigidity and inequality of the established institutions and attitudes' and 'the economic and social power relations embodied in this framework of institution and attitudes' are considered the primary reason for inability to get economic plan effectively implemented. The preconditions, therefore, are that before the development takes place, the old religious attitudes, cultural patterns and social structures must be destroyed (Frankel, 2008).

The hierarchy was a part of India's socio-cultural history. 'Hereditary caste group, each placed in a position of ritual superiority or inferiority to others and all governed in their mutual relationship by customary norms of reciprocal, non-symmetrical rights and obligations, continued to provide building blocks of social organisation in the hundreds of thousands of India's villages' (ibid., 5). In the premodern time, the criteria of pollution and purity had qualified only to the upper twenty per cent or so (of population) for access to education. And those twenty per cent belonging to the elite *jatis* (sub-castes) of the higher varna of Brahmins, Kshatriyas and Vaishas – the priest-scholars, warrior-rulers and traders.[24] Sudras is the fourth one in varna hierarchy, which includes the peasantry and members of artisan caste, etc. Apart from this, at the bottom of the religious hierarchy, a group (one in five persons) stand who have no varna at all. They were considered so polluted that they can defile other caste person merely by touch or even by sight. These pariahs were known as outcast or untouchables, whose hovels were segregated in the edges of the villages. The middle caste Hindu rarely visits the street of the untouchables but depends on them for menial, and indispensable services. The only source of livelihood for these *avarna* or the untouchable groups was their service to the four upper caste groups. The varna hierarchy is paramount even in the modern times and the avarna are still socially excluded (ibid., 6). Therefore, the incidence of hunger and deprivation can be more due to social exclusion. When these avarna shows concentration, hunger and deprivation may show a regional pattern as well.

Therefore, a transformed social and cultural system and an active, reformative and constructive political party can also be a necessary part to remove hunger and deprivation.

Hunger and deprivation at the societal level may also arise due to external shocks to the state or to the local economy: either through natural calamities which can affect on the societal output, employment and put the seeds of hunger and deprivation at the societal, households and individual level or through displacement in the name of development, which also bring equally undesirable impacts and contributes to hunger and deprivation. The external sources i.e. various natural calamities, are – flood (Satapathy, 1993), drought/famine (Sen, 1981), cyclone/tsunami and earthquake. Keeping Odisha situation in mind, two crucial natural calamities that occur frequently are flood and drought. Though cyclone and earthquake occur occasionally, they leave a huge negative impact in the economy. Therefore, preparedness for natural calamities is also necessary to avoid hunger and deprivation. Also, with the implementation of export promotion zone and special economic zone, external shocks come to the local community through displacement without a proper rehabilitation and resettlement policy, which has a huge social cost by producing hunger and deprivation. The following section deals with the political economy in the context of Odisha.

Political Economy Framework in the Context of Odisha

In a broader sense, political economy indicates a longstanding institutional and structural process that brings changes in a society. The process may occur mainly due to the pattern of the ownership on means of production and the appropriations of the output from production by various groups. In the Marxian political economy, these groups are the property-holding and property-less classes. In India, the ownership of the means of production and appropriations of the output can also be seen by various social groups. However, the class and social groups often show positive relations, i.e. higher the caste, higher is the class and vice versa. Hence, both class and caste dimensions affect the changes in a society. In addition, regional disparities, socio-cultural history, geographical specificity etc. can also affect the changes in a society.

In states like Odisha, the political economy can include a number of variables, such as property distribution across class and across social groups, regional biases of politics and policies, geographical specificity and socio-cultural history. But the crucial factor is the social hierarchy and structure, based on which a group of people have been segregated from the state affairs as well as from state's development process. Incidentally, these groups are also found in particular region of the state. Both these groups together, show a pattern across social groups and regions. The study looks through these specific features of the Odisha's economy and society.

Methodology of the Study

Area of study: The study was carried on in Odisha, which is located in the eastern part of the country between the parallels of 17.49° N and 23.34° N latitudes and meridians of 81.27° E and 87.29° E longitudes. As can be seen from the Map 1.1 (see at the end of the chapter) Odisha is bounded by Bay of Bengal (with about 450 kms of coastline) in the east, West Bengal in the north-east, Jharkhand in the north, Chattisgarh in the west; and Andhra Pradesh in the south. In 2001, the state covers an area of 1,55,707 sq km, accounting for 4.87 per cent of total area of India and 3.6 per cent of country's population (around 36 millions).[25]

Geographically, the state is mixture of both hilly and plain areas. While the eastern region of the state comprises of fertile plain areas, the south-western part is dense with hilly areas. The secondary data analysis in this study is done for all Odisha and the primary surveys were carried on in three districts such as Boudh, Kalahandi and Sambalpur. Boudh district has been chosen for the study of social exclusion as there is higher percentage of SC population in the district (21.88%) compared to 16.53 per cent for the state as a whole. Scheduled Caste population face the maximum social exclusion based on caste. Kalahandi district has been

Map 1.1 *Districts map of Odisha*

chosen for the study of the functioning of NREGA, as Kalahandi is the most backward district. The district also is chosen to see the impact of displacement from a recent development induced displacement (by the Vedanta project). Sambalpur district has been chosen for the study of the functioning of the various institutions that provide social security to reduce hunger and deprivation.[26]

Methods of study: The study followed comparative paradigm to explore and compare hunger, poverty and deprivation. While growth and distribution were examined as causal factors in bringing about hunger and deprivation, social exclusion, displacement and social insecurity aspects are seen as major impediments to alleviation of hunger and deprivation.

For achieving the objectives of the study, both qualitative and quantitative data were utilised.

Data Sources

The study also uses data from both secondary and primary sources. The secondary literature includes books, journals and published magazines in English, Hindi, and Odia languages and different Odia novels, narrating the extent and consequences of hunger in the state.

The major secondary data sources include data and reports published by the Census of India, National Sample Survey Organisation (NSSO), National Family Health Survey (NFHS) and the data provided by the Central Statistical Organization (CSO).

1. Data from the NSSO used are calorie, protein and fat intakes, availability of two square meals and income poverty. Per capita monthly consumption

expenditure, employment, urban-rural residence, region, landholding are also obtained from NSSO.

2. Data from NFHS 1992–93, 1998–99, 2005–06 included various anthropometric indicators, such as body mass index, underweight, stunting and wasting.
3. Data for life expectancy and infant mortality rate are obtained from Sample Registration System. The data sources for housing and educational deprivation are NSSO and Census[27] and the source for assessing clothing and morbidity is NSSO.
4. Data for growth aspects are obtained from Central Statistical Organisation.
5. The Census data and reports were used to obtain information on aspects of education, occupation, geographical and demographic indicators. The old census reports also provide different livelihood patterns of people at different times. Some of the important variables that have been used from the census data are population, gender, social group, location, region, work participation rate, livelihood or occupation, literacy and housing.
6. Data on social exclusion are obtained from the Gazetteer of Odisha of various years. Recent evidences are also considered as sources of data on social exclusion found from different published materials (Praxis, 2001; de Haan, 2005).

Tools of data collection: Keeping in view the objectives in mind, standardised questionnaires as well as open-ended interview schedules were adopted to collect the data. The available secondary data do not provide enough scope for addressing questions on social exclusion. Further, it is those regions where maximum share of the SC and ST population of the state reside that indicate high proportions of hunger and deprivation. For obtaining qualitative data on social exclusion, interview method was used. About 20 interviews were conducted to understand social exclusion in Odisha and its association with hunger and deprivation. Kundamal village (Kantamal block, Palasagora gram panchayat) in Boudh district was chosen as site for the study. Boudh district was chosen since 21.88 per cent of the population is Scheduled Caste according to the Census of India, 2001. The Ghansi community, one of the 'untouchable' communities in Odisha, is examined in detail.

Interview method was used to understand the impact of development-induced displacement and effectiveness of the resettlement and rehabilitation policy on hunger and deprivation in Kalahandi district of Odisha. Kalahandi is one of the most backward districts of Odisha in many respects (50.3% agricultural labour and about 50 per cent SC/ST population). The Vedanta project in the Lanjigarh block of Kalahandi district has led to displacement of about 300 households covering four villages in the region. It is significant to observe this because Kalahandi is reeling from drought in the previous years and there is a prevailing condition of hunger and deprivation.

Also, a survey using a questionnaire method was conducted in the Sason Block of the Sambalpur district to examine social security aspects including functioning of the various institutions.

The functioning of the NREGA was examined through interviews conducted in the Balangi village (Badachatrang Gram Panchayat of Thuamul Rampur block) in Kalahandi district. About 15 households were interviewed.

Preparation of Questionnaire and Interview Schedule

All the questionnaires were prepared and developed based on need and issues. In addition to this, interview schedule was prepared for the household and individual separately. Officials and personnel who are involved and associated with poverty alleviation, displacement and social security programmes were administered the interview schedules.

Data Analysis

As the data analysis is concerned, simple bivariate and multivariate tabulations (descriptive statistics) have been used. For trend analysis, the compound growth rate has been used.

Issues in Data Collection

While collecting data, we have confronted with some of the problems. For information on historical aspects of hunger and deprivation, historical data for the entire state was not available. With respect to NSSO data also, there were some limitations. The sample size is not large enough for a disaggregated analysis. Given the constraint, the analysis was done by merging some subsectors.

In primary survey too, we have confronted some problems, such as interview time (either in morning or evening), lack of communication and lack of transport to certain villages. The respondents were also afraid to provide information especially about government programmes and aspects of social exclusion. In most of the cases, the respondents provided approximate figures than exact figures.

Organisation of the Study

The chapters in this study are organised as follows: The first chapter presents the introduction and association of poverty, hunger and deprivation from various perspectives including measurement aspects. This chapter also presents objectives, theoretical framework, methodology and sources of data utilised in the study. Chapter two, titled 'Structure of Hunger and Deprivation' presents growth and structure of hunger and deprivation in Odisha. It also provides insight into the background factors (the social, political, economic and geographical factors) that might have contributed to the existing hunger and deprivation. It also traces the political economy framework in the context of Odisha.

Chapter three, titled 'Slow and Lopsided Structural Transformation and its Impact', presents the nature and rate of growth (output and employment) and their association with hunger and deprivation in Odisha. Chapter four, titled 'Absence of Distributive Justice: Access to Land and Related Resources in a Comparative Perspective' presents the trend and pattern of agrarian structure and its relation with hunger and deprivation.

While, the political economy aspects of hunger and deprivation in the state is examined by looking into the growth and distribution questions, the social and cultural aspects, which also contribute to the hunger and deprivation in Odisha, has been examined by looking into the social exclusion phenomenon in the state in Chapter five.

The sixth Chapter titled 'Public Provisioning to Reduce Hunger and Deprivation' examines the failure of public actions (public provisioning, i.e. distribution of the essential commodities) to reduce hunger and deprivation in Odisha.

Chapter seven, presents the external sources of hunger and deprivation in the state, where it tries to cover drought, flood and cyclone aspects.

Chapter eight, 'Coping with External Shocks: Displacement in the Name of Development' looks into the impacts of displacement as a reason of hunger and deprivation in the state. Chapter nine, concludes the study.

Significance and Limitations of the Study

The significance of the study lies more on its empirical contributions than its methodological improvements. The methodology adopted here is a revised and refined one. The study locates itself in Odisha, where incidence of hunger and deprivation is high. Though, there have been research studies in hunger and deprivation in the all-India context and in some Indian states, yet Odisha presents an entirely different picture. Therefore, there is a need to explore some of the pathways to improve conditions to prevent hunger and deprivation.

Secondly, the study adopted an interdisciplinary approach to study the problem of hunger. Further, theoretical framework adopted is not merely a derived one from the existing literature, but rather the socio-cultural, political and the economic history of Odisha is incorporated in the framework.

Thirdly, the use of the term 'value', although has been used frequently in other fields of economics, in poverty literature it has been used rarely. In the present study, the value of the labour power of certain communities has been always dependent on the demand of the some others' better turnover. Hence, the application and interpretations of such terms (value, derived demand, elasticity of demand and specialisation of occupations, etc.) in the study of hunger and deprivation provide a newer dimension of interpretation and possibly helps to understand the processes better.

Despite this significance, the study also has some implicit limitations. Hunger and deprivation itself is the result of many factors prevailing in a society such as economic, socio-cultural, topological and political history. Hence, the problem itself requires attention from other disciplines as well. Therefore, an interdisciplinary approach has been adopted here. Understanding a society and economy, especially for which not enough literature and historical facts are available is not an easy task. Given the limited literature, facts and figures, we have taken utmost care to prevent erroneous inferences. On the one hand, time, space and money constraints and on the other hand, the interest to understand the problem in all its manifestations have brought about a rich but complex perspective. Both the internal and external sources of hunger and deprivation have been discussed to an extent here, which is not the end but perhaps the beginning.

Figure 1.1 *Flowchart of hunger and deprivation*

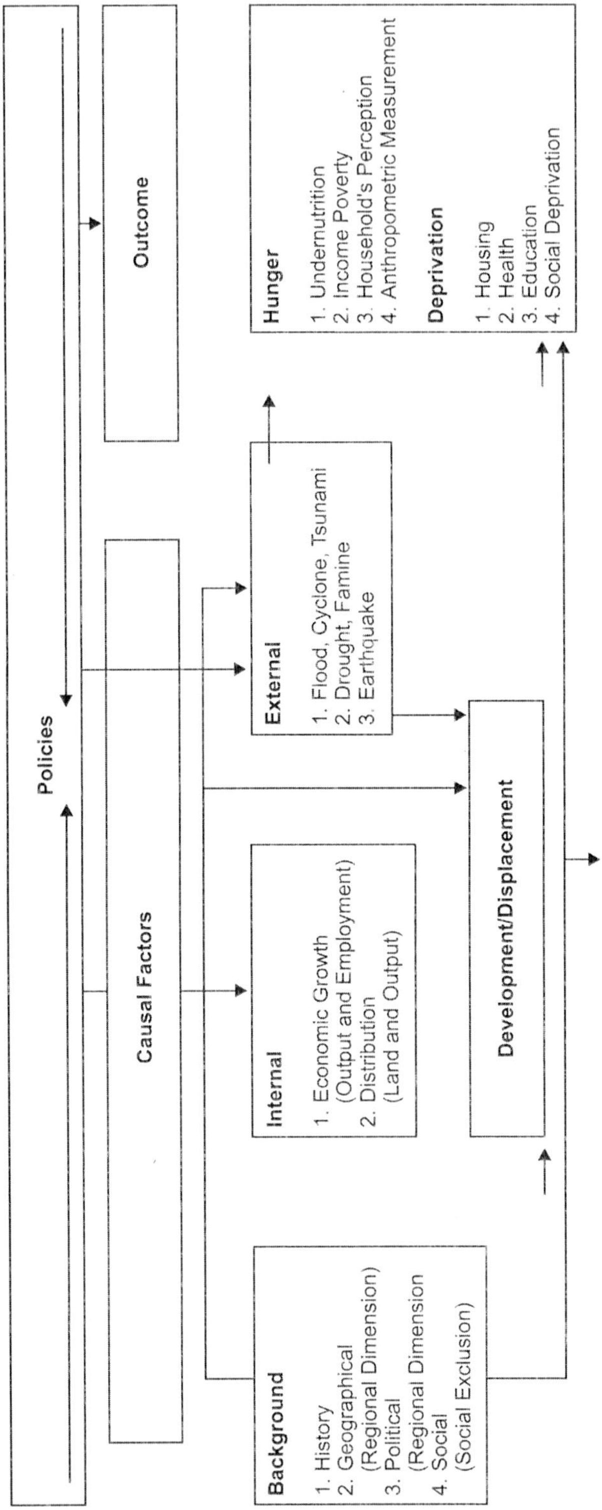

Notes

1. The study is limited to involuntary hunger, i.e. those who remain hungry.
2. As Sen (1989) has specified chronic hunger involves sustained nutritional deprivation on a persistent basis.
3. Anthropometry is the science of measuring the human body and its parts. As the literature shows, (Foster and Leather, 1999: 55), human physical growth (increase in size) and development (differentiation) are both responsive to variations in calorie and protein intakes. If a particular human body does not achieve a satisfactory growth, then one reason could be the deficiency of adequate protein and calorie, for example, the low Body Mass Index (BMI, weight in kilograms divided by the height in metres square).
4. But the poor BMI also can be due to poor health status, which again is mostly due to inadequate calories intake.
5. Sen considered the work of the classical political economy of Adam Smith, Robert Malthus, David Ricardo, Karl Marx, John Stuart Mill and others are interdisciplinary because of their larger scope and coverage of the economy compared to the dominant traditions of the economy of later part of 20th century. However, it may be helpful to note that in economic literature, there are two distinct schools of thought, first the classical political economic, which includes the thoughts of Smith, Malthus, Ricardo, Marx, Mill and others. And the other is neo-classical political economy, where the demand and supply play dominant role for any economic analysis.
6. The Nobel prize winner Robert Fogel (in 1994) pointed out, food shortages were so severe in Europe in the eighteenth and early nineteenth century that on one hand the bottom 20 per cent subsisted on such poor diets that they were effectively excluded from the labour force and on the other hand he estimated the improved nutrition accounts for 30 per cent of the growth in income per capita in Britain between 1790 and 1980.
7. Details figures on per capita SDP, growth rates and trends, etc. have been shown in the chapter three.
8. During 2004–05, the poverty ratios in the coastal, northern and southern Odisha are 29.82, 56.59 and 71.27 per cent respectively.
9. Dreze and Sen (1989) stated lack of adequate food might lead to higher morbidity. Kannan et al., (1991) found the inverse relationship between the rate of morbidity and socio-economic status within Kerala. However, for detailed examination the present study is confined to first four indicators as the anthropometrics indicators can also capture the health aspect to an extent.
10. The kind of deprivation discussed here is common to all but social exclusion (of castes, class, gender) is another aspect of deprivation (specific to a group), which can also cause hunger has been discussed later.
11. Malnutrition could be caused either by underconsumption or by overconsumption of any essential nutrient. The underconsumption of essential nutrient could be considered as hunger. There are four types of malnutrition, such as overnutrition, secondary malnutrition, dietary deficiency or micronutrient malnutrition and protein-calorie malnutrition. Overnutrition occurred out of overconsumption, which is mainly the problem of high-income countries or high-income group. Whereas, undernutrition occur mainly out of under consumption of calorie, protein and micronutrient. However, overnutrition is not a bigger problem for the developing countries rather, the most common types of malnutrition faced by the developing countries like India and especially by developing state like Odisha is undernutrition, which arises out of inadequate consumption of specified diet.

12. Consumer unit is the rate of equivalence of a normal person determined on the basis of age-sex compositions of a person. From the 26th round, NSSO has been using a level of 2,700 kcal per consumer unit per diem as a standard.
13. Same observations have been made by Planning Commission, 2004; Deaton and Dreze, 2004.
14. There are four kinds of malnutrition (see Foster and Leather, 1999), such as: undernutrition, Proteins Calorie malnutrition and micronutrients malnutrition and overnutrition. Among these, the first three are our concern.
15. BMI data was not collected in the first round of NFHS, 1992–93.
16. The free education aspects in this regard may not make a special case for Odisha as it is same for all other Indian states.
17. A flowchart is presented (Figure 1, given at the end of the chapter) to indicate the linkages of hunger and deprivation with other factors.
18. For agricultural growth and poverty reduction see, Ahluwalia, 1978; Chadha and Sharma (1997). However, while bringing the growth dimensions, the displacement effects often arises, but given the scope of the study the displacement issues have not been addressed here adequately.
19. The net per capita availability also included the food grains received by the state from different social security schemes.
20. However, in the long run why an economy fails to achieve a sound structural transformation is a question of public expenditure, which requires a special attention.
21. For arguments in favour of land distributions on poverty reduction please see Raj, 1974; Kumbhar 2006.
22. Kumbhar's study is based on the functioning of only one scheme – the public distribution system.
23. In the Keynesian thought, the situation of hunger and deprivation may occur due to failure of effective demand. That means, there may be adequate food available in a region but people may die out of hunger due to lack of purchasing power with them. According to Sen, the situation of hunger and starvation death may occur when an individual lacks exchange entitlements (Sen, 1981: 154). Exchange entitlements may be in the form of physical assets or may be in the form of knowledge or skill, which can enhance person's capability to earn higher income and reduce his/her critical poverty.
24. Varna refers to the four universal unchanging categories of the Hindu hierarchy – Brahmin, Kshtriya, Vaisha and Sudra. Jati refers to smaller, hereditary, endogamous groups associated with a traditional occupation and related to one another in terms of ritual pollution or purity.
25. However, in the pre-independence period, the geographical boundary of the state is not that distinct as it was fragmented into different parts of India when different rulers ruled over the state, especially after 1568 A.D. when Odisha lost her last independent king. A brief history of the state has been presented in Annexure A, B and C.
26. Sambalpur district was chosen because it represents medium level of development in Odisha.
27. Data on housing is available from both the sources but while Census data is based on decadal survey the NSSO data is based on quinquennial rounds survey. Hence, depending on different purpose, both the data sources will be used.

2

Structure of Hunger and Deprivation in Odisha

This chapter presents the structure of hunger and deprivation within the state of Odisha. It found a high level of hunger and deprivation with a large variation across different social groups, regions and location of the state. These variations are explained through various historical factors of the state. In Odisha, the political economy aspects of the problem has been seen through the history of the accumulating and de-accumulating groups; through higher and lower strata of the society and their access to wealth, health, education, housing; and through gender dimension. The chapter comprises of three sections. The first section examines the structure of hunger and deprivation within Odisha. The second section explains the findings through various historical factors of the state such as the economy, society and topography. Section three, summarises the chapter.

Structure of Hunger in Odisha

Hunger in Terms of Income Poverty

As the aim is to examine the structure and magnitude of hunger, we have explored the recent poverty estimates from 61st NSSO Round Monthly Per Capita Consumer Expenditure (MPCE). The estimated head count ratio across different variables such as region, social groups, types of household, types of occupation, landholding classes have been presented in Table 2.1. If we compare various groups, the incidence of poverty is the highest among casual labourers residing in urban areas (89%) followed by STs population (75%), Southern region (71%) and agricultural labour residing in rural areas (65%). At the other extreme, the incidence of poverty is the lowest among regular/salaried people in urban areas (21%), higher castes (25%), and coastal areas and households owning more than 4 acres of land (30% each).

Table 2.1 shows that there is a huge difference in the head count ratio of poverty across the regions of Odisha. The incidence of poverty is the highest in Southern Odisha (71%) followed by Northern Odisha (56%) and coastal Odisha (29%). The difference between rural and urban head count ratio at the aggregate level is minor (46% in rural area and 44% in urban area).

Table 2.1 *Distributions of below poverty line population by their socio-economic characteristics across the regions of Odisha during 2004–05*

	Coastal	*Northern*	*Southern*	*Total*
Based on place of location				
Rural	27.39	59.07	72.66	46.9
Urban	44.11	42.9	55.03	44.72
Total	29.82	56.59	71.27	46.61
Based on social group				
ST	66.76	71.99	82.92	75.25
SC	36.91	66.38	67.46	52.63
OBC	27.23	47.09	65.53	38.32
Other	22.45	30.28	37.53	25.37
Total	29.82	56.59	71.27	46.61
Based on types of households (Rural)				
Self employed in Non-agriculture	19.94	44.84	57.65	32.78
Agricultural Labour	34.02	78.45	88.06	64.54
Other Labour	33.1	57.05	76	51.95
SE in Agl	30.83	55.32	69.58	46.11
Other	15.42	22.31	34.89	19.78
Total	27.36	59.07	72.66	46.9
Based on types of households (Urban)				
Self-employed	50.66	47.42	56.99	50.16
Regular/Salaried	18.85	21.8	24.57	20.9
Casual Labour	85.43	89.89	97.53	89.25
Other	19.05	23.94	61.86	26.38
Total	44.01	42.9	55.03	44.67
Number of Observations				
Total	2,268	1,755	1,000	5,023
Based on land possession (hectares)				
Land less	30.4	32.56	34.48	31.85
0.001 to .4	30.78	54.95	65.25	44.48
0.41 to 1.0	29.65	57.42	82.81	47.8
1.001 to 2.0	30.79	64.82	70.6	51.13
2.001 to 4.0	17.15	45.34	81.18	42.14
> 4.001	11.97	30.24	63.24	29.65
Total	29.59	55.95	70.22	45.81
Number of observations for land possession				
Total	2,220	1,694	953	4,867

Notes: In land possession data 156 cases are excluded in the data itself. Coastal-Balasore, Bhadrak, Cuttack, Gajpati, Ganjam, Jagatsinghpur, Jajpur, Khurda, Nayagarh, Puri, Kendrapada, (Total 11); Northern – Angul, Bargarh, Balangir, Deogarh, Dhenkanal, Jharsuguda, Keonjhar, Mayurbhanj, Sambalpur, Sonepur, Sundargarh; (Total 11); Southern – Boudh, Kalahandi, Koraput, Malkangiri, Nuapada, Nawarangpur, Kandhamal, Rayagada (Total 8), Grand Total – 30.

Source: NSSO unit level data, 1993–94 and 2004–05.

At the regional level, there exists substantial differences. The magnitude of rural poverty remained higher in South (72%) and North (59%) Odisha compared to their urban counterpart (55% and 42% respectively). The situation is reverse in coastal Odisha. While the magnitudes of rural head count ratio was 27 per cent, the same for urban area is 44 per cent.

Across the social groups also, there exists substantial differences in the incidence of poverty. The head count ratio across social groups varies from 25 per cent among 'other' category of social group (higher caste) to 75 per cent among STs and 52 per cent among SCs and 38 per cent among OBC. These variations also pointed out that, irrespective of the place of living, the SC/ST are more prone towards poverty than OBC and 'other' category of social groups. On the whole, irrespective of region, as we move in the social ladder from STs to SCs to OBC to other category of social groups, the incidence of poverty declined sharply.

Across occupational groups in rural areas, the magnitude of head count ratio remains highest among the agricultural labour (64%), followed by 51 per cent among 'other' category labour, 46 per cent among self-employed in agriculture, 32 per cent among self-employed in non-agriculture and 19 per cent among others. A point to be noted here is that the magnitude of head count ratio, among the agricultural labour households of southern Odisha and northern Odisha is very high, 88 per cent and 78 per cent respectively.

In urban areas too, there exists substantial differences. The head count ratio varies from 89 per cent among the casual labour to 50 per cent among the self-employed to 26 per cent among 'other' and 20 per cent among the regular and salaried types of households. Here too, the incidence of poverty among the casual labour remains very high in southern part of Odisha (97%), compared to 89 per cent in northern Odisha and 85 per cent in coastal Odisha.

As the land possession and head count ratio is concerned, the relationship is not that sharp, except that large landowners (>4 acres) in the coastal and northern region exhibit relatively lower incidence of poverty. Households are below the poverty line even with adequate land possession with them, which might be due to low land productivity (discussed in Chapter 4).

Hunger in Terms of Anthropometrical Indicators

Boxes 2.1 to 2.3 present the nutritional status of children using anthropometric indicators in Odisha across rural and urban location, regions, caste, standard of living, occupations and education (see appendix A2.1 to A2.3 for detailed results). The unit level data has been extracted and analysis has been carried out.

Box 2.1
Stunting (height per age)

	Moderate Nutritional Status	*Severe Nutritional Status*
Rural/Urban	Children from urban Odisha are more disadvantaged with 44.8 per cent of them are moderately stunted compared to rural Odisha (37%).	Children from rural Odisha are more disadvantaged with 18 per cent of them are severely stunted compared to urban Odisha (14%).
Subregions	Southern region is more disadvantaged with 47 per cent of the children are moderately stunted compared to around 43 percentage in northern and coastal Odisha.	Coastal Odisha is relatively disadvantaged with 19.43 per cent of the children are severely stunted compared to 17.1 and 15.3 per cent in southern and northern Odisha respectively.
Caste	SC children are most disadvantaged with 50 per cent of them are moderately stunted, which is followed by children from STs (49.6%), OBC (44.28%) and other caste (32.24%).	SC children are most disadvantaged with 22.84 per cent of them are stunted, which is followed by children from ST, (19.71%) OBC (18.29%) and other caste (9.84%).
	In both the cases (moderate and severe) as we move from SCs to STs to OBC to other category of social group, the nutritional status of the children becomes better.	
Standard of living	Children from low standard of living households are most disadvantaged with 50.6 per cent of them are moderately stunted, which is followed by children from medium standard of living households (38.7%), and high (21.2%) standard of living households.	Children from low standard of living households are most disadvantaged with 20.4 per cent of them are severely stunted, which is followed by medium (16.4%), and then high (4.5%) standard of living.
Based on occupation of mother	Children, whose mother works in non-agricultural occupations are most disadvantaged with 54.2 per cent of them are moderately stunted, which is followed by children whose mothers are non-workers with 42.1 per cent, and that is followed by children, whose mother works in agriculture (38.2%).	Children, whose mother works in non-agricultural occupations are most disadvantaged with 23.1 per cent of them are severely stunted, which is followed by children whose mothers are non-workers with 16.8 per cent, and that is followed by children whose mother works in agriculture with 12.5 per cent.
Based on education of mother	Children from illiterate mothers are most disadvantaged with 50.1 per cent of them are moderately stunted, which is followed by children from primary school educated mother (42.1%) and children from high school educated mother (31.1%) and the least moderated stunted children was found among mothers with secondary or more educated (12.8%).	Children from illiterate mothers are most disadvantaged with 22.5 per cent of them are severely stunted, which is followed by children from primary school educated mother (12.8%) and children from high school educated mother (9.9%) and the least moderated stunted children was found among mothers with secondary or more educated (2.1%).
Based on occupation and education of father	Child's nutritional status based on his/her father's education or occupational status followed the similar pattern as it was in the case of mother's education and occupation.	

Box 2.2

Wasting (Weight per age)

	Moderate Nutritional Status	*Severe Nutritional Status*
Rural/Urban	Children from rural Odisha are more disadvantaged with 55.5 per cent of them are moderately wasted compared to urban Odisha (45.3%).	Children from rural Odisha are more disadvantaged with 21.2 per cent of them are severely wasted compared to urban Odisha (16.6%).
Subregions	Northern region is more disadvantaged with 60.2 per cent of her children are moderately wasted compared around 57 per cent in southern and 49 per cent in coastal Odisha.	Southern Odisha is relatively disadvantaged with 26.3 per cent of her children are severe wasted compared to 23 and 16.6 per cent for northern and coastal Odisha respectively.
Caste	SC and STs children are most disadvantaged with 59 per cent of them are moderately wasted, which is followed by OBC (56.59%) and other (42.95%) caste.	ST children are most disadvantaged with 26.43 per cent are severely wasted, which is followed by SCs (24.31%) OBC (20.73%) and other castes (12.17%).
Standard of living	Children from low standard of living households are most disadvantaged with 62.01 per cent of them are moderately wasted, which is followed by medium 48.88 per cent, and high (27.93%) standard of living.	Children from low standard of living households are most disadvantaged with 25.98 percentages of them are severely wasted, which is followed by medium 15.47 per cent, and then high (9.91%) standard of living.
Based on occupation of mother	Children, whose mother works in non-agricultural occupations are most disadvantaged with 63.23 per cent of them are moderately wasted, which is followed by children whose mothers are non-workers with 53.15 per cent, and that is followed by children, whose mother works in agriculture (47.25%).	Children, whose mother works in non-agricultural occupations are most disadvantaged with 26.01 per cent of them are moderately wasted, which is followed by children whose mothers are non-workers with 19.81 per cent, and that is followed by children, whose mother works in agriculture, 17.58 per cent.
Based on education of mother	Children from illiterate mothers are most disadvantaged with 60.9 per cent of them are moderately wasted, which is followed by children from primary school educated mother (50.6%) and children from high school educated mother (45.1%) and the least moderated wasted children was found among mothers with secondary or more educated (17.0%).	Children from illiterate mothers are most disadvantaged with 26.0 per cent of them are severely wasted, which is followed by children from primary school educated mother (15.3%) and children from high school educated mother (13.1%) and the least moderated wasted children was found among mothers with secondary or more educated (8.5%).
Based on education of father	Child's nutritional status based on his/her father's educational status followed the similar pattern as it was in the case of mother's education. But in case of occupation, it is slightly different.	
Based on occupation of father	Children, whose father works in non-agricultural occupation are most disadvantaged with 60 per cent of them are moderately wasted, which is followed by children whose father works in agricultural occupations with 48.1 per cent, and that is followed by children whose father does not work, with 28.6 per cent.	Children, whose father works in non-agricultural occupation are most disadvantaged with 24.7 per cent of them are moderately wasted, which is followed by children whose father works in agricultural occupations with 15.4 per cent, and that is followed by children whose father does not work, with 10.7 per cent.

Box 2.3
Under weight (height per age)

	Moderate Nutritional Status	*Severe Nutritional Status*
Rural/Urban	Children from rural Odisha are more disadvantaged with 24.4 per cent of them are moderately under weight compared to urban Odisha (23.9%).	Children from rural Odisha are more disadvantaged with 4 per cent of them are severely underweight compared to urban Odisha (3.6%).
Subregions	Northern region is more disadvantaged with 31 per cent of her children are moderately under weight compared to around 27.8 per cent in southern and 17.65 per cent in coastal Odisha.	No or negligible sample size.
Caste	ST children are most disadvantaged with 30.47 per cent are moderately underweight, which is followed by OBC (24.57%) SCs (23.10%) and other castes (19.8%).	ST children are most disadvantaged with 5.71 per cent are severely underweight, which is followed by OBC (4.63%), SCs (3.11%) and other castes (2.61%).
Standard of living	Children from low standard of living households are most disadvantaged with 28.73 per cent of them are moderately underweight, which is followed by medium (19.73%), and high standard of living (16.22%).	Children from low standard of living households are most disadvantaged with 5.03 per cent of them are severely underweight, which is followed by medium (2.69%), and then high (2.68%) standard of living.
Based on occupation of mother	Children, whose mother works in non-agricultural occupations are most disadvantaged with 32.1 per cent of them are moderately under weight, which is followed by children whose mothers are non-workers with 23.1 per cent, and that is followed by children, whose mother works in agriculture, 18 per cent.	Children, whose mother works in non-agricultural occupations are most disadvantaged with 4.4 per cent of them are severely underweight, which is followed by children whose mothers are non-workers, with 4.3 per cent.
Based on education of mother	Children from illiterate mothers are most disadvantaged with 26.9 per cent of them are moderately underweight, which is followed by children from high school educated mother (23.1%) and children from primary school educated mother (21.1%) and the least moderate wasted children was found among mothers who are secondary or more educated (14.9%).	Children from illiterate mothers are most disadvantaged with 4.8 per cent of them are severely underweight, which is followed by children from primary school educated mother (3.1%) and children from high school educated mother (2.2%) and the least moderate wasted children was found among mothers who are secondary or more educated (2.1%).
Based on education of father	Children from illiterate father are most disadvantaged with 30.4 per cent of them are moderately under weight, which is followed by children from high school educated mother (23.2%) and children from primary school educated mother (20.8%) and the least moderated wasted children was found among mothers with secondary or more educated (17.6%).	Children from illiterate father are most disadvantaged with 5.1 per cent of them are severely underweight, which is followed by children from high school educated mother (3.3%) and children from primary school educated mother (3.9%) and the least moderated wasted children was found among mothers who are secondary or more educated (2.9%).

Cont'd...

...Cont'd

Based on occupation of father	Children, whose father works in non-agricultural occupation are most disadvantaged with 28.7 per cent of them are moderately underweight, which is followed by children whose father works in agricultural occupations with 19.3 per cent, and that is followed by children whose father does not work, with 3.6 per cent.	Children, whose father works in non-agricultural occupation are most disadvantaged with 5.1 per cent of them are severely underweight, which is followed by children whose father works in agricultural occupations, with 2.7 per cent.

Structure of Deprivation in Odisha

Three essential types of deprivation have been dealt with here, such as housing, education and health. The aggregate trends of these variables have been seen in Chapter 1, here we have examined the structures of the same.

Housing Deprivation

For housing conditions, we have used the 2001 census data. Table 2.2 presents the percentages of houses by their conditions across the regions of Odisha (districts level data has been given in Appendix A2.4). Table 2.2 shows that percentages of 'good houses' are higher in coastal region than northern region and then followed by southern region of the state. The district level data depicts that the following: the highest percentages of good houses are found in Khorda (38.17%) followed by Ganjam (33.74%) and then followed by Cuttack (31.78%), whereas the lowest share of good houses has been found in Deogarh (15.04%) followed by Sonepur (19.61%). In terms of dilapidated houses, the highest percentages have been found in Jagatsinghpur (21.98%) followed by Jajpur (18.33%).

Table 2.2 *Percentages distribution of the conditions of houses across the regions of Odisha, 2001*

Regions	*Total*	*Good*	*Livable*	*Dilapidated*
Odisha	75,56,918	25.64	64.41	9.95
Southern	14,75,417	22.91	72.23	4.86
Northern	26,66,755	24.10	68.03	7.87
Coastal	34,14,746	25.97	60.42	13.60

Source: Census of India, 2001.

Educational Deprivation

Table 2.3 presents the level of literacy rate across the regions of Odisha (districts level data has been given in Appendix A2.5) to show the educational deprivation.

Both female and male literacy rates are the lowest in southern Odisha (28 and 56% respectively) and the corresponding rates are the highest in coastal Odisha (60 and 82% respectively). It is also interesting to note that the gender-gap in literacy is the highest in southern Odisha and the lowest in coastal Odisha. Rural literacy rate is much lower than the urban literacy rate, regardless of regions. However, female literacy is the lowest in the rural areas of southern Odisha (24%).

Table 2.3 *Literacy rates across the regions of Odisha, 2001*

	Total			*Rural*			*Urban*		
Regions	*P*	*M*	*F*	*M*	*F*	*T*	*P*	*M*	*F*
Coastal	70.8	81.7	59.6	69.0	80.6	57.3	80.5	87.4	72.9
Northern	63.2	76.1	49.8	60.1	73.9	46.1	79.6	87.6	70.6
Southern	41.8	56.2	27.5	38.3	53.0	23.9	74.8	84.5	64.5

Source: Census of India, 2001.

The district level analysis shows the following: The overall literacy is the highest in Khorda (80%) followed by Jagatsinghpur (79%), Puri (78%) and Cuttack (77%). The lowest level of literacy is seen in Malkanagiri (30%), Nabarangapur (34%), Koraput (36%) and Rayagada (36%). The level of literacy is low in rural areas and among females. The female literacy in rural areas is the lowest in Nabarangpur (20.7%) followed by Malkanagiri, (20.9).

Health Deprivation

An indicator that has been considered to analyse health deprivation is the infant mortality rate. The district-level Infant Mortality Rate (IMR) for 2001 is presented in Table 2.4. The Table shows that IMR is the highest in Phulbani district (186) followed by Gajapati (165), Rayagada (163), Malkanagiri (158), Kalahandi (157), Nabarangapur (156), and Koraput (153) and the IMR is low in Mayurbhanj (91), Jharsuguda (96), Angul (109), Bargarh (110) and Sonepur (110). A noticeable point that emerges is that the IMR has remained high among the districts of southern Odisha.

Table 2.4 *Infant mortality rate (under five years) across the districts of Odisha, 2001*

Districts	*IMR 2001*	*Districts*	*IMR 2001*	*Districts*	*IMR 2001*
Phulbani	186.5	Puri	135.9	Bolangir	119.6
Gajapati	165.1	Kendrapara	134	Jagatsinghpur	117.4
Rayagada	162.9	Bhadrak	130.6	Balasore	116.2
Malkanagiri	158.3	Nayagarh	130.3	Sundergarh	114.9
Kalahandi	157.4	Dhenkanal	130	Khurda	114
Nabarangapur	156.2	Keonjhar	129	Sonepur	109.9
Koraput	153.4	Cuttack	128.1	Bargarh	109.6
Deogarh	148.5	Ganjam	127	Angul	109
Boudh	143.3	Jajpur	125	Jharsuguda	95.6
Nuapada	135.9	Sambalpur	122.2	Mayurbhanj	90.6

Source: www.iipsindia.org.

As discussed above regarding the various indicators of hunger and deprivation, here are some salient overall findings:

The incidence of hunger and deprivation is more prominent among SC/ST. It is found that irrespective of the place of living, the SC/ST population is more prone towards poverty than OBC and 'other' category of social groups. Hence, study of the social structure of the state is necessary to understand the dynamics of the problem.

The region-wise analysis shows that the coastal region is much better-off in terms of hunger and deprivation and the southern region is worse-off. This can also be due to less proportion of SC/ST population in coastal region and more proportion of SC/ST in the southern region.

Given the high initial level of hunger and deprivation, there is a tardy progress in reducing the same. Hence, role of the state is more essential.

To explain the higher incidence and variations, we have examined below certain historical factors of the state.

Explaining Through Historical Factors

The existing hunger and deprivation in the state is seen from four broad historical factors: (a) geography and topology of the state, (b) economic structure in the state, (c) social structure and (d) political structure. The following sections examine these factors in Odisha's economy.

Geography and Topology of the State and the Regional Dimension of the Hunger and Deprivation in the State

Geographically and topologically, the state as a whole is a mixture of both hilly and plain areas. While the coastal region of the state comprises of fertile plain areas, the south-western and some of northern part is dense with hilly areas. Due to this type of topology, while the south-western and some of northern Odisha are affected with drought, the coastal Odisha gets affected with flood. As the coastal Odisha is close to sea, it also gets affected with cyclone. Such topological situation might be leading to more natural calamities in the state.[1] This could be also the reason that the total cropped area in the state remained low even during pre-independence era and hence might have contributed to the existing hunger and deprivation in the state.

Table 2.5 presents the land utilisation pattern in Odisha, vis-à-vis all-India for 1945–46 and 1955–56. Two major points can be drawn from the Table: (i) at the initial period (i.e., 1945–46 to 1955–56) itself, the total cropped area of India has remained roughly 7 to 10 percentage points higher than Odisha, (ii) though the area sown more than once is low in both cases, the level of all-India remained almost always higher than Odisha, (iii) there is no substantial changes in the land utilisation pattern over the period of ten years.

There also prevailed variation across the regions of Odisha in the land utilisation pattern. As Table 2.6 shows, the gross cropped area remained higher in coastal part of Odisha (52.21%) compared to southern Odisha (37%) and northern Odisha (33.94%), which might be the reason why hunger and deprivation in southern Odisha is higher compared to coastal Odisha (for district level variation see Appendix A2.6).

Table 2.5 *Land utilisation pattern in Odisha and India (figures are in percentages)*

Year	*Total area**		*Under forest*		*NAC*		*CWOFL*		*Fallow land*		*Net sown area*		*Sown more than once*		*Total cropped area*	
	ORI	*IND*	*ORI*	*IND*	*ORI*	*IND*	*ORI*	*IND*	*ORI*	*IND*	*ORI*	*IND*	*ORI*	*IND*	*ORI*	*IND*
1945–46	100	100	12.94	14.44	32.82	17.37	16.01	15.77	6.21	9.69	32.03	42.57	3.98	6.07	36.01	48.64
1955–56	100	100	23.21	17.45	15.43	16.45	17.00	13.48	7.58	8.40	36.96	44.22	2.97	6.16	39.92	50.39

Notes: Total Area is according to village papers, NAC – Not available for Cultivation, CWOFL – Cultivable Waste other than fallow land, NSA – Net sown area.

Source: Statistical Abstract of India, 1955–56.

Table 2.6 *Land utilisation pattern across the districts of Odisha, 1955–56*

Districts	*Total area**	*Under forest*	*NAC*	*CWOFL*	*Fallow land*	*NSA*	*Sown more than once*	*Total cropped area*
Coastal	100	16.21	12.34	17.55	9.02	44.92	7.29	52.21
Northern	100	29.84	11.47	17.98	7.69	32.81	1.13	33.94
Southern	100	19.87	23.51	14.38	6.28	35.93	1.83	37.74

Source: Statistical Abstract of Odisha, 1955–56.

The agro-climatic conditions were neither conducive for agriculture, nor consistent in the state and across the districts of the state. Figure 2.1 presents the annual rainfall and its variability from 1931 to 1950 in Odisha. The figures show a high fluctuation over the time. It ranges from 47.98 inches in 1935 to 74.36 in 1936. The coefficient of variation of rainfall (across districts) was 7.61 per cent in 1932 and this increased to 19.59 per cent in 1941 (in relation to mean rainfall). Such fluctuations might have affected state's economic situations.

Figure 2.1 *Annual rainfall and its variability in Odisha, 1931–1950*

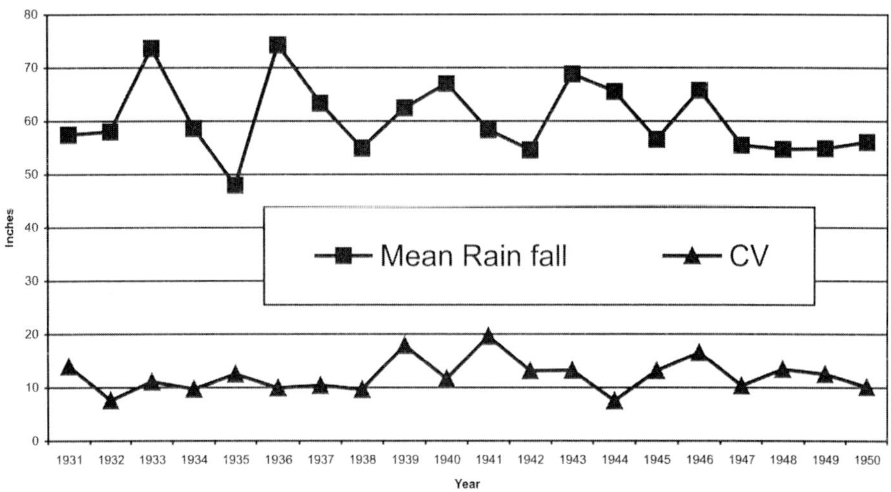

CV – Coefficient of Variation

Source: Census of India, 1951, Volume XI, Odisha, Part-I Report.

The district level situation is presented in Appendix A2.7 and A2.8. It shows, as the level of normal rainfall was concerned, during the periods, Ganjam received least rainfall, i.e. 50.22 inches followed by Bolangir, Kalahandi and Dhenkanal. As the variability of rainfall is concerned, the coefficient of variations of rainfall remained highest in Phulbani, Bolangir and Kalahandi. The low rainfall with high fluctuations might have caused drought or flood in the districts.

As the agro-climatic conditions from 1931 to 1950 depicts, three districts (Kalahandi, Phulbani and Balangir) were prone towards the conditions of either drought or flood (according to variability and mean annual rainfall). On the other hand, Cuttack, Puri and Balasore, where the variability of rainfall is around

15 per cent of the 'mean' rainfall and the 'mean' rainfall is around 61, 62 and 60 inches respectively may be more prone to flood as these districts are having plain area and geographically located in the coastal area. In either case, lack of preventive measures, may lead to non-conducive agrarian climate with regards to rainfall and hence may affect employment and output and may lead to a situation of hunger and deprivation.

This geographical and topological disadvantage might have been the perennial features and might have affected the economic base of the state in the absence of effective policy. The following section looks into the economic structure of the state.

Economic Structure and the Hunger and Deprivation in the State

To explain the existing hunger and deprivation in Odisha (in a historical perspective), the land revenue system which deals with the growth and distribution of wealth in the economy is important. The pre-independence Odisha was governed by four kinds of land revenue system: These are:

(i) the coastal Odisha was under Zamindari land revenue system;
(ii) the western Odisha was under Mahalwari system;
(iii) the southern Odisha was under mainly Zamindari but partly Ryotwari system as well; and
(iv) the feudatory states were initially having separate kinds of land revenue system[2] (since, 1905).

These situations arose in the state after Odisha lost her last independent king during 1568 and especially after the fuller occupation of Odisha by the British. These different forms of land revenue systems have their own consequences in producing hunger and deprivation in the state.

One of the crucial reasons behind the deterioration of the land revenue system in the state is that the Britishers, who took over Odisha had least knowledge on the previous land revenue system that operated in Odisha (see Annexure B for more). As a result, the mixture of several systems led to different consequences, which apparently produced hunger and deprivation.

Economic Consequences of Different Land Revenue Systems

Some of the economic consequences that different land revenue systems in Odisha brought about were (i) Indebtedness, (ii) Uncertainty, (iii) Force commerce and distress migration, (iv) Dadan system, (v) Absentee landlordism, (vi) Cultivable fallow land, (vii) Bethi and Gothi system, (viii) Share cropping system, and (ix) Low yield rate.

Indebtedness

'One of the most important consequences of the British system of land administration in Odisha was that the bulk of the peasantry was more or less perpetually indebted by the end of 19th century' (Chaudhury, 1981). Estimates made during 1890–1930 showed that, 75 to 80 per cent of the total agricultural population was in debt. Not only that they were compelled into market relations through the debt mechanism, but finally they had to escape from the state to far distant places.

Uncertainty

Uncertainties had been brought into the position of ryots with the introduction of British land revenue system. In the earlier system, the rent was paid in kind and proportionately related to output, which was replaced by the cash and a fixed quantity of payment of land revenue. This has brought two types of uncertainties for the ryots. First, there were uncertainties in the output due to natural calamity and its impacts on the ryots. Even if there would be failure in crops, then the ryots have to pay a fixed amount of rent. Otherwise, their status as a ryot may be more volatile. Secondly, if there is fluctuation in the commodity market then also the situations of some ryots, especially those who are relatively less susceptible to market forces may become volatile and weak. The price fluctuations of rice within a year have been presented in Figure 2.2. In both the cases, the marginal ryots may get more affected. Such distress situations brought two outcomes in the Odisha's society such as force commerce and Dadan system.

Figure 2.2 *Price fluctuation of rice, 1919–1930*

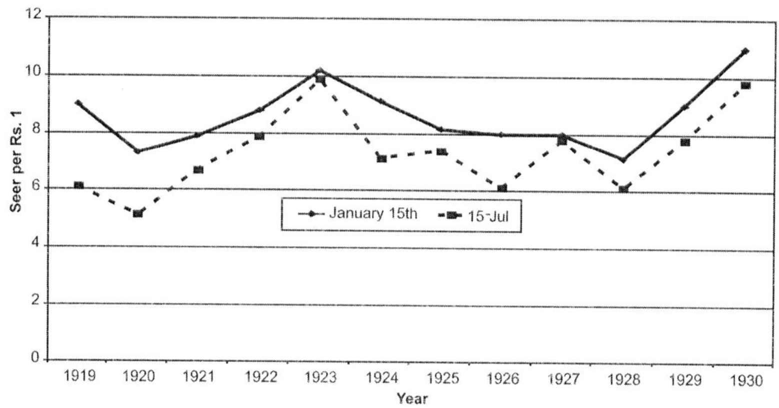

Source: Babu, 1984.

During the time of 1890, Bengal-Nagpur railway also opened and it went through Sambalpur. By the time of 1894, Sambalpur town was linked with a branch of railway line. With the introduction of railway system the pace of development might have increased but there also started an outflow of grains from Sambalpur. Figure 2.3 shows the steady increase in export of rice from Sambalpur both in quantity terms and in its share in total export. Not only that, even at the time of intense famine, there was export (1899–1900). As a consequence of such exports, the price of grains has increased steadily (see Figure 2.4). As it is evident from the Figure, roughly after 1894–98 onwards, there was a steady increase till 1919–23. This rise in price was mainly due to the movement of commodity. However along with commodity if there would have been movement of labour (i.e. migrated) then the migrated labour would have earned higher wages in the destination place. Further because of migration, supply of labour might have fall in Sambalpur and wage rate might have risen. But the movement of labour is less easy than commodity. As a result of which the supply of labour in Sambalpur might have increased compared to its demand hence low wage and resultant hunger.

Figure 2.3 *Quantity of rice exported*

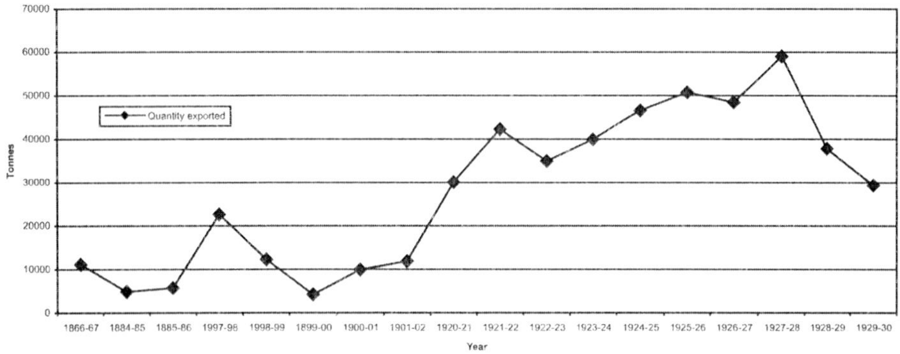

Source: Complied from, Babu, 1984, Padhi, 1999.

Figure 2.4 *Annual average price of rice*

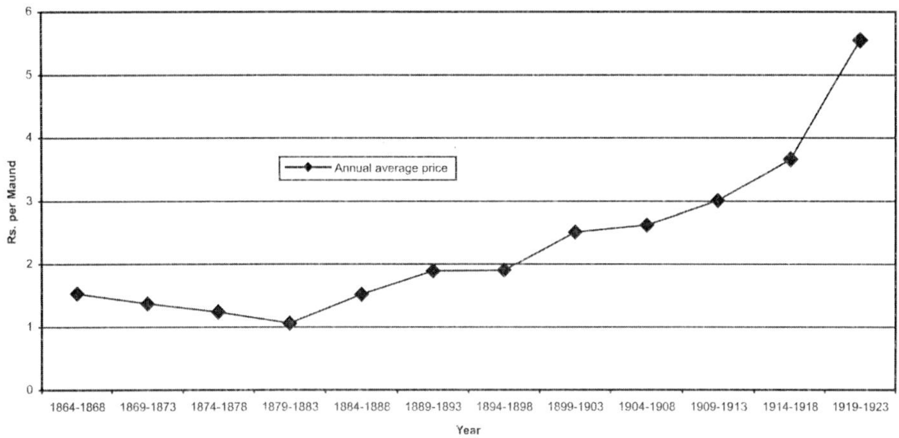

Source: Complied, Babu, 1984.

The Forced Commercialisation

At the first stage, to repay the previous debt and to meet family's current cash requirements the farmer has to go for a distress sale (a larger proportion of his agricultural output at a lower price). After the distress sale, the amount of paddy left with him was insufficient to meet the consumption requirements of his family till the next harvest. Hence, at the second stage again he was forced by the situation to borrow and go for distress purchase of rice (when price was high). This distress sale and purchase became a regular practice for almost the entire small peasantry. Here, the source of the informal credit was Mahajans (money-lender), Zamindars (landlords) and rich peasants. And they are the agents to buy the output of the peasants also. Over a period, when the market for credit and output was seriously interlocked, these agents (moneylenders, landlords and rich peasants) taking the advantage of the poor conditions of the small peasants set the pre-harvest price of rice at a low level. Sometimes, the price received by the small peasants from these agents was less than the free market harvest price of rice. Hence, the small peasants were continuing cultivation (commerce) not for profit motive but they were forced to continue the commerce just to get rid of debt.

Ultimately, the debt burdens were increased to such an extent that, the peasants had to mortgage the 'occupancy right'. That means along with the output and credit market the land and labour market also got interlocked over the short period of time. The Bihar-Odisha provincial Banking Enquiry Committee of 1929 showed that 45 per cent of the total agricultural indebtedness in Odisha was covered by registered simple mortgages of ryot land. Again between 1901–1902 to 1920, the total number of cases of registrations of mortgages, mainly of land, increased by 303 per cent in Cuttack, 168 per cent in Balasore and 127 per cent in Puri. On the other hand, in case of registration of sale of land (immovable property), during 1901–1902 to 1920, it increased by 256 per cent in Cuttack, 147 per cent in Balasore and 105 per cent in Puri. The committee also pointed out that both the mortgages and selling of land were inversely related with fluctuations.

Ultimately, the forced commerce led to a situation of massive migration (from Odisha) of both woman and man to the distant places like Assam and Bengal. It is interesting to see that although the real wage was stagnant in Assam, Bengal and Odisha; and the exploitation of jobbers; during later part of 19th century and earlier part of 20th century; still there were out-migrations from Odisha to Assam and Bengal. This might be due to the fact that even at the low wage rate there was no regular employment available in Odisha. Taking into account caste distribution of migrants as a supportive explanation, Chauduri found 'vast majority of emigrant workers belonged to caste, which were traditionally agriculturists or strongly attached to land'. Whereas Ranajit Dasgupta found, the majority of jute mill operators in Bengal belonged to other than landholding and cultivating caste. He found, around one-fourth of the jute labour force are the most downtrodden, lowest in social hierarchy and the untouchable caste.

Although findings differ across studies, the gist of both the cases could be the same; it is those group of people that migrated, whose total entitlements were not enough for their survival or whose outputs were less preferred by the society than the outputs of others.

Dadan System

The *dadan* labour system is a form of labour contract under which people are recruited by intermediaries on payment of advance to work in different parts of India. But the negative element in this system was that it has all the ingredients of bonded labour system. This has developed in many parts of Odisha. Below the experiences of Sambalpur has been described.

Sambalpur was considered as the 'Garden of Eden' or the land of promises till 1899/1900. In that year, the districts experienced one of the worst famines, the death per mile of population increased from 29.17 in 1899 to 108.18 in 1900 (Babu, 1984). The worst burden of the famines fell on small farmers and labour. The large farmers, however, managed to get at least half of their crops. The impact of this famine reduced the demand for the agricultural labour also. Labour started migrating and there were agents to export labour to Assam and Bengal. In such situation the small farmers also started migrating. Taking the advantage of the situation the trader class started giving advance to the small cultivators as they were in need of money to pay cash rent and to buy food. In that system, the cultivators were receiving interest free advances by agreeing to supply a certain quantity of produce at predetermined prices, immediately after harvest. If the cultivators (borrower) were unable to keep the commitment, then a heavy rate of interest of fifty per cent was charged. Hence, when the cultivators failed to repay the debt, then also they were forced to continue their cultivation, hence it might be known as forced commerce. And the benefit was going to the traders, who used to buy when the price was low and sell when it was high. Slowly, the traders started controlling both credit market as well as output market over the time. The indebtedness has increased mainly among the bottom two types of ryots although it was worse for the lowest one. These two together constituted around 50 per cent of total holdings. As a result of high indebtedness the cultivators escaped from the place as *dadan* labour.

It is worthy to mention a sentence regarding the sources of debt. As the Dewar Report points out 'throughout the Bihar-Odisha provinces the money-lender was generally Bania caste. Hence, the question of caste arises again. Banias are generally shopkeepers and moneylenders, from higher caste and generally found in larger villages. The high interest rate charged by them and the harsh crude methods of collection has given them notorious prominence.

Over a period of time, the small farmers emerged as landless class and the rich and substantial ryots and moneylenders became surplus landholding class. As it was pointed out, the average size of holdings in Odisha increased from around 3 acres in 1898 to 4–5 acres in 1929, i.e. the number of landholders came down. And in the whole process, not only the landless labour has increased the employment opportunities in agriculture but also decreased due to the greedy nature of the landholders and desire for less number of hired labour for cultivation.

Absentee Landlordism

During the early British period many defaulting estates were sold at Cuttack, which were bought by the speculative Bengal who expected a permanent settlement to be made in Odisha. 'Most of these Bengali landlords were absent landlords, who performed no useful functions. They were mainly rent receivers, who stayed outside the state of Odisha. They left the management of their estates in the hands of Bengali amalas, who exploited the tenants to the maximum limit without taking interest in the improvement of the land (Samal, 2000).

Cultivable Fallow Land

The British land revenue system brought such havoc that even the landlords, scared of the probable increase in the land revenue over their landholdings, often threw the major portion of tillage out of cultivation before the commencement of the new settlements. This caused a great misery to the cultivator tenants and created a food scarcity situation in the state.

Practice of Bethi and Gothi

Bethi is a kind of forced unpaid labour that a tenant has to perform for his master, who provided him the tenantship. Whenever a landlord needs he can call any tenant for free service and the master was not obliged to pay for the tenant's work. Gothi is a kind of bonded labour. In this system, the labourer was pledged to his master to be attached to him for a span of years or for his entire life like a slave. For his work he would be given food, boarding facilities at master's house and some annual remuneration. When a particular Gothi died, his son might be inducted into the service to fulfil the unredeemed pledge.

Practice of Share Cropping

Another feature that got developed in the rental system was that of share cropping. Mostly, a big farmer leased out a piece of land to small farmers and the small farmers have to pay some fixed amounts of produce to the big farmers. However, the amount used to be fixed at such a level that very less was left for the small farmers.

The lesson can be drawn from here that due to the peculiar kind of British land revenue system, complemented with natural calamity (output fluctuations) and absence of formal credit sources; the small peasants were forced to borrow from informal sources. Due to the uncooperative debt mechanism, ultimately the small peasants fell into the net of debt and subsequently slipped into the group of hunger.

Table 2.7 *Percentage distributions of households across their landholdings in Odisha 1951*

Land	*Coastal*	*Northern*	*Southern*	*Odisha*
Per Household Land Cultivated (PHHLC)	2.84	5.69	5.63	4.20
0 Acres	39.67	28.66	35.68	36.46
1 Acres	17.60	10.67	11.76	13.99
2 Acres	11.33	10.02	8.72	10.26
3 Acres	8.37	10.14	7.09	8.49
4 Acres	5.79	7.96	6.26	6.44
5 Acres	4.63	6.66	5.45	5.29
6–10 Acres	6.02	13.06	11.03	8.89
11–32 Acre	3.04	9.44	6.92	5.76
33 acres	0.27	0.27	0.52	0.31
34–79 Acres	0.56	1.32	1.42	0.96
80 and over	0.19	0.29	0.33	0.26
Acres not stated	2.54	1.53	4.83	2.89
Total	100	100	100.00	100

Source: Census, 1951.

Hence, some of the features of Odisha's economy prior to independence could be – low land productivity, credit unavailability (high interest rate), emergence of landless class and man made hunger. Along with that, in the first quarter of the 19th century, there were unemployed labour from manufacturing sector due to the salt monopoly and during the early period of 20th century, there were unemployment in agriculture due to landlessness of small peasants.

Finally, during 1951, till then not much reforms have been implemented in the land revenue system shows an agrarian structure (see Table 2.7) where (i) The average cultivated area per household in coastal, northern and southern regions are 2.84, 5.69 and 5.63 per cent respectively, compared to 4.2 acres for Odisha. (ii) The percentages of landless households in the coastal, northern and southern regions were 40, 29 and 36 per cent compared to 37 per cent for Odisha. (iii) The percentages of large farmers (more than 10 acres) in coastal, northern and southern regions were 7 per cent, 14 per cent and 15 per cent, compared to 10 acres for Odisha. The district level analysis which can be seen from Appendix A2.9 shows that the average cultivable land per household varied in the range of 1.91 in Ganjam district to more than 6 acres in Phulbani, Balangir, Sundargarh and Mayurbhanj. As far as the percentage of landless households are concerned, almost all the districts have sizeable shares of landless households, which ranged from 13 per cent in Keonjhar, 25 per cent in Mayurbhanj and 26 per cent in Sundargarh to 48 per cent in Ganjam, 45 per cent in Cuttack and 41 per cent in Kalahandi. As far as the percentages of large farmers or landlords are concerned (here we have taken more than 10 acres), the shares remained higher in Balangir, Sundargarh and Kalahandi.

Table 2.8 *Livelihood categories and their characteristics, 1951*

Livelihood classes	*Characteristics*
Class I	Cultivators of land, wholly or mainly owned and their dependants. For example, occupancy tenants, ryots, tenants with heritable right but without transfer right (Sambalpur), village servants; Zamindar or thikadar who had large extent of nijchas land which he/she cultivated himself/herself and lesser extent of other lands from which he/she collects rent.
Class II	Cultivators of land, wholly or mainly unowned and their dependants. Varieties of cultivators doing cultivation under lease fall in this category. Such as *bhag* (share) and Thicka.
Class III	Cultivating labours and theirs dependants. Persons doing agricultural work on daily or on periodical basis fall in this category.
Class IV	Non-cultivating owners of land; agricultural rent receivers and their dependants.
Class V	People involved in production other than cultivation and their dependants. Such production relates to primary industry such as livestocks raising, forestry, plantation, hunting, fishing as well as employment in other types of production such as mining and quarrying, processing and manufacture of food stuffs, textiles, leather goods, metal, chemicals, etc.
Class VI	Persons involved in various kinds of trades (commerce) and theirs dependants. This relates to moneylending, banking, insurance, etc.
Class VII	Persons involved in transport work and theirs dependants.
Class VIII	Persons involved in other services (miscellaneous) and theirs dependants.

Notes: First four types are agricultural classes. And others are non-agricultural classes.

Source: Census, 1951.

The economic structure of the state also can be examined from the livelihood information of the Census data of 1951. That census provides information based on economic classifications in place of classification based on religion prior to 1951. Tables 2.8 to 2.12 attempt to examine the livelihood patterns of people of Odisha during 1951. In the 1951 census the population has been divided into two broad livelihood categories, (i) agricultural and (ii) non-agricultural. Both the categories were again subdivided into four each. Table 2.8 presents the characteristics of each category.

Table 2.9 presents the distribution of population across different livelihood classes during 1951. The proportion of population who depended on agriculture in Odisha was roughly 80 per cent, compared 70 per cent for India and 55 per cent for Travancore-Cochin (the least proportion dependent on agriculture). The region-wise analysis shows that the proportion of workers who depend on agriculture for their livelihood in coastal, northern and southern regions are 76 per cent, 85 per cent and 82 per cent respectively, compared to 79 per cent for Odisha. Across the districts of Odisha (see Appendix A2.10 and A2.11). Keonjhar, Mayurbhanj, Sundargarh, Balasore, Kalahandi and Dhenkanal remained above the state average with more than 80 per cent dependent on agriculture, whereas, the same has remained at 66 per cent for Ganjam.

Table 2.9 *Livelihood patterns of Odisha and other Indian states, 1951*

	Agricultural classes	*Non-agricultural classes*	*Livelihood classes*							
			I	*II*	*III*	*IV*	*V*	*VI*	*VII*	*VIII*
India	69.84	30.16	46.92	8.87	12.56	1.49	10.56	5.98	1.58	12.04
Vindhya Pradesh	87.12	12.88	62.61	6.36	17.62	0.53	4.6	2.8	0.43	5.05
Bihar	86.05	13.95	55.29	8.27	21.87	0.62	3.94	3.4	0.72	5.89
Odisha	79.28	20.72	59.53	5.94	12.31	1.5	6.33	2.91	0.53	10.95
Madhya Pradesh	76	24	49.5	4.47	20.41	1.62	10.6	4.39	1.47	7.54
Uttar Pradesh	74.18	25.82	62.26	5.15	5.71	1.06	8.39	5.03	1.36	11.04
Assam	73.34	26.66	57.89	12.81	1.74	0.9	14.68	3.9	1.28	6.8
Madhya Bharat	72.22	27.78	50.43	10.22	10.67	0.9	9.96	5.61	1.07	11.14
Rajasthan	70.87	29.13	43.31	22.86	3.1	1.6	8.88	6.58	0.94	12.73
Mysore	69.9	30.1	55.46	4.76	6.79	2.89	10.24	5.57	1.16	13.13
Madras	64.93	35.07	34.95	9.58	18.23	2.17	12.35	6.69	1.68	14.35
Punjab	64.73	35.27	38.76	16.09	7.74	2.14	7.36	9.15	1.04	17.72
Bombay	61.46	38.54	40.74	9.69	9.05	1.98	13.76	7.61	2.23	14.94
West Bengal	57.21	42.79	32.34	12.01	12.26	0.6	15.36	9.32	3.05	15.06
Travancore-Cochin	54.85	45.15	26.34	7.1	20.17	1.24	21.19	6.8	3.41	13.75

Source: Census of India, 1951.

Table 2.10 *Livelihood patterns across the districts of Odisha, 1951*

Regions	*Agricultural*	*Non-agricultural*
Coastal	76.2	23.8
Northern	84.9	15.1
Southern	80.2	19.8
Odisha	79.28	20.72

Source: Census of India, 1951.

Yield Rate

The land revenue system and the agrarian structure of the colonial Odisha might also have affected the yield rate in the state, which might have contributed to the existing hunger and deprivation. Tables 2.11 and Table 2.12 depict the following: (i) yield rate of rice in Odisha always remained lower than that of all-India during 1945–46 to 1956–57, (ii) in case of wheat, the yield rate was higher for Odisha prior to 1948–49 but after that it has started declining or stagnant, whereas it has shown increase for all-India. This shows that there was a gap in the yield rate between Odisha and India during this period.

Table 2.11 *Yield rate of rice and wheat in Odisha during 1945–46 to 1956–57 (in kg per acre)*

	Yield rate of rice		*Yield rate of Wheat*		*Percentages of gross irrigated area*	
Year	*Odisha*	*India*	*Odisha*	*India*	*Odisha*	*India*
1945–46	241	—	318	—	22.94	18.83
4946–47	268	335	317	199	22.36	19.03
1947–48	244	330	319	267	22.35	18.03
1948–49	141	312	170	253	22.47	18.08
1949–50	214	307	254	261	15.09	16.72
1950–51	207	266	250	264	16.91	17.11
1951–52	222	284	238	260	16.71	17.36
1952–53	232	304	241	304	18.74	16.98
1953–54	235	359	240	299	11.54	17.07
1954–55	229	323	216	319	12.75	17.24
1955–56	221	334	252	286	16.07	17.45
1956–57	236	357	258	277	—	—

Source: Statistical Abstract of India and Statistical Abstract of Odisha, 1956–57.

Within Odisha, as Table 2.12 shows, the yield rate of rice remained higher for Koraput (309 kg), Ganjam (276 kg) and Sundargarh (272 kg) districts; and it remained at lower levels for Balangir (97 kg), Keonjhar (186 kg), Kalahandi (186 kg), Phulbani (190 kg) and Dhenkanal (193 kg) districts. In case of wheat, the yield rate remained low in Phulbani (162 kg), Mayurbhanj (193 kg), Kalahandi (193 kg) and Puri (193 kg) districts and remained higher for Cuttack (308 kg), Sambalpur (289 kg) and Koraput (289 kg) districts.

However, the exploitative land revenue system was not without any resistance from below. There have been resistance by the Odia people before the modern Odisha was formed under the leadership of different parties and organisations. The upper social groups, who were not so concerned about benefit of the common people, mainly led such political parties.

Table 2.12 *Yield rate of rice and wheat across the districts of Odisha during 1956–57 (in kg per acre)*

Districts	*Yield rate of rice*	*Yield rate of wheat*	*% gross irrigated area*
Balasore	202	Negligible	9.19
Balangir	97	243	28.57
Cuttack	262	308	19.73
Dhenkanal	193	210	35.03
Ganjam	276	Negligible	39.07
Kalahandi	186	193	14.28
Keonjhar	186	217	5.38
Koraput	309	289	32.61
Mayurbhanj	233	193	7.33
Phulbani	190	162	0.26
Puri	230	193	17.91
Sambalpur	247	289	29.00
Sundargarh	272	243	2.93

Notes: Percentages of Irrigation for 1951, 1941, 1931 and 1921 are based on average net area sown and average net area irrigated, but for 1956–57 it is based on gross irrigated area and total cropped area.

Source: Statistical Abstract of Odisha, 1957 and Census of India, 1951.

Social Structure

Social structure refers to a definite relation among certain entities or groups, where the behaviour and relationship within social systems, or social institutions is based on the social systems and social hierarchy. Hence, the notion of social structure as relationships between different entities or groups indicates the idea that society is grouped into structurally related groups or sets of roles, with different functions, meanings or objectives or purposes. Such stratifications and functional relationship may put certain groups into a disadvantaged position when the system may force them to specialise with some profession which has less exchange value compared to other professions. In the Indian context, due to the prevalence of 'caste system' the structure of the society became such that people from SCs, STs social groups are more backward than other social groups. Similarly, certain socio-cultural practices in the Indian society also had some barriers for the women and people of disadvantaged caste to be educated in the educational institution. That might not only create more illiterate people in the society, it also may create more dependents. Certain social factors that have been looked here are – population, low level female work participation and high dependents, higher SC/ST population, low level literacy etc.

Population

There are many theories, which have established the relationship between population and poverty. It appears that in Odisha, even if there is not much population pressure and having the reputation of surplus rice producers for decades, people suffered from food insecurity, acute economic distress and even death due to starvation (for details on population growth, see Appendix A2.16).

Low Level of Female Work Participation and High Dependents

During 1951 and prior, participation of women in the labour market remained very low. Table 2.13 presents the dependent females per thousand males in agricultural and non-agricultural sector. The Table depicts that in agriculture, the non-earning dependent females remained highest in coastal region (2,060 per thousand) followed by northern (1,823) and southern (1,925) regions. Whereas, earning female dependents is highest in northern region (1,509) followed by southern region (901) and coastal region (511). In the non-agricultural sector the non-earning dependents remained highest in southern region (1,931) followed by coastal region (1,919) and northern region (1,724). But there are highest number of earning dependents in northern region (1,973) compared to 870 in southern region and 849 in coastal region.

On the whole, the results suggest that in Odisha, women as compared to men are twice likely to be non-earning dependents in both agricultural and non-agricultural sectors. Also, women are only one-fifth to one-third as likely as men to be self-supporting persons in both the sectors. This clearly reflects the very high incidence of dependency among females in Odisha. The district level picture can be seen from Appendix A 2.12.

Table 2.13 *Females per 1000 males in agricultural non-agricultural classes, 1951*

	Agriculture				*Non-agriculture*			
	Total	*SSP*	*NED*	*ED*	*Total*	*SSP*	*NED*	*ED*
Coastal	1,038	184	2,060	511	1,047	321	1,919	849
Northern	1,015	195	1,823	1,509	963	263	1,724	1,973
Southern	1,004	177	1,925	901	1,028	222	1,931	870
Odisha	1,021	182	1,936	1,081	1,025	290	1,868	1,135

Note: SSP – self-supporting person, NED – Non-earning dependent, ED – Earning dependent.

Source: Census of India, 1951.

Higher SC/ST Population

In caste-based ladder of the society, Brahmins were in the top rank and accorded the highest ritual and social status in Odisha. The second rank holder in caste-based ladder were Rajputs, which represent Kshatriya status with a group of castes (Khandayat, Paiks, Binjhals (tribal), Gond (tribal), Rana, Bhuiyan (tribal), etc. Next to them, come the Karan or Mahanti castes, whose profession was writing. They are prosperous and highly educated. Hence, although ritually they ranked below Kshatriya, socially they ranked next to Brahmin. In the third place there are a number of intermediate castes who are artisans, Kulta (cultivators), Patara (traders), Pathuria (stone carvers), Sundhi (distillers), Gouda (cowherd), Guria (sweetmakers), Teli (oilman), etc. At the lowest place of the ladder come the then untouchable castes (formerly untouchable officially, but in reality now also untouchables in many parts of Odisha). But here one needs to be careful with regard to the Indian planning that – all scheduled castes are not untouchables but almost all the untouchables are scheduled castes. For example, though Dhoba (washerman) is a SCs community, they are allowed to the upper caste festival, or to the educational institute or the temple without any hurdles. Whereas, the Hadi (scavenger) were/are one of the scheduled castes, who fulfill very essential functions of cleaning the latrines, were/are not allowed to many social institutions without restrictions. Similarly, the Ganda, who play music or the Ghusurias or ghansi (the pig-rearers) comes under similar category. Hence, traditionally, these communities are excluded from the mainstream society and hence are likely to face high deprivation.

The untouchable castes were also having almost no direct relation with land (production) and the only exchangeable output they were having was labour power, demand for which fluctuates so much at the time of natural calamity (drought or flood). Hence, the intensity of hunger would be more among those people. But, although the above-mentioned occupations were their main occupation, due to lack of regularity in those caste-based occupations, in the up season (of their occupation) they remained mainly as agricultural or non-agricultural labour. However, the situation of these communities remains backward even in the subsequent era.

As it was well known in the Odisha history, there have been consistent number of movements (whether social or political or economical), and Odisha also was a part of the Sanskritisation, which was a movement for the transition of the castes. In the Sanskritisation process, many subcastes have been given the status of upper group caste, but as it was pointed out in the *Odisha Gazetteer* (1991), it was almost impossible for a traditionally untouchable caste to cross the pollution barrier and become recognised as a non-polluting caste. Therefore, these groups of populations (who constituted around 18% during 1951) may be more deprived and hungry compared to other individuals of the society. As Table 2.14 shows, the percentage of SCs and STs population was highest in southern and northern region with 56 per cent and 50 per cent respectively compared to 23 per cent in the coastal region (the district-wise distribution of population can be seen in Appendix A2.13).

However, in spite of such social deprivation, prior to independence there was no reservation policy. Observing the condition of untouchable castes, during 1949–50, Babasaheb Ambedkar demanded for a separate electorate with separate settlement. But opposing this, Gandhi had gone for 'fast unto death'. And ultimately the reservations for backward classes came up (which was weak targeting) for the development of those lower castes people. The demand posed by Babasaheb was for the hungry and deprived people, who was opposed by Gandhi and the solution that came out was indirectly for the wealthy and upper caste people as well. The policy had helped more to touchable SC, who often stay in the vicinity of the upper castes. Hence, they would be better than the untouchable SC. There could have been a significant improvement of those communities had the reservation policy been more scientific and specific.

But as far as scheduled tribes and hunger and deprivation are concerned, although generally, they do not encounter untouchability, they have locational disadvantage. In many villages, tribals were the landlords as they were the indigenous population of the locality. However, they may require more provision of physical infrastructures, such as educational building, road transport and vocational centre near to them. Since one group is material resource poor and other is knowledge poor, both the groups might face hunger and deprivation, but still it might be more among untouchable castes. This is due to the earlier institutional rigidity and present inefficient enforcement of the law.

However, by raising the caste issue here, the study does not exclude the occupational issue like the problems of landless labour or agricultural labour. But traditionally, occupations were based on castes. The people belonging to low caste have low earnings and, therefore, inherently suffer from hunger and deprivation. The distribution of occupations across the castes group in agricultural and non-agricultural class in Odisha is examined below in Table 2.15A. It shows that among the cultivators, most of them are either from OBC (30.11%) or Non-backward Castes (NBC) (33.68%) compared to STs (22.63%) and SCs (13.58%). This may imply that STs and SCs (particularly SC) were landless at the initial level itself. As regards agricultural labour occupation, the incidence was highest among the SCs (28.45%) followed by OBC (26.31%), and STs (25.68%), compared to Non-backward Castes (19.57%). In the non-agricultural sector, the OBC and NBC have dominated the shares in all the occupations, compared to SCs and STs category.

Table 2.15B reveals that SCs social groups is the single most disadvantaged social groups, which is having only 45 per cent of population as cultivators class compared to all other social groups put together at 60 per cent. Further SCs social group is having highest share of population as agricultural labour class (19.51%) compared to 15.61 per cent of ST, 11.37 per cent of OBCs and 7.24 per cent of NBC.

Table 2.14 *Distribution of backward classes population across the districts of Odisha during 1951 categories of castes*

	NBC	*TBC*	*SC*	*ST*	*SC/ST*	*OBC*
Coastal	50.58	49.43	4.95	18.48	23.43	26.15
Northern	18.98	81.02	34.03	16.62	50.65	30.37
Southern	13.67	86.33	37.43	18.67	56.10	30.23
Odisha	33.3	66.7	20.2	18	38.2	28.5

Notes: NBC – Non-backward classes, TBC – Total backward classes, SC-Scheduled Tribes, ST – Scheduled Castes, OBC – Other Backward Classes.

Source: Census of India, 1951, Volume XI, Odisha, Part-I Report, pp. 7.

Table 2.15A *Distribution of occupations across social groups*

	ST	*SC*	*OBC*	*NBC*	*Total*
Shares to total population	20.26	17.96	28.49	33.28	100
Category			Agricultural Class		
I	22.63	13.58	30.11	33.68	100
II	19.09	23.99	23.40	33.53	100
III	25.68	28.45	26.31	19.57	100
IV	5.97	9.20	24.08	60.75	100
			Non-Agricultural Class		
V	10.30	26.03	34.25	29.43	100
VI	5.64	17.70	32.35	44.31	100
VII	5.92	12.31	21.10	60.67	100
VIII	14.22	23.63	21.52	40.63	100

Notes: Generally I –Cultivator class; II – Tenant class; III – Agricultural labour class; IV – Landlord class; V – Engaged in primary product producing industrial class; VI – Traders class; VII – Engaged in transport and their dependants; VIII – Engaged in service and their dependents.

Source: Census of India, 1951.

Table 2.15B *Distribution of occupation across castes*

Category	*Odisha*	*ST*	*SC*	*OBC*	*NBC*
I	59.53	66.50	44.99	62.92	60.24
II	5.94	5.59	7.93	4.88	5.98
III	12.32	15.61	19.51	11.37	7.24
IV	1.50	0.44	0.77	1.27	2.74
Agl. Total	79.29	88.14	73.20	80.43	76.20
V	6.33	3.22	9.17	7.61	5.59
VI	2.91	0.81	2.87	3.30	3.87
VII	0.53	0.15	0.36	0.39	0.97
VIII	10.95	7.67	14.40	8.27	13.37
Non Agl. Total	20.71	11.86	26.80	19.57	23.80
Grand Total	100	100	100	100	100

Source and Notes: same as Table 2.15A.

Table 2.16 presents the percentage shares of landless households (based on land ownership) by their social background in different regions of Odisha, which depicts: (i) the percentages of landless household in coastal, northern and southern regions are 37 per cent, 29 per cent and 36 per cent respectively, (ii) same pattern has followed in the agricultural classes but among the non-agricultural classes, the percentages of landless household in coastal, northern and southern regions are 54 per cent, 67 per cent and 79 per cent respectively, (iii) across the districts (see Appendix A2.14), in agriculture, the incidence of landlessness has remained higher in Dhenkanal (20.93%), Koraput (17.60%) and Phulbani (13.62%), and the same remained lower in Balangir (0.78%). In Kalahandi, Mayurbhanj and Sambalpur, it remained negative, which shows that in those parts, the incidence of landless households was higher among non-backward class. In non-agriculture, the difference remained highest in Ganjam.

However, the Table 2.16, as such, does not show much difference in landlessness between the backward and non-backward castes. Within the backward caste, the positions of the scheduled castes is not clear here as most of the socially excluded groups are from this social category.

Table 2.16 *Percentages distributions of landless households by their social groups across districts in Odisha, 1951*

	Coastal	*Northern*	*Southern*	*Odisha*
Total Avg	36.83	28.66	35.68	36.46
Ag.classes	28.40	19.13	24.98	25.43
Backward	33.28	19.75	25.07	25.96
Non-backward	24.20	20.86	22.50	24.42
Difference	9.08	–1.10	2.57	1.54
Non-agr	54.37	67.48	78.91	65.14
Backward	55.93	68.17	80.02	67.31
Non-backward	52.36	66.56	75.31	61.83
Difference	3.57	1.60	4.71	5.47

Source: Census, 1951.

Literacy

Literacy is another aspect, which could have enhanced the earning capacity of the people. But it has remained at low level in Odisha and hence might have contributed to the existing hunger and deprivation. By 1951, the literacy rate in the state was just 15.8 per cent (see Table 2.17). The percentage of literacy rate among male was 13.5 per cent, whereas the same among female was 2.3 per cent. Broadly, the level of literacy remained low among those regions (e.g. the southern Odisha), where the percentage of SC/ST population was high. The district level picture can be seen from Appendix A2.15.

Table 2.17 *Level of literacy across the districts of Odisha during 1951*

	Total	*Males*	*Females*	*SC/ST*
Coastal	21.3	17.9	3.4	23.4
Northern	12.9	11.3	1.5	50.7
Southern	7.2	6.57	0.6	56.1
Odisha	15.8	13.5	2.3	38.2

Source: Census of India, 1951.

Political Structures and Hunger and Deprivation in Odisha

Rise and Growth of Political Consciousness and Social Change in Odisha[3]

The political consciousness in Odisha started with when an elite group of educated Odias and Bengalis settled in Odisha took leadership in the socio-political activities of the post-1870 period. Then, the journalists, literature and historical writings have projected the movement of Odia nationalism. The spread of English education in Odisha and the participation of Odias in various government services have impeded the movement. The Odias employed in Government services became the leaders of the society and being educated, they made constant efforts to create political consciousness among the masses. The *Utkal Dipika* edited by Girija Sankar Ray; the printing press of Fakir Mohan Senapati; *The Sambalpur Hitesini* published from Bamara; *The Ganjam and Oriya Hitabadini* and the *Cuttack Chronicle* were important newspapers, which contributed valuable services for the growth of Odia nationalism during 19th century.

Language Movement

The language movement was another aspect that added into the growth of the Odia nationalism while the people of Ganjam, Koraput and Sambalpur felt the Telugu and Hindi speaking intolerance in them. In the coastal Odisha, there was attempt to eliminate Odia from the official language and replace with Bengali. The Odias strongly opposed to the situations through different newspapers. During the last quarter of the 19th century, 'the 'Utkal Sabha' or the Odisha Association was formed which played a major role in this respect. Further, the movement of Odia nationalism became pervasive when an outstanding leader, Madhu Sudan Das (1848–1934) became the symbol of Odia nationalism. 'The Utkal Sabha put forth before the Government the grievances of Odia peasantry, the necessity of improving communication and the need for using Odia in the Courts of Ganjam and Visakhapatanam agency where Odia population formed a majority' (Tiwari, 1994). The Utkal Union Conference in 1903 attempted to find out the ways and means to solve the problem of the Odia-speaking people. The main objectives were to accelerate political progress to unite the Odia-speaking people, to achieve quick economic progress by developing the industrial potentialities of Odisha and to promote social progress to ensure the spread of general, technical and female education. The emphasis, however, has remained more on the amalgamation of Odia-speaking tracts. At the later period, the Utkal Union Conference has merged with the mainstream of national movement under the agencies of the Congress party. The movement of Odia nationalism has continued and it was in 1905, the initial success towards unifications was achieved, when Sambalpur was amalgamated with the Odisha division. The province of Bihar-Odisha was formed during 1912, which was yet another major step in the direction. But the Odia-speaking people of Ganjam and Visakhapatnam agency were still under Madras presidency. Meanwhile, the people of Odisha felt that their interest would not be safeguarded under the new government dominated by the Biharis. So the struggle continued and finally in 1936, British Parliaments' Constitution of Odisha Order got approval of emperor of England on 3 March 1936. It was on 1 April 1936, Odisha was announced as a separate province. During 1948–49, all the princely states finally merged with Odisha but these princely states emerged as a single political party named as Ganatantra Parishad Party, who dominated the politics for two years. However, as far as the constitutional culture in Odisha was concerned, it dates back to 1936–37. During 1936, Constitution of Odisha Order provided for a legislature and the election to the legislative assembly was held in January 1937. During that elections though Congress had the majority, there was division within the party, hence could not form the government. As different authors pointed out from time to time, the political instability due to conflicting interest within the group and among the elite continued to be a permanent feature in the Odisha politics (Banerjee, 1984; Tiwari, 1992). The general election in Odisha was held during 1952, where all the feudatory states played an active role but the result was replacement of feudal leaders in the so-called democratic elections. All the local institutions, the Zamindar, the gountia, the raja, etc. became the democratic representative (Bailey, 1963). Hence, there was just a change by name from feudalism to democracy, but there was no change in real sense and probably therefore the social changes have been obstructed.

Regional Politics

On the leadership pattern of the regional political parties in Odisha, it was pointed out that 'a considerable sub-cultural differences on the basis of caste and region continued for long time among the legislators of the state' (Banerjee, 1984). The politics of the state was guided in the same manner as the traditional Odia society was. In the traditional Odia society, all socio-cultural and economic domination were influenced by two upper castes – Brahmins and Karanas. Later on, the Kshyatriyas and the Khandayats had emerged increasingly powerful and occupied vital position in politics as well as in governments. The SCs and STs became conscious about their political right and power very late. Yet, despite 40 per cent constitutional reservation for them, they remained unorganised and have failed to emerge as an independent political force in the state. Rather, all other political parties have own their loyalty for forming the government.

Political system of Odisha also got affected by deep regional differences, which occurred due to the centuries old differences in the climate, topography, history and traditions. In Odisha, mainly two kinds of political parties were formed, one is geographical and the other is factional. The geographical political party emerged out of regional support, leadership and ideology. The later one are factions from the National Congress and formed state-specific party like Jana Congress or Utkal Congress. In Odisha, the regional party like Ganatantra Parishad emerged due to feudal structure of the Odisha society. In the post-merger period the ex-rulers, mostly from western part of Odisha, had to lost their kingdoms, which might be the reason that they were looking for political space. Tiwari (1992) pointed out that, 'the post-merger anti-Congress psychology of few powerful ex-rulers paved the way for the formations of a political organisation on regional lines'. When the ex-rulers came to the assembly as elected legislators, the Congress ruling party of the state, instead of filling the gaps (regional, historical, traditional), antagonised the regional feeling by giving berths to a few insignificant ex-rulers of the coastal region in the ministries and other important posts. That continued in the subsequent period and the regional imbalances in all spheres including hunger and deprivations aggravated in the state.

Therefore, as the history indicates, hunger and deprivation in the state have been rooted from the combined effects of geographical specificity, social structure and cultural practices, economic structure (agrarian structure) and weak political will for the development of some regions of the state.

Summary

The chapter examined the structure of hunger and deprivation within the state of Odisha and explored relation of the same with the economic history of Odisha. It found a higher variation of hunger and deprivation across social groups, geographical regions, occupations, education status, landholding and dependency ratio. However, a single most important variable that can explain the higher incidence of hunger and deprivation in Odisha to a greater extent is the social structure and social hierarchy prevailing in the state, where the SCs/STs, women, children and old dependents faced most disadvantages. Thirty-eight per cent of the people of the state belong to SCs/STs, who were traditionally excluded from the state affairs and continued to face various exclusions. The findings here suggest that irrespective of place of living, incidence of hunger and deprivation among SCs/STs are higher. Historically, these communities do not possess adequate wealth, neither they had access to any valuable socio-cultural institutions like school, temple and other public places, which might have been the reason of higher hunger and deprivation among them. Therefore, these communities are in need of special attention from a developmental state but as it will be seen in a later chapter, the problem of social exclusion is pervasive in the state.

Geographically too, these communities concentrate in specific regions (southern and part of northern Odisha). Being away from the focus of the state, the regional disparities might have increased. Lack of awareness among the people of this region on one hand, and lack of interest of the state on the other, together failed to bring development in a sustainable manner to reduce hunger and deprivation. Instead, certain investments were made (mostly in mines and coal), which are highly extractive and unsustainable in nature. That added to the hunger and deprivation through displacement and destroying the source of livelihood of the people.

Where (79.28) per cent of the state manpower was engaged in uncertain agriculture in 1951, though it has reduced to 61.80 per cent in 2011, yet there is higher chances of unemployment, disguised unemployment, low labour productivity, less wages and hence a low overall economic growth to reduce hunger and deprivation. In the absence of adequate and balanced economic growth, the state may not only fail to cope with the natural calamities like drought, flood and cyclone, it may also fail to provide adequate social security to her people, especially to the disadvantaged.

In brief, it seems the strong social hierarchy, which was there in the socio-cultural history of Odisha might be still persisting in different forms and hence remains an important barrier to reduce hunger and deprivation in the state.

Given these identifications, the following chapters attempt to explain the causal factors in detail. Chapter three presents the structural transformations in the state (in output and employment) and its impact on hunger and deprivation.

Notes

1. For elaborate description on the topology and geography of the state see Annexure A.
2. Arrival of each of these system in the state can be seen from Annexure B.
3. The history of Odisha as an administrative unit can be seen from Annexure C.

3

Slow and Lopsided Structural Transformation and its Impacts

It has been observed from the review of literature and secondary data that hunger and deprivation can occur in a country or a state due to the specific structure of her economy. If the structure of an economy is shaped with a lopsided growth or if the economy is at a very low level and economic growth also growing at a very slow rate, then, it may lead to mass hunger and deprivation, as the gross available wealth is insufficient for the requirements. As Dantwala put forth, growth through a systematic structural transformation, where both output and employment can expand adequately is required for poverty reduction.[1]

In most of the cases, economic development has historically been associated with structural transformations of an economy.[2] The most common form of structural transformation that has been observed historically had shown a sequence of shift of sectors from agriculture to industry, and then to services. Therefore, in an underdeveloped economy, there is a higher chance that agriculture may have a predominant share in the total produced output and as the development progresses, the share of industry increases and that of agriculture declines. Subsequently, after reaching a reasonably high level of development, the share of service sector increases and becomes a major component of the output of the economy. The transformation is not only confined to output but it has also been followed by employment. For more on structural transformation see Annexure D.

As the relations of the structural transformations with hunger and deprivation (poverty) are concerned, the first kind of shift of labour force, i.e. agriculture to industry or from rural to urban is more essential because in most of the developing countries, rural sector accounts for a substantially higher share of absolute poverty than urban sector; therefore a rural resident is more likely to be poor by any standard (Lipton and Ravalion, 1994). When the rural poor shift from agricultural sector to urban industrial sector, it may reduce poverty in two ways by increasing productivity both in agriculture and industry (Lewis, 1954; Fai and Ranis, 1964). Therefore, the rural-urban shift of labour or the sectoral transformation of output and employment has an impact on hunger and deprivation.[3] The following section looks into the structural transformations exercise, where it starts with output transformations.

This chapter looks into the growth and structure of the Odisha economy and attempts to explore the association of hunger and deprivation with the output and employment growth, labour productivity especially in agriculture in the state. The chapter comprises of four sections. Section I examines the output/product transformations; Section II examines the employment transformations; Section III expands the employment aspects further and tries to understand the characteristics of labour and its relationship with poverty; and Section IV discusses the output and employment transformations with hunger and deprivation followed with final conclusion.

Section I

Product Transformations in Odisha Economy

Sectoral Shares of NSDP and NDP

This section presents the structural transformation of output and seeks to examine the association with existing hunger and deprivation in Odisha. The sources for the output data are the Central Statistical Organisation and Directorate of Economics and Statistics. Tables 3.1 and 3.2 show the percentage distribution of Net State Domestic Product (NSDP)[4] of Odisha and percentage change of NSDP across different sectors from 1950–51 to 2005–06 in 1993–94 prices. Similarly, Tables 3.3 and 3.4 show the same for National Domestic Product (NDP) of India (see Appendix A3.1 and Appendix A3.2 for subsectoral contribution to NSDP of Odisha and NDP of India respectively).

Table 3.1 *Percentage distribution of NSDP of Odisha across different sectors, 1950–51 to 2005–06 (1993–94 prices)*

Sectors	*50s*	*60s*	*70s*	*80s*	*90s*	*2000–01 to 2005–06*
Primary	65.39	62.11	60.31	57.55	45.43	39.18
Secondary	10.39	15.38	15.76	14.19	16.64	13.87
Service	24.22	22.51	23.93	28.26	37.94	46.95
Total NSDP	100	100	100	100	100	100

Notes: 50s – 1950–51 to 1959–60, 60s – 1960–61 to 1969–70, 70s – 1970–71 to 1979–80, 80s – 1980–81 to 1989–90, 90s – 1990–91 to 1999–2000. Output of all variables has been converted to constant prices of 1993–94 by using the splicing method.

Sources: Compiled from the estimates of State Domestic Product, 1950–51 to 1993–94; Directorate of Economics and Statistics and Central Statistical Organization.

Table 3.1 shows that during 1950s the primary sector has contributed highest percentage to NSDP (65.39%) compared to service sector (24.22%) and secondary sector (10.39%). In recent period (i.e. 2000–01 to 2005–06), the share of primary sector came down to 39.18 per cent and that of service sector increased to 46.95 per cent. However, the share of the secondary sector to NSDP has increased marginally from 10.39 per cent during the 1950's to 13.87 in the recent period (2000–01 to 2005–06). In other words the service sector plays the leading role followed by the primary sector in recent years in Odisha. As most of the theories on structural changes suggest, the share of secondary sector too increases to a higher level[5] as development progresses, but this has not yet happened in Odisha.

Table 3.2 *Percentage change in each sector of NSDP of Odisha over different decades, 1950–51 to 2005–06*

Sectors	*50s to 60s*	*60s to 70s*	*70s to 80s*	*80s to 90s*	*2000–01 to 2005–06*	*1950–51 to 2005–06*
Primary	–5.02	–2.90	–4.57	–21.06	–13.76	–40.08
Secondary	48.02	2.52	–9.97	17.22	–16.63	33.51
Service	–7.05	6.27	18.10	34.26	23.77	93.83

Source and Notes: Same as Table 3.1.

The all-India product transformation shows that during 1950s, the primary sector has contributed the highest percentage to NDP (60%) compared to service sector (26%) and secondary sector (14%). In recent period (i.e. 2000–01 to 2005–06), the share of primary sector came down to 25 per cent and that of service sector increased to 53 per cent and the secondary sector has shown only a marginal increase to 22 per cent (Table 3.3 and Appendix A3.4, A3.5). Almost a similar trend is found in the NDP as it was in NSDP – the share of primary sector to NDP has shown persistence decline and the share of service sector has shown a persistent increase.

Table 3.3 *Percentages distribution of NDP of India across different sectors, 1950–51 to 2005–06*

Years	*50s*	*60s*	*70s*	*80s*	*90s*	*2000–01 to 2005–06*
Primary	59.65	51.82	46.92	40.59	33.15	24.94
Secondary	14.17	18.24	19.53	21.03	22.29	22.29
Service	26.17	29.94	33.55	38.38	44.55	52.76
NDP	100	100	100	100	100	100

Source and Notes: Same as Table 3.1.

Table 3.4 *Percentages change in each sector of NDP of India over different decades, 1950–51 to 2005–06*

Years	*50s to 60s*	*60s to 70s*	*70s to 80s*	*80s to 90s*	*2000–01 to 2005–06*	*1950–51 to 2005–06*
Primary	–13.13	–9.45	–13.49	–18.33	–24.76	–58.19
Secondary	28.69	7.07	7.68	6.01	–0.01	57.27
Tertiary	20.59	12.26	11.77	9.00	22.84	102.59
Services	14.39	12.05	14.40	16.09	18.43	101.61

Source and Notes: Same as Table 3.1.

Similarities and Dissimilarities

Considering the all-India pattern as a reference point, Odisha is much behind in its contribution to NSDP, especially from secondary sector. If we look at the share of the sub-sectors (see appendix A3.3), some striking difference between all-India and Odisha are found. For instance, in Odisha, the contribution of 'forestry and logging' to NSDP has declined from 8.45 per cent during 1950s to 3.18 per cent in 2000–01 to 2005–06. On the contrary, in India as a whole, the share of the corresponding subsector to NDP has increased from 6.8 per cent to 11.08 per cent during the same time period. Similarly, in case of unregistered manufacturing also, its share to NSDP has persistently declined in Odisha. On the contrary, the share of unregistered manufacturing to NDP has shown a marginal increase in all-India. Similar trend is found in case of 'real estate, ownership of dwellings and business services'. In all other subsectors, though there is a difference in the magnitude between Odisha and India, the pattern has remained the same.

The last column of Tables 3.2 and 3.4 shows the percentage change in the share of various sectors to NSDP of Odisha and NDP of India during 1950s and 2000–01–2005–06.

During the 55 years, there has been a 40 per cent decline in the share of primary sector to NSDP of Odisha as compared to 58 per cent decline in the share of the corresponding sector for all-India. In the secondary sector, the increase in the share to NSDP in Odisha is much lower than the increase in the share to NDP in India during the same period (33% and 57% respectively). In the service sector, both all-India and Odisha have experienced substantial increase in the share to NSDP and NDP respectively from 1950s to the recent period (2000–01 to 2005–06, these are 94% and 102% respectively). Another distinct feature is that in Odisha there is a marked shift between the performance of 1980s and 1990s in the sectoral share to NSDP (Table 3.1) and rate of change in the sectoral share to NSDP (Table 3.2). An analysis of growth rate of each sector, as given below, will provide clarification to this point.

Annual Compound Growth Rate of Output

Growth either in NSDP or NDP is another indicator which can have a close association with hunger and deprivation via structural transformation. It is observed that the NSDP of Odisha at constant prices increased from ₹ 5,16,559 lakh in 1950–51 to ₹ 28,67,097 lakh in 2005–06, which shows a 5.5 times increase. During the same time period, the NDP of India has increased 13.6 times. This reflects a higher extent of deterioration of Odisha's position in comparison to all-India. Here, the growth exercise has been discussed.

The growth rate of each sector is calculated by fitting semi-log model. The results (including growth rate in per capita income) are presented in Tables 3.5 and 3.6 respectively for Odisha and India (for detailed results, see Appendix A3.6 and A3.7).

The results in Tables 3.5 and 3.6 shows that both Odisha and India have experienced significant increase in the growth rate of the economy. However, the growth rate in Odisha continues to remain lower than the all-India growth rate. The results suggest that during 1950s, while the annual compound growth rate of

NSDP was 1.20 per cent, the annual compound growth rate of NDP was much higher (3.53%). This shows that during the initial stage itself, there persists a gap in the growth rate of the domestic product of India and Odisha. In the recent period (2000–2001 to 2005–06), while the annual compound growth rate of NSDP remained at 7.84 per cent, the same for NDP was 10.75 per cent. For the whole time period (1950–51 to 2005–06), while the NSDP was growing at an annual rate of 2.81 per cent, the NDP was growing at 4.4 per cent per annum.

Table 3.5 *Annual compound growth rates of NSDP of different sectors and of per capita income of Odisha, 1950–51 to 2005–06 (%)*

Sectors	*50s*	*60s*	*70s*	*80s*	*90s*	*2000–01 to 2005–06*	*1950–51 to 2005–06*
Total Primary	–1.25	4.27 *	1.07	2.64 *	2.75 *	6.54 *	1.81 $
Total Secondary	10.30 $	0.40	6.58 $	7.14 $	1.55	10.63 *	3.49 $
Service	4.54 *	3.16 $	2.96 $	6.92 $	6.29 $	8.15 $	4.25 $
Total NSDP	1.20 *	3.46 *	2.39 *	4.48 $	3.91 $	7.84 $	2.81 $
Per Capita Income	–0.32	1.16	0.47	2.92 *	2.44 $	6.64 $	0.99 $

Notes: 50s – 1950–51 to 1959–60, 60s – 1960–61 to 1969–70, 70s – 1970–71 to 1979–80, 80s – 1980–81 to 1989–90, 90s – 1990–91 to 1999–2000. All sectors' output and per capita income have been converted to constant prices of 1993–94 by using splicing method. Statistical significance has been shown for major sector only. * and $ implies 't' statistics significant at 5 and 10 per cent level respectively.

Source: Same as Table 3.1.

Table 3.6 *Annual compound growth rates of NDP of different sectors and per capita NDP of all-India, 1950–51 to 2005–06 (%)*

Years	*50s*	*60s*	*70s*	*80s*	*90s*	*2000–01 to 2005–06*	*1950–51 to 2005–06*
Primary	2.64$	1.51$	1.67$	3.10$	3.33$	6.09$	2.69$
Secondary	5.62$	5.20$	4.36$	6.53$	5.84$	12.06$	5.37$
Services	4.43$	4.61$	4.73$	6.99$	8.14$	12.45$	5.86$
NDP	3.53$	3.07$	3.23$	5.30$	6.03$	10.75$	4.40$
Per Capita NDP	1.57$	0.84	0.93	3.08$	3.96$	—	1.81$ ^

^ 1950–51 to 1999–2000.

Source and all other notes: Same as Table 3.5.

By normalising this NSDP and NDP with population, the growth rate in the per capita NSDP and NDP has been examined. It shows that during 1950s while the per capita NSDP was growing negatively (–0.32%), the per capita NDP was growing at 1.57 per cent per annum. By 1990s, the per capita NSDP has grown at 2.44 per cent per annum and NDP has grown at 3.96 per cent. For the whole period (1950–51 to 2005–06), while the per capita NSDP has grown at 0.99 per cent per annum, the per capita NDP has grown at 1.81 per cent per annum.

As regards sectoral growth rates, as the Tables depict, during 1950s, the annual compound growth rate of primary sector of Odisha was lower than the growth rate of primary sector of all-India (1.25% and 2.64% respectively). Since then, the growth rates have increased in both: by 1990s, the annual primary sector growth rate in Odisha and India were 2.75 per cent and 3.33 per cent respectively. During the last 6 years subperiod, the annual growth rate of primary sector for Odisha and India is 6.54 per cent and 6.09 per cent respectively. For the whole period, the annual growth rate for Odisha and India remained at 1.81 per cent and 2.69 per cent respectively. A point can be noted here is that in the recent period, there is a sign of convergence in the growth rate of primary sector.

The compound annual growth rate of secondary sector in Odisha remained at 10.30 per cent during 1950s, which was almost double than that of India (5.62%) at that time. But Odisha economy could not maintain this pace and by 1990s it fell to 1.55 per cent per annum. The same for all-India was 5.84 per cent. In the recent times (2000–01 to 2005–06), the secondary sector has grown at 10.63 per cent and 12.06 per cent per annum for Odisha and India respectively. For the whole period (1950–51 to 2005–06), the secondary sector of Odisha and India has grown at 3.49 and 5.37 per cent per annum respectively.

The growth rate of service sector has remained better than other sectors, both in Odisha as well as in all-India. In Odisha, during 1950s the annual compound growth of the service sector was 4.54 per cent; the same for India during that time was 4.43 per cent. Over time, the growth rates have shown persistent increase and during 1990s; the annual compound growth of the service sector of Odisha was 6.29 per cent while the same for India during that time was 8.14 per cent. The growth rates have further increased in the recent period (2000–01 to 2005–06). The annual compound growth rate of the service sector during this period was 7.84 per cent and 12.45 per cent for Odisha and all-India respectively. In the whole period, 1950–51 to 2005–06, the annual growth rate was 2.81 per cent and 5.86 per cent for Odisha and all-India respectively.

Sectoral Contribution to NSDP Growth

A sector's contribution to overall growth depends not only on its growth rate, but also on its relative size. A sector may grow faster than others, but its contribution to overall growth may not be large if the sector's absolute size is small and vice versa. Hence, here we have combined both the growth rate of each sector and their respective share to NSDP (or relative size), which will provide better insight into the structural change and composition of the NSDP.

Chenery and Syrquin (1986) had shown the relation between the aggregate and sectoral growth by differentiating with respect to time. By definition, the total output growth (the sum of sectoral output), $V = \sum V_i$, and expressing the result in growth rate terms:

$$gv = \sum p_i gv_i$$

where gv_i and gv are the growth rates of Vi and V respectively, and the weights are the sectoral output shares, $p_i = Vi/V$. By definition, $p_i gv_i$ is the contribution of ith sector to the overall growth. This methodology has been followed here and results are present in Table 3.7.[6]

Table 3.7 *Sectoral compositions of the NSDP growth*

Periods	*Total NSDP*	*Sectoral contributions to NSDP growth*			*Share of each sector in the growth of NSDP*		
		Primary	*Secondary*	*Service*	*Primary*	*Secondary*	*Service*
1950s	1.35	–0.82	1.07	1.10	–60.50	79.15	81.34
1960s	3.43	2.65	0.06	0.71	77.47	1.79	20.75
1970s	2.39	0.64	1.04	0.71	26.97	43.37	29.66
1980s	4.49	1.52	1.01	1.96	33.89	22.57	43.53
1990s	3.90	1.25	0.26	2.39	32.10	6.64	61.26
2001 to 2006	7.86	2.56	1.47	3.83	32.60	18.74	48.66

Source: Same as Table 3.1.

Table 3.7 depicts a sharp declining contribution of the primary sector towards the overall NSDP growth over the time period. During 1960s, the primary sector has contributed 77 per cent to the NSDP growth, which fell to 33 per cent during 2001–06.[7] The contribution of the secondary sector towards the overall NSDP growth over the time period has shown a lot of variations. During 1960s, while the secondary sector has contributed 2 per cent to the NSDP growth during 2001–2006 it increased to 19 per cent. The service sector has shown a sharp increase in its share to the growth of NSDP. The contribution was 21 per cent in 1960s, which increased to 49 per cent during 2001–2006.

In short, during the last fifteen years (1990–91 to 2005–06), the primary, secondary and service sectors contributed on an average 32 per cent, 12 per cent and 56 per cent to the NSDP growth of Odisha.

The results are summarised below:

(a) There is very slow progress in the rate of growth of per capita income, perhaps due to less demand for the product of secondary sector.
(b) The service sector has shown a tremendous growth and within the service sector, it is the hotel and restaurant, finance and related sectors, public administration and other community services that have shown an impressive growth.
(c) However, this kind of output transformation (service sector's supremacy over secondary sector) may be less sensitive to the existing hunger and deprivation in Odisha, as the people affected with hunger and deprivation primarily depend on the primary sector. In this context, it may be useful to see the employment scenario in the economy.

Section II

Employment Transformations in Odisha Economy

Previous analysis shows that transformations have occurred in the sectoral composition of NSDP in Odisha. The share of primary sector has declined and that of service sector has increased and the share of secondary sector remained almost stationary or has experienced a marginal improvement. The per capita income has shown an increase but has grown at a tardy pace. Given this, as discussed earlier, theoretically the shift in product is expected to be accompanied by a shift in the employment structure as well and that can reduce poverty. This section examines the growth and structure of employment in Odisha. But before that, we have looked into the growth of population in the state very briefly.

Table 3.8 *Population growth of Odisha since 1951*

			Pt/p0		*Ln(Pt/PO)*		*Avg. Annual Exp. Gth**	
Years	*India*	*Odisha*	*India*	*Odisha*	*India*	*Odisha*	*India*	*Odisha*
1951	35,66,28,312	1,46,45,946	1.23	1.20	0.21	0.18	2.08	1.81
1961	43,89,36,918	1,75,48,846	1.25	1.25	0.22	0.22	2.22	2.24
1971	54,79,49,809	2,19,44,615	1.21	1.20	0.19	0.18	1.94	1.84
1981	66,52,87,849	2,63,70,271	1.26	1.20	0.23	0.18	2.31	1.83
1991	83,85,67,936	3,16,59,736	1.23	1.16	0.20	0.15	2.04	1.51
2001	1,02,86,10,328	3,68,04,660						
2005	1,11,30,00,823	3,91,14,167	—	—	—	—	—	—

Note: The growth rate has been calculated by using exponential growth rate, i.e. $Pt = P0e^{rt}$, where p0 is the initial population, pt is the population at the end of the year,'t' is the time at the end of the period and 'r' is the exponential growth rate. *The average annual growth rate of 1951 implies the growth rate of that decade, i.e. 1951 to 1961. Similarly other decades' growth rate.

Source: Census of India, Various Issues.

Table 3.8, presents the average annual exponential growth rate of population in Odisha vis-à-vis India. It shows that the average annual growth rate of population in Odisha has always remained lower than all-India during 1951 to 2001 (except similar growth rate during 1961–71). This may imply that any level of higher Worker Population Ratio in Odisha (or even equal WPR) should put Odisha in a better-off position given all other things being constant.

Structure of Employment

There are two major data sources, which provide information on the economic characteristics of the population in India, namely Census of India and NSSO. The various quinquennial rounds of the NSSO data are used to examine the growth and structure of employment.[8]

Worker Population Ratio or Work Participation Rate

Work participation rate is one of the best indicators for informing on employment scenario. As per the NSSO definition, the work-force participation rates (WFPR) or worker-population ratio (WPR) refers to the number of persons or person-days employed per thousand persons or person-days. Table 3.9 presents the WPR of India and Odisha.[9] The results are presented below (also see Appendix A3.8):

(a) Overall, the WPR of Odisha was either at par with or marginally lower than that of all-India in the initial level, but has shown an improvement in the subsequent period and has remained higher than all-India. This is due to the higher percentage of rural employment in Odisha. During 1977–78 the

WPR of Odisha was 41.1 per cent, which increased to 43.6 per cent during 2004–05. For all-India, these figures are 42.3 per cent during 1977–78, which marginally fell to 42 per cent during 2004–05.

(b) The WPR across the location shows that it has remained higher in urban India than urban Odisha in both the cases (males and females), though in case of urban females the gap is higher than males. During 1977–78, the WPR in rural Odisha was 44.9 per cent, which fell to 42.2 per cent during 2004–05. For rural India, these figures are 44.4 during 1977–78 and 43.9 per cent during 2004–05. In urban Odisha, the WPR during 1977–78 was 34.1 per cent, which fell to 33.4 during 2004–05. Similarly, figures for urban India are 34.1 per cent during 1977–78 and 36.5 per cent during 2004–05.

Table 3.9 *Trend in percentages distribution of workers in the rural and urban areas of India and Odisha from 1977–78 to 2004–05*

NSS Round (year)	*All-India*			*Rural India*			*Urban India*		
	Male	*Female*	*Total*	*Male*	*Female*	*Total*	*Male*	*Female*	*Total*
32 (1977–78)	54.31	29.7	42.3	55.2	33.1	44.4	50.8	15.6	34.1
38 (1983)	53.8	29.51	42	54.7	34	44.5	51.2	15.1	34
50 (1993–94)	54.5	28.6	42	55.3	32.8	44.4	52.1	15.5	34.7
61 (2004–05)	54.7	28.7	42	54.6	32.7	43.9	54.9	16.6	36.5
NSS Round (year)	*All Odisha*			*Rural Odisha*			*Urban Odisha*		
	Male	*Female*	*Total*	*Male*	*Female*	*Total*	*Male*	*Female*	*Total*
32 (1977–78)	55.7	26.2	41.1	56.4	27.4	44.9	50.4	14.9	34.1
38 (1983)	56.5	27.3	42.1	57.5	29.4	43.4	50.1	11.1	32.1
50 (1993–94)	55.7	29.6	42.8	56.6	31.7	44.2	51	15.7	34.1
61 (2004–05)	57.2	29.5	43.6	58.6	32.2	42.2	50.4	14.8	33.4

Source: Sarvekshana Reports of the various quinquennial NSSO rounds.

Given the lower population growth and higher WPR in Odisha compared to all-India, the conditions of the people in the state may be expected to be better but this has not happened in reality. Though, in the post-1990 phase, the WPR of Odisha remained higher than the WPR of India, the magnitude of poverty in the former has remained much higher. Hence, magnitude of employment alone may not explain poverty better; rather where these workers work, how much they earn etc. may explain poverty better.

Hence, we have looked into the sectoral distribution of these workers in Table 3.11. But before that, we have examined the distribution of total employment of Odisha by their location (i.e. in terms of rural and urban composition) in Table 3.10 and Appendix A3.9.

The results show that:

(a) Overall, even in 2004–05, as high as 88 per cent of the employed people are located in the rural sector and only 12 per cent are located in the urban sector.

(b) Sectorwise, the proportion of rural workers in the primary, secondary and service sector are 97 per cent, 83 per cent and 63 per cent respectively.

(c) Over the last 27 years, there is very little change in the share of employment in urban and rural areas. Probably, that might be the reason why the rural wage rate in Odisha remained at low level and perhaps the reason for poverty in the state.

Not only the transfer of workers from rural to urban area is very less in Odisha, but as Table 3.11 (also see Appendix A3.9) shows the transfer of employed persons from agriculture to non-agriculture also remained less impressive in Odisha compared to all-India.

Table 3.10 *Percentages distributions of workers by rural urban location in Odisha from 1977–78 to 2004–05*

		Primary	*Secondary*	*Service*	*Total*
Rural	1977–78	97.75	74.29	59.98	91.22
	1983	97.47	78.47	66.00	90.78
	1993–94	97.35	73.16	57.33	88.91
	2004–05	97.16	82.68	62.76	88.01
Urban	1977–78	2.25	25.71	40.02	8.78
	1983	2.53	21.53	34.00	9.22
	1993–94	2.65	26.84	42.67	11.09
	2004–05	2.84	17.32	37.24	11.99

Source: Sarvekshana Reports of the various quinquennial NSSO rounds.

Table 3.11 presents the sectoral distribution of employment at one digit level[10] in Odisha. The main results are presented below:

(a) In rural areas, primary sector has been the major employment provider (85.73% during 1977–78 and 69.67% during 2004–05) and within primary sector, agriculture provides almost all employment. The secondary and service sectors have shown some increase during this period (secondary sector increased from 5.85% during 1977–78 to 16.69% during 2004–05 and service sector increased from 8.43% to 13.64% during this period), yet the share of employment in the non-agricultural sector has remained only one-third of the total employment.

(b) In the urban sector, though service sector has been the major share of employment provider, it did not show significant improvement over time. The share of service sector employment in 1977–78 was 58.47 per cent, which remained almost at the same level in 2004–05 (59.39%). As the primary sector employment in the urban sector of Odisha is concerned, there has been around 5 percentage point decline (from 20.53% in 1977–78 and 14.95% in 2004–05). This decline has been mostly substituted by the secondary sector's increase in employment from 21.01 per cent in 1977–78 to 25.66 per cent in 2004–05.

(c) Overall (both in rural and urban areas together), during 1977–78 and 2004–05, there has been around 17 percentage point decline in the employment share of the primary sector (from 80% to 63%); 10 percentage point increase in the secondary sector (7.17% to 17.76%) and 6 percentage point increase in the service sector (from 12.82% to 19.12%).

(d) The employment diversification for all-India, during 1977–78 and 2004–05 is presented below: 10 percentage point decline in primary sector (66.25 to 56.94%); 3 percentage point increase in the secondary sector (14.87 to 18.10%) and around 6 percentage point increase in service sector (18.88 to 24.96%).

Overall, this shows that for all-India the magnitude of diversification of employment remained lower than Odisha and the level of employment in the primary sector for India remains much lower than that of Odisha (around 6 percentage point difference). This might be a reason for the disadvantaged position faced by the workers of Odisha compared to all-India.

Table 3.11 *Trend in percentage distribution of workers across the sectors by rural urban location in Odisha and India from 1977–78 to 2004–05*

		Odisha				*India*			
	Years	*Primary*	*Secondary*	*Service*	*Total*	*Primary*	*Secondary*	*Service*	*Total*
Rural	1977–78	85.73	5.84	8.43	100	83.92	7.70	8.60	100
	1983	79.71	9.83	10.47	100	81.76	8.58	9.39	100
	1993–94	82.03	8.28	9.69	100	79.00	9.60	11.40	100
	2004–05	69.67	16.69	13.64	100	73.20	13.20	13.60	100
Urban	1977–78	20.53	21.01	58.47	100	*			
	1983	20.36	26.55	53.09	100	15.7	32.3	51.1	100
	1993–94	17.87	24.35	57.78	100	13.5	30.9	55.5	100
	2004–05	14.95	25.66	59.39	100	9.6	33.3	57.2	100
Total	1977–78	80.01	7.17	12.82	100	66.25	14.87	18.88	100
	1983	74.23	11.37	14.40	100	69.24	13.12	17.64	100
	1993–94	74.91	10.06	15.02	100	64.67	14.26	21.07	100
	2004–05	63.11	17.76	19.12	100	56.94	18.10	24.96	100

Source: Sarvekshana Reports of the various quinquennial NSSO rounds.

Table 3.12 *Percentages distributions of workers by gender (male and female) in Odisha from 1977–78 to 2004–05*

		Primary	*Secondary*	*Service*	*Total*
Male	1977–78	66.94	71.24	76.05	68.41
	1983	65.71	66.28	80.30	67.88
	1993–94	62.65	67.17	81.50	65.94
	2004–05	61.97	64.17	83.67	66.51
Female	1977–78	33.06	28.76	23.95	31.59
	1983	34.29	33.72	19.70	32.12
	1993–94	37.35	32.83	18.50	34.06
	2004–05	38.03	35.83	16.33	33.49

Source: Sarvekshana Reports of the various quinquennial NSSO rounds.

Table 3.12 (also see Appendix A3.11) presents the percentage distribution of total workers by their gender. The table shows that:

(a) There has been a small increase in the share of female workers to total workers from 1977–78 to 2004–05 (31.5% in 1977–78 to 33.5% in 2004–05). During the same period, the share of male workers to total workers has shown a small decline (68.5% to 66.5%). A point can be made here is that such underrepresentation of females in the total employment may put them in a vulnerable position.

(b) During 1977–78 to 2004–05, while there is a decline in the share of male workers in the primary sector (by 5 percentage point) and in the secondary sector (7 percentage point), there is an increase in the share of female workers in these sectors. But in case of service sector, during the same time period, there is 7 percentage point increase in the share of male workers. This implies that service sector is predominantly occupied by male workers.

Table 3.13 *Trend in percentage distribution of workers across the sectors by gender in Odisha from 1977–78 to 2004–05*

		Primary	*Secondary*	*Service*	*Total*
Male	1977–78	78.28	7.47	14.25	100
	1983	71.86	11.10	17.03	100
	1993–94	71.18	10.25	18.57	100
	2004–05	58.80	17.14	24.06	100
Female	1977–78	83.75	6.53	9.72	100
	1983	79.24	11.93	8.83	100
	1993–94	82.14	9.70	8.16	100
	2004–05	71.67	19.00	9.33	100
Total	1977–78	80.01	7.17	12.82	100
	1983	74.23	11.37	14.40	100
	1993–94	74.91	10.06	15.02	100
	2004–05	63.11	17.76	19.12	100

Source: Sarvekshana Reports of the various quinquennial NSSO rounds.

Given the discussion on the structure of employment in Odisha, Table 3.14 (also refer Appendix A3.13) presents the growth rate of employment across different sectors. The Table shows that there are three time points where the annual compound employment growth rate has grown beyond 2%. Those are (i) during 1977–78 to 1983 (2nd to 3rd quinquennial round), where the annual compound employment growth was 2.53%, (ii) during 1987–88 to 1993–94 (4th to 5th quinquennial round), where it has grown at 3.01%[11] and (iii) during 1999–00 to 2004–05 (6th to 7th quinquennial round), where it has grown at 2.78%. In the case of (i) and (ii), the primary sector has grown at an annual rate of 1% and 4.09% respectively; in the case of (iii) it has grown at an annual compound growth rate of 0.33%.

Further, as Table 3.13 (see Appendix A3.12) shows, more female employees have concentrated in the primary sector (84% in 1977–78 to 71% in 2004–05) compared to their male counterparts (78% in 1977–78 to 59% in 2004–05). However, such bias was not found in the secondary sector. But in the service sector, the share of male workers remained higher than females.

Moreover, in the service sector, the share of male workers has increased while the share of female workers has declined during 1977–78 and 2004–05 (for males, 14.25% during 1977–78 to 24% during 2004–05 and for females, 9.72% during 1977–78 to 9.33% 2004–05).

Table 3.14 *Trends in the employment Growth in Odisha*

		Primary	*Secondary*	*Service*	*Total*
Male	2nd to 3rd	0.63	10.82	6.08	2.37
	3rd to 4th	0.52	2.65	4.01	1.39
	4th to 5th	2.80	–0.75	1.27	2.11
	3rd to 5th	1.65	0.94	2.63	1.75
	5th to 6th	0.47	4.69	1.80	1.19
	6th to 7th	–0.44	10.03	7.52	2.71
	5th to 7th	0.02	7.33	4.62	1.95
Female	2nd to 3rd	1.74	16.06	0.91	2.87
	3rd to 4th	–0.36	5.09	0.61	0.44
	4th to 5th	6.49	–3.82	3.08	4.89
	3rd to 5th	3.01	0.54	1.84	2.64
	5th to 6th	–1.02	11.81	–1.72	0.49
	6th to 7th	1.66	5.79	8.07	2.91
	5th to 7th	0.31	8.76	3.06	1.69
Total	2nd to 3rd	1.00	12.42	4.93	2.53
	3rd to 4th	0.22	3.50	3.37	1.09
	4th to 5th	4.09	–1.82	1.59	3.01
	3rd to 5th	2.13	0.80	2.48	2.04
	5th to 6th	–0.08	7.24	1.19	0.95
	6th to 7th	0.33	8.39	7.61	2.78
	5th to 7th	0.13	7.82	4.35	1.86

Source: Sarvekshana Reports of the various quinquennial NSSO rounds.

The head count ratio of Odisha has shown substantial decline during 1983 to 1993–94 (i.e. 3rd to 5th quinquennial rounds). This may imply that the overall employment growth induced by the primary sector may have better impact on poverty reduction than the employment growth rate induced by service or secondary sector. During 1999–00 to 2004–05, however, the overall employment growth induced by service sector has not resulted in poverty reduction.

We have calculated the annual compound growth rate of employment from 1983 to 1993–94 and from 1993–94 to 2004–05 along with the other quinquennial round's employment growth. That will allow us for a clear comparison of employment growth and poverty reduction. The result shows that there is not much difference in the overall employment growth rate between the two periods: (i) 1983 to 1993–94 (around 2%) and (ii) 1993–94 to 2004–05 (1.9%). Yet there is around 17 percentage point poverty reduction in the former period compared to only 2 percentage point decline in poverty in the latter. The reason may lie in the primary sector's growth in recent years.

Further, we have examined the sensitivity effects of output on employment, which has been presented in Table 3.15.

Table 3.15 *Employment elasticity in Odisha*

Rounds Q	*Agri*	*M and Q*	*Primary*	*Manf.*	*EGWC*	*Sec.*	*T and H*	*Trans*	*FIPAD*	*Service*	*Total*
2nd to 3rd	–0.47	6.89	–0.52	–1.42	4.73	–3.31	–7.57	3.61	1.03	2.13	–2.28
3rd to 4th	0.01	1.59	0.10	–0.07	11.71	0.58	1.19	0.55	0.14	0.47	0.25
4th to 5th	2.69	–0.13	1.57	–0.05	–0.64	–0.29	0.05	0.95	0.30	0.25	0.68
3rd to 5th	1.16	0.37	0.88	–0.07	1.16	0.13	0.46	0.67	0.21	0.37	0.47
5th to 6th	–0.09	–0.82	–0.07	0.51	3.38	1.02	0.83	0.23	–0.19	0.13	0.19
6th to 7th	0.09	0.62	0.06	1.66	19.11	3.33	1.08	0.66	1.18	0.98	0.47
5th to 7th	0.12	–0.29	0.05	3.12	37.16	6.24	1.69	0.81	0.75	1.14	0.64

Notes: Agri. – Agriculture, M and Q – Mining and quarrying, Manf. – Manufacturing, EGWC – Electricity, Gas, Water supply and Construction, T and H – Trade and Hotel, Trans – Transport, FIPAD – Finance, Insurance and public administration etc. Q – Quinquennial rounds.

Source: Sarvekshana Reports of the various quinquennial NSSO rounds.

Section III

Linking up NSDP Transformation with Employment Transformation

The previous section shows a moderate degree of output transformation in Odisha, but these output transformation failed to generate corresponding employment growth and absorption of workers. In Table 3.15 we have tried to find out the employment absorptive capacity of different sectors by calculating the employment elasticity for each sector.

Employment elasticity of output (Eo) measures the sensitivity of employment to output. It is the percentage change in employment to percentage change in income. Employment Elasticity (Eo) = (% change in employment/% change in output). The analysis shows:

(a) In the primary sector, only once the employment elasticity is greater than unity is during the 4th and 5th quinquennial rounds, where the Eo is 1.57, i.e. the output growth has a positive effect on employment. For all other rounds, the Eo is either negative or less than unity, i.e. employment is less sensitive to output.

(b) In the secondary sector, since 1990s the Eo is greater than 1, which shows higher sensitivity of the output on employment.

(c) In the service sector such situation (Eo greater than 1) arises twice during the 2nd to 3rd quinquennial rounds and during 6th to 7th quinquennial rounds. In all other rounds the Eo has remained low. Overall, for all sectors as a whole, the Eo has never been greater than 1. Such low Eo may be the reason behind both lopsided structural transformations and hunger and deprivation in the state. The comparison of employment analysis, shows during 3rd to 5th quinquennial rounds the overall Eo was 0.47 but during that time period the Eo in primary sector was 0.88, which is not only higher than the overall Eo but also higher than the Eo of primary sector during 5th to 7th quinquennial rounds.

(d) The Eo of both service and secondary sector have remained much higher than primary sector in the later period yet it has very less impact on poverty reduction (only by 2 percentage points). In the former period the Eo has remained higher only in primary sector yet poverty has declined by 17 percentage points. Hence, Eo of primary sector may be more sensitive to poverty than other two sectors.

Differential Labour Productivity and Poverty

Ideally, when a dual model successfully works out in an economy, then the labour productivity of agricultural sector should increase and the gap of labour productivity among different sectors should decline (the structural transformation also assumes so) but as can be seen from Table 3.16, it has happened only to a limited extent. Initially, i.e. during 1977–78, the labour productivity of the secondary sector was 2.86 times higher than that of primary sector, which has come down to 1.39 times higher during 2004–05. This is due to the decline in the labour productivity of secondary sector and the increase of the same in the primary sector. Similarly, in the initial period the labour productivity of service sector was 2.42 times higher than that of primary sector, which has increased to almost 4 times higher during 2004–05. The labour productivity of service sector compared to secondary sector has increased during 1977–78, the labour productivity of service sector was 0.84 times of the secondary sector, which has reached to 2.84 times during 2004–05. Hence, as the above analysis shows, there was a structural transformation both in output and employment in the state but it has remained lopsided by the service sector. Since the poor primarily depends on primary sector the tardy growth of primary sector both in terms of employment and output and the declining share in NSDP might have failed to reduce the existing hunger and deprivation in the state.

The changes in each kind of work to explain its links to hunger and deprivation in Odisha are also examined.

Table 3.16 *Labour productivity and compound annual growth rate of labour productivity in Odisha*

	Agri	*M and Q*	*Primary*	*Manf.*	*EGWC*	*Sec.*	*T and H*	*Trans*	*FIPAD*	*Service*	*Total*
1977–78	8,186	26,112	8,339	16,049	57,805	23,815	20,323	27,459	19,180	20,236	10,974
1983	7,063	19,046	7,204	5,236	42,707	10,951	14,447	24,846	19,075	17,836	9,161
1987–88	7,812	15,220	7,973	9,319	18,904	12,342	13,826	34,193	25,115	21,437	10,700
1993–94	6,869	41,958	7,422	11,038	39,461	18,310	20,298	34,690	30,045	26,904	11,444
1999–00	6,564	1,61,879	7,839	14,186	27,323	18,213	21,211	49,966	54,073	38,895	13,972
2004–05	7,402	2,07,577	9,923	12,500	16,077	13,784	20,582	64,613	51,888	39,209	16,209
					Annual compound growth rate of labour productivity						
2nd to 3rd	2.91	−6.12	−2.88	−20.07	−5.87	−14.39	−6.60	−1.98	−0.11	−2.49	−3.55
3rd to 4th	2.03	−4.39	2.05	12.22	−15.04	2.42	−0.87	6.59	5.66	3.75	3.15
4th to 5th	−2.54	22.49	−1.42	3.44	15.86	8.21	7.98	0.29	3.65	4.65	1.35
3rd to 5th	−0.28	8.22	0.30	7.74	−0.79	5.27	3.46	3.39	4.65	4.20	2.25
5th to 6th	−0.90	31.00	1.10	5.15	−7.09	−0.11	0.88	7.57	12.47	7.65	4.07
6th to 7th	2.43	5.10	4.83	−2.50	−10.06	−5.42	−0.60	5.28	−0.82	0.16	3.01
5th to 7th	0.75	17.34	2.95	1.25	−8.59	−2.80	0.14	6.42	5.62	3.84	3.54

Source: Sarvekshana Reports of the various quinquennial NSSO rounds and CSO.

Transformations in the Nature of Employment

The above analysis shows a less impressive employment growth in Odisha and that might be a causal factor behind the higher magnitude of hunger and deprivation in the state. Further, we have examined the employment transformation by the nature of employment. Tables 3.17 and 3.18 show the details of the share and growth of different types of employment in Odisha.[12] However, due to definitional changes the trend comparison may be inappropriate, hence the Table 3.17 has been remade (Table 3.17A) by clubbing some of the types of job together, which may allow us for trend comparison in a meaningful way.

Table 3.17 *Percentages distribution of usual status workers (UPSS) by their place of location and gender in Odisha from 1983 to 2004–05*

	Years	*T_M*	*T_F*	*T_M_F*	*R_M*	*R_F*	*R_T*	*U_M*	*U_F*	*U_T*
Own Account workers	1983	37.91	11.84	29.45	39.21	11.81	29.83	29.00	12.35	26.19
	1993–94	37.28	10.78	28.24	38.40	10.35	28.42	29.16	17.31	26.70
	2004–05	38.86	10.16	28.99	39.02	10.08	28.63	37.70	11.27	32.10
Employers	1983	—	—	—	—	—	—	—	—	—
	1993–94	1.81	0.88	1.49	1.78	0.92	1.48	1.99	0.36	1.65
	2004–05	0.75	0.23	0.57	0.80	0.18	0.57	0.47	0.99	0.58
Unpaid Family workers	1983	13.80	40.52	22.47	14.61	41.09	23.68	8.26	30.53	12.02
	1993–94	14.70	43.22	24.44	15.92	44.73	26.19	5.86	20.26	8.87
	2004–05	15.54	49.99	27.38	16.61	51.65	29.19	8.02	25.68	11.76
Regular/salaried/wage employee	1983	15.81	2.76	11.58	11.28	1.77	8.02	46.85	20.35	42.38
	1993–94	11.29	3.14	8.51	6.19	1.41	4.49	48.12	29.23	44.20
	2004–05	11.01	4.26	8.69	7.50	2.15	5.58	35.69	35.20	35.59
Casual wage labour in public works	1983	0.75	0.35	0.62	0.55	0.28	0.46	2.11	1.59	2.02
	1993–94	0.66	0.39	0.57	0.73	0.41	0.62	0.16	0.00	0.12
	2004–05	0.16	0.13	0.15	0.18	0.14	0.17	0	0	0
Other type of work	1983	31.72	44.53	35.88	34.34	45.06	38.01	13.78	35.18	17.39
	1993–94	34.27	41.57	36.77	36.98	42.15	38.84	14.69	32.84	18.49
	2004–05	33.68	35.23	34.21	35.89	35.80	35.86	18.12	26.87	19.98

Note: During 1983, there were .30 per cent of the employees attached to either moneylenders or landlords and have been excluded here.

Source: NSSO, Unit level data various quinquennial rounds.

Table 3.17 reveals the following:

(a) The share of total own account workers has remained almost stagnant at around 29 per cent though there is a little variation around this figure during 1983 to 2004–05. In case of female workers, it has marginally declined from 11.84 per cent in 1983 to 10.16 per cent in 2004–05, whereas in case of male workers the share has shown a marginal increase from 37.91 per cent in 1983 to 38.86 per cent in 2004–05. In case of rural area, the share has declined from 29.83 per cent in 1983 to 26.59 per cent in 2000 and then shown a marginal increase to 28.63 per cent in 2004–05. In urban areas, it has shown a steady increase from 26.19 per cent in 1983 to 32.10 per cent in 2004–05.

(b) As the share of 'employers'[13] category is concerned, it constitutes only around one per cent of the total workforce. Overall this category has shown a decline from 1.49 per cent in 1993–94 to 0.57 per cent in 2004–05.

(c) The share of total unpaid family labour has shown an increase from 22.47 per cent in 1983 to 27.38 per cent in 2004–05. While the share of male workers has increased from 13.80 per cent in 1983 to 15.54 per cent in 2004–05, the share of female workers has increased from 40.52 per cent in 1983 to 49.99 per cent in 2004–05. In case of rural area, the share has increased from 23.68 per cent in 1983 to 29.19 per cent in 2004–05, whereas in case of urban area, it has slide from 12.02 per cent in 1983 to 11.76 per cent in 2004–05. However, here a general observation is that in almost all cases, there has been some increase in 1993–94 from 1983 and then it again fell in 1999–00 from 1993–94 and then it has again increased in 2004–05.

(d) The share of total salaried/wage employee has shown a decline from 11.58 per cent in 1983 to 8.25 per cent in 1999–00 and then it has shown a slight increase to 8.69 per cent in 2004–05. While the share of male workers has declined from 15.81 per cent in 1983 to 10.92 per cent in 1999–00 and then increased to 11.01 per cent in 2004–05, the share of female workers has shown an overall increase (with a little variations) from 2.76 per cent in 1983 to 4.26 per cent in 2004–05. In case of rural area, the share has declined from 8.02 per cent in 1983 to 4.19 per cent in 1999–00 and further increased to 5.58 per cent in 2004–05, whereas in case of urban area, it has decreased from 42.38 per cent in 1983 to 35 per cent in 2004–05.

(e) As the share of total casual wage labour in public works category is concerned, it has never gone beyond 1 per cent of the total workers. Only in urban Odisha it was around 2 per cent in 1983 and it has come down to zero in 2004–05.

(f) The share of total other types of workers has shown an increase from 35.88 per cent in 1983 to 43.57 per cent in 1999–00 but recently it fell to 34.21 per cent in 2004–05. While the share of male workers has increased from 31.72 per cent in 1983 to 41.33 per cent in 1999–00 and then fell to 33.68 per cent in 2004–05, the share of female workers has decreased from 44.53 per cent in 1983 to 41.57 per cent in 1993–94 and then increased to 47.93 per cent in 1999–00 and then again fell to 35.23 per cent in 2004–05. In case of rural area, the share has increased from 38.01 per cent in 1983 to 46.93 per cent in 1999–00 but then it fell to 35.86 per cent in 2004–05, whereas in case of urban area, the share has increased from 17.39 per cent in 1983 to 20.69 per cent in 1999–00 but then it has marginally declined to 19.98 per cent in 2004–05.

However, due to certain definitional changes in the NSSO data and due to non-capturing of certain category of job separately, at the disaggregated level, the trend comparison has certain limitations. Hence, we have clubbed certain categories of employee into a broad category of self-employed (own account workers, employers and unpaid family labour), regular labour and casual labour (casual labour in the public work and other types of workers) in Table 3.17(A).

Table 3.17(A) *Percentage distribution of usual status workers in broad employment group (UPSS) by their place of location and gender in Odisha from 1983 to 2004–05*

		T_M	T_F	T_M_F	R_M	R_F	R_T	U_M	U_F	U_T
Own account workers, employers and unpaid family labour	1983	51.72	52.36	51.92	53.82	52.90	53.51	37.26	42.88	38.21
	1993–94	53.78	54.89	54.16	56.10	56.01	56.07	37.02	37.93	37.21
	2004–05	55.15	60.38	56.95	56.43	61.91	58.40	46.18	37.93	44.44
Regular salaried/wage employee	1983	15.81	2.76	11.58	11.28	1.77	8.02	46.85	20.35	42.38
	1993–94	11.29	3.14	8.51	6.19	1.41	4.49	48.14	29.23	44.19
	2004–05	11.01	4.26	8.69	7.50	2.15	5.58	35.69	35.20	35.59
Casual wage labour in public work and other types of Work	1983	32.47	44.88	36.50	34.89	45.34	38.47	15.88	36.77	19.41
	1993–94	34.93	41.97	37.33	37.71	42.57	39.44	14.85	32.84	18.61
	2004–05	33.84	35.36	34.36	36.08	35.94	36.03	18.12	26.87	19.98

Source: NSSO, Unit level data various quinquennial rounds.

Table 3.17(A) shows the following:

(a) The share of workers in the self-employed category has increased from 51.92 per cent in 1983 to 56.95 per cent in 2004–05 (however, the share has declined from 54.16% in 1993–94 to 47.92% in 1999–00). The share of male workers has increased from 51.72% in 1983 to 55.15% in 2004–05 (here too, the share has declined from 53.78% in 1993–94 to 47.40% in 1999–00). The share of female workers has increased from 52.26% in 1983 to 60.38% in 2004–05 (here too there was a decline from 54.89% in 1993–94 to 48.94% in 1999–00). In case of rural area, the share has increased from 53.51% in 1983 to 58.40% in 2004–05 (which also has shown a decline from 56.07% in 1993–94 to 48.67% in 1999–00). In urban area, the share has increased from 38.21 per cent in 1983 to 44.44 per cent in 2004–05 (which also has shown a decline in 1993–94 over 1983).

(b) The share of total casual labour has shown an increase from 36.50 per cent in 1983 to 43.85 per cent in 1999–00 and then has fallen to 34.36 per cent in 2004–05. In case of male, it has increased from 32.47 per cent in 1983 to 41.68 per cent in 1999–00 and then has fallen to 33.84 per cent in 2004–05. In case of female, the share has increased from 44.88 per cent in 1983 to 48.08 per cent in 1999–00 and then it has fallen to 35.36 per cent in 2004–05. In case of rural area, it has increased from 38.47 per cent in 1983 to 47.14 per cent in 1999–00 and then has fallen to 36.03 per cent in 2004–05. In case of urban area, the share has increased from 19.49 per cent in 1983 to 21.42 per cent in 1999–00 and then it has fallen to 19.98 per cent in 2004–05.

Overall, the nature of employment has shown a transformation, in that the share of casual labour has increased and that of self-employed and regular labour have declined over time (1983 to 1999–00). The 2004–05 round has shown a reverse change – the share of self-employment has increased and that of casual labour has decreased. During 1999–00 to 2004–05, while the share of self employment and regular labour has increased that of casual labour has decreased. During 1999–00 and 2004–05, the incidence of poverty has remained stagnant. This suggests that the benefit of increase in the share of regular labour or self-employed might have gone to the non-poor section. Below, we have explored the growth of employment in the said categories.

The employment growth rate in the state has been calculated by applying the simple compound average annual growth rate formula. Table 3.18 shows the following:

(a) The annual compound growth rate of self-employed workers has grown at 2.28 per cent during 1983 to 1993–94, which has fallen to – 0.64 per cent during 1993–94 to 1999–00 and then has increased to 5.10 per cent during 1999–00 to 2004–05. The trend has remained similar for male, female and rural area. However, in urban area, there was no negative growth. In urban area, while the employment has grown at 2.89 per cent during 1983 to 1993–94, during 1993–94 to 1999–00, it has grown at 6.15 per cent per annum and then it has shown a decline to 3.46 per cent during 1999–00 to 2004–05. The growth rate of regular labour was negative during 1983 to 1993–94 (–1.24% per annum), but it has improved in the subsequent period (1.15% during 1993–94 to 1999–00 and 2.64% during 1999–00 to 2004–05).

(b) In the case of casual labour, while the growth rate during 1983 to 1993–94 was 2.08 per cent and during 1993–94 to 1999–00 it was 5.16 per cent, the rate has fallen to 3.30 per cent per annum during 1999–00 to 2004–05.

Table 3.18 *Growth rate of employment across different types of workers in Odisha*

	Q rounds	*T_M*	*T_F*	*T_M_F*	*R_M*	*R_F*	*R_T*	*U_M*	*U_F*	*U_T*
Own account workers, Employers and unpaid family labour	3rd to 5th	2.29	2.29	2.28	2.12	2.19	2.13	3.02	2.00	2.89
	5th to 6th	–0.73	–0.47	–0.64	–1.36	–1.02	–1.24	5.61	7.53	6.15
	6th to 7th	4.64	5.90	5.10	4.50	6.08	5.09	4.55	–1.06	3.46
	5th to 7th	1.53	2.47	1.87	5.08	3.15	4.80	1.92	2.66	2.19
Regular salaried/wage employee	3rd to 5th	–1.49	3.11	–1.24	–4.24	–0.63	–4.09	3.37	7.06	3.60
	5th to 6th	1.14	0.76	1.15	0.17	0.42	0.19	–0.78	–2.99	–1.04
	6th to 7th	1.69	9.09	2.64	6.81	11.45	7.32	0.28	13.83	2.57
	5th to 7th	3.44	5.79	3.69	–0.25	5.08	0.75	1.42	4.84	1.89
Casual wage labour in public work and other types of Work	3rd to 5th	2.63	1.13	2.08	2.49	0.97	1.91	2.40	2.09	2.74
	5th to 6th	5.48	4.64	5.16	5.66	4.68	5.28	7.39	3.37	6.18
	6th to 7th	–2.63	–4.51	–3.30	–3.41	–4.95	–3.98	2.37	–1.12	1.25
	5th to 7th	1.02	–0.25	0.54	4.85	1.10	3.69	1.35	–0.04	0.84

Source: NSSO, Unit level data various quinquennial rounds.

The following sub-section discusses the intra-regional aspect of employment patterns in Odisha.

Intra Regional Aspects of Employment in Odisha[14]

As discussed in Chapter 2, due to the existence of the geographical diversity and other socio-economic disparities in the state, there are also variations in the magnitudes of hunger and deprivation across different regions of the state. The variations in the incidence of poverty of different regions of Odisha were presented in previous chapter. Here we examined the same with respect to employment.

First, the population growth across various NSS regions (Coastal, Northern and Central) in Odisha has been presented. The Table 3.19 depicts, there is not much difference among these three regions as far as the average annual rate of population growth is concerned.[15] Given this population growth, we have also examined the pattern of worker population ratio in theses regions.

Table 3.19 *Population growth of different NSS regions of Odisha since 1951*

Regions		*R_M*	*R_F*	*U_M*	*U_F*	*Total*	*Avg. Annual Expn. Gth**
Coastal	1981	54,84,640	55,04,488	7,97,136	6,77,288	1,24,63,552	1.9
	1991	65,19,512	64,35,216	11,43,905	9,75,233	1,50,73,866	1.55
	2001	74,15,708	72,81,501	15,47,153	13,63,029	1,76,07,391	—
	2005	78,07,781	76,50,398	17,68,854	15,23,116	1,87,50,149	—
Northern	1981	40,78,563	40,42,080	6,69,044	5,67,555	93,57,242	1.72
	1991	47,86,360	46,99,455	8,70,698	7,52,523	1,11,09,036	1.39
	2001	54,30,566	53,30,298	10,56,063	9,50,875	1,27,67,802	—
	2005	57,11,905	56,05,742	11,40,824	10,44,155	1,35,02,626	—
Southern	1981	20,64,201	20,77,012	2,07,202	1,92,062	45,40,477	1.87
	1991	24,89,083	24,95,127	2,54,588	2,38,036	54,76,834	1.6
	2001	29,02,696	29,26,653	3,08,384	2,91,734	64,29,467	—
	2005	30,86,785	31,19,483	3,32,961	3,25,998	68,65,227.7	—

Note: The growth rate has been calculated by using exponential growth rate, i.e. $Pt = P0e^{rt}$, where p0 is the initial population, pt is the population at the end of the year, t is the time at the end of the period and r is the exponential growth rate. *The average annual growth rate of 1981 implies the growth rate of that decade, i.e. 1981 to 1991. Similarly other decade' growth rate implies. R_M – Rural Male, U_M – Urban Male, R_F – Rural Female, U_F – Urban Female;
Coastal-Balasore, Bhadrak, Cuttack, Gajpati, Ganjam, Jagatsinghpur, Jajpur, Khurda, Nayagarh, Puri, Kendrapada, (Total 11); Northern – Angul, Bargarh, Balangir, Deogarh, Dhenkanal, Jharsuguda, Keonjhar, Mayurbhanj, Sambalpur, Sonepur, Sundargarh; (Total 11); Southern – Boudh, Kalahandi, Koraput, Malkangiri, Nuapada, Nawarangpur, Kandhamal, Rayagada (Total 8) Grand Total – 30.

Source: Census of India, Various Issues.

Worker Population Ratio or Work Participation Rate

Table 3.20 presents the WPR of different regions of Odisha since 1983. The regional analysis reveals the following:

(a) The WPR has continued to remain higher in southern part of Odisha compared to northern and coastal part of Odisha. And for this matter the

WPR of southern Odisha has remained higher than Odisha and all-India during the time period considered here. The WPR of southern Odisha in 1983, was 55.87 per cent, which fell to 48.90 per cent in 2004–05. Similarly, figures for the northern part of Odisha are 44.55 per cent in 1983 and 48.31 per cent in 2004–05; and in coastal Odisha 33.75 per cent in 1983 and 37.97 per cent in 2004–05.

(b) As the trends of WPR across the regions are concerned, it has persistently declined for southern Odisha; whereas it has increased for the coastal and northern Odisha, where the increase of coastal Odisha is more noticeable.

Table 3.20 *Trend in percentages distribution of worker population ratio in the rural and urban areas among different regions of Odisha from 1983 to 2004–05*

NSS Round (year)	*Coastal Odisha*								
	Coastal			*Rural Coastal*			*Urban Coastal*		
	Male	*Female*	*Total*	*Male*	*Female*	*Total*	*Male*	*Female*	*Total*
38 (1983)	51.96	15.64	33.75	52.32	17.03	34.56	49.61	6.17	28.32
50 (1993–94)	51.89	17.55	34.92	51.92	18.10	35.03	51.70	13.38	34.21
61 (2004–05)	55.87	20.85	37.97	56.84	21.90	38.81	50.01	13.52	32.52
NSS Round (year)	*Northern Odisha*								
	Northern			*Rural Northern*			*Urban Northern*		
	Male	*Female*	*Total*	*Male*	*Female*	*Total*	*Male*	*Female*	*Total*
38 (1983)	58.27	30.47	44.55	59.72	33.55	46.68	51.48	14.25	34.01
50 (1993–94)	58.0	37.77	48.06	59.46	41.73	50.75	49.27	14.17	32.03
61 (2004–05)	58.23	38.37	48.31	59.76	42.35	50.98	50.83	16.52	34.62
NSS Round (year)	*Southern Odisha*								
	Southern			*Rural Southern*			*Urban Southern*		
	Male	*Female*	*Total*	*Male*	*Female*	*Total*	*Male*	*Female*	*Total*
38 (1983)	64.23	47.82	55.87	66.06	50.68	58.17	47.63	17.75	33.43
50 (1993–94)	62.39	45.79	54.15	63.56	48.13	55.83	52.98	23.54	39.44
61 (2004–05)	59.47	37.39	48.90	49.95	37.77	32.40	60.34	39.59	50.42

Source: NSSO, Unit level data various quinquennial rounds.

The analysis across gender shows the following points:

(a) The WPR of male has remained higher than female.

(b) In coastal Odisha, trends of both male and female WPR have increased but the level of male WPR is almost three times higher than female WPR. In 1983 the share of male WPR in coastal Odisha was 51.96 per cent, which increased to 55.87 per cent in 2004–05; the corresponding figures of female WPR in the same period were 15.64 per cent and 20.28 per cent respectively.

(c) Unlike coastal Odisha, the gap of WPR between male and female is less in northern Odisha. In 1983, the share of male WPR in coastal Odisha was 58.27 per cent, which remained at 58.23 per cent in 2004–05; the corresponding figures of female WPR in the same time period were 30.47 per cent and 38.37 per cent.

(d) Unlike the coastal and northern part of Odisha, which have shown an increasing trend in WPR, in the southern Odisha, the trend of WPR has declined both in the case of male and female yet the level of WPR in both cases (male and female) have remained higher in southern Odisha compared to other two regions. In 1983 the share of male WPR in southern Odisha was 64.23 per cent, which fell to 59.47 per cent in 2004–05; the corresponding figures of female WPR in the same time period were 47.82 per cent and 37.39 per cent.

Across the rural and urban area, the analysis reveals the following:

(a) In all the three regions, the WPR in rural area has remained higher than their urban counterparts except in southern Odisha in 2004–05, where there is sudden decline of rural WPR by 17 percentage point. In the rural areas of coastal and northern Odisha, the WPR has shown a 4 percentage point increase in each, where as in the rural areas of southern Odisha it has declined by 26 percentage point in 1983 to 2004–05.

(b) The WPR in rural areas of coastal Odisha in 1983 was 34.56 per cent, which increased to 38.81 per cent in 2004–05; the corresponding figures for the rural areas of northern Odisha in the same time period were 46.68 per cent and 50.98 per cent; and for the rural areas of southern Odisha were 58.17 per cent and 32.40 per cent.

(c) In urban coastal Odisha while the WPR has shown a 4 percentage point increase, in the urban areas of northern Odisha it has remained almost stagnant (with minor fluctuations) but in the urban areas of the southern Odisha it has increased very rapidly by 17 percentage point and that increase has happened during 1999–00 to 2004–05. The WPR in urban coastal in 1983 was 28.32 per cent, which increased to 32.52 per cent in 2004–05; the corresponding figures for the urban area of northern Odisha in the same time period were 34.01 per cent, and 34.62 per cent; and for the urban area of southern Odisha were 33.43 per cent and 50.42 per cent.

The analysis of the Table 3.20 shows, there is a huge decline in the percentage of WPR in southern Odisha, where the magnitudes of hunger and deprivation are rampant. Such decline can be related to the building of Indravati dam in Kalahandi in early 1980s, which might have pushed the WPR so high and lack of subsequent investment in the regions might have shown the gradual decline of the WPR.

Table 3.21 and Appendix A3.14 present the sectoral distribution of employment at one digit level, which depicts the following:

Table 3.21 *Trend in percentages distribution of workers across different sectors by rural urban location in different regions of Odisha from 1983 to 2004–05*

		Primary	*Secondary*	*Service*	*Total*
Coastal	1983	68.3	11.0	20.70	100
	1993–94	72.56	7.52	19.91	100
	1999–00	61.90	13.77	24.33	100
	2004–05	59.21	17.01	23.78	100

Cont'd...

...*Cont'd*

Northern	1983	73.44	13.77	12.78	100
	1993–94	74.83	12.44	12.73	100
	1999–00	70.91	16.13	12.97	100
	2004–05	64.31	20.38	15.31	100
Southern	1983	81.38	8.90	9.72	100
	1993–94	82.23	9.20	8.58	100
	1999–00	83.89	10.27	5.84	100
	2004–05	73.53	13.50	12.97	100

Source: Same as previous table.

(a) In all the regions, primary sector (especially agriculture) provides employment to most people, though the magnitude differs across the regions. In 1983, the share of employment provided by primary sector in coastal Odisha was 68.3 per cent, which fell to 59.21 per cent in 2004–05; Similar figures in northern Odisha were 73.44 per cent and 64.31 per cent respectively; and for the southern Odisha were 81.31 per cent, and 73.53 per cent respectively.

(b) Across the rural urban location, in all these three regions, while the primary sector is the major employment provider in rural areas, the service sector is the major employment provider in urban sector. In 1983, the share of employment in the rural primary sector in coastal Odisha was 74.98 per cent, which fell to 64.84 per cent in 2004–05; similar figures in the northern Odisha were 80.43 per cent and 70.91 per cent respectively and in southern Odisha were 85.01 per cent and 77.05 per cent respectively. Similarly, in urban location, the share of employment in the service sector in coastal Odisha was 65.35 per cent in 1983, which fell to 62.54 per cent in 2004–05; Similar figures in the northern Odisha were 39.94 per cent in 1983 and 58.31 per cent in 2004–05 and for the southern Odisha these were 66.92 per cent in 1983 and 51.32 per cent in 2004–05.

(c) As the transformation of employment from agricultural sector to non-agricultural sector is concerned, there is hardly any difference among the regions in the rate of transformation, i.e. over the time period in all regions there are around 8 to 9 percentage point of the employee transfer from agriculture to non agriculture. As shown already, in the southern Odisha, there is higher dependency in agricultural sector compared to northern and coastal regions. Such overdependency might be the reason of the higher incidence of hunger and deprivation in southern part of Odisha compared to the northern and coastal part of Odisha.

Table 3.22 and Appendix A3.15 presents the genderwise distribution of employment.

In the primary sector, the share of female employee is always higher than male workers in all the regions; whereas in the service sector, reverse is the situation. Such biases do not exist in secondary sector.

Table 3.22 *Trend in percentages distribution of workers across different sectors by gender in different regions of Odisha from 1983 to 2004–05*

Region	Gender	Year	Primary	Secondary	Service	Total
Coastal	Male	1983	67.62	9.71	22.68	100
		1993–94	70.18	7.60	22.22	100
		2004–05	55.28	16.64	28.08	100
	Female	1983	70.54	15.30	14.17	100
		1993–94	79.80	7.29	12.95	100
		2004–05	69.27	17.96	12.77	100
Northern	Male	1983	70.76	14.07	15.16	100
		1993–94	70.77	12.21	17.02	100
		2004–05	61.30	18.11	20.59	100
	Female	1983	78.71	13.18	8.11	100
		1993–94	81.32	12.80	5.89	100
		2004–05	68.89	23.83	7.28	100
Southern	Male	1983	78.41	9.41	12.18	100
		1993–94	78.08	11.23	10.72	100
		2004–05	67.37	15.49	17.14	100
	Female	1983	85.22	8.25	6.53	100
		1993–94	87.87	6.43	5.66	100
		2004–05	84.21	10.04	5.74	100

Source: NSSO, Unit level data various quinquennial rounds.

Transformations in the Nature of Employment at the Regional Level

Given the distribution of workers, across different regions of Odisha, below we have examined the employment transformation by the nature of employment. Tables 3.23 and 3.23 (A) show the details of the changing share of different types of employment across different regions of Odisha by their place of location and gender respectively.[16] As due to definitional changes, the trend comparison of all kinds of job over time may be inappropriate, we have grouped some types of job that may better facilitate trend comparison.

Table 3.23 *Percentages distribution of usual status workers in broad group (UPSS) by their place of location in Odisha from 1983 to 2004–05*

		Total			Rural			Urban		
		C	N	S	C	N	S	C	N	S
Own account workers, Employers and unpaid family labour	1983	55.98	46.04	55.12	57.89	47.42	56.18	40.53	36.66	37.08
	1993–94	57.97	52.57	50.36	59.84	54.98	51.74	44.58	29.76	33.28
	1999–00	54.03	45.13	42.51	55.05	46.39	42.71	48.74	36.28	40.23
	2004–05	63.34	53.05	51.83	64.78	55.03	52.64	52.31	38.09	38.13
Regular salaried/wage employee	1983	11.71	13.05	9.10	7.66	9.12	6.96	44.58	39.81	45.22
	1993–94	10.11	8.56	5.67	5.62	4.11	3.29	42.29	50.59	35.14
	1999–00	11.03	8.57	2.93	6.13	4.37	0.90	36.48	38.15	26.80
	2004–05	9.69	9.16	5.78	6.72	5.23	4.07	32.48	38.92	34.65
Casual wage labour in public work and other types of Work	1983	32.31	40.91	35.79	34.45	43.47	36.86	14.89	23.53	17.70
	1993–94	31.92	38.87	43.97	34.54	40.91	44.97	13.13	19.65	31.59
	1999–00	34.94	46.30	54.55	38.82	49.24	56.39	14.78	25.57	32.97
	2004–05	26.97	37.79	42.40	28.50	39.75	43.29	15.20	23.00	27.22

Source: NSSO, Unit level data various quinquennial rounds.

The results presented in Table 3.23 show the following:

(a) The percentage of workers involved in the self-employment and regular job has always remained higher in coastal Odisha than northern Odisha and southern Odisha. On the other hand, the percentage of workers involved in casual labour has remained higher in southern Odisha than northern Odisha and coastal Odisha. This suggests that the earning opportunities are likely to be higher in coastal and northern Odisha compared to southern Odisha.

(b) In 1983, the share of self-employed in coastal Odisha is 55.98 per cent, which has increased to 63.34 per cent in 2004–05. Similar figures for north Odisha in the same time span were 46.04 per cent and 53.05 per cent respectively and for southern Odisha these figures were 55.12 per cent and 51.83 per cent respectively.

(c) In case of regular job, in 1983, the share in coastal Odisha is 11.71 per cent, which has marginally declined to 9.69 per cent. Similar figures for north Odisha in the same time span were 13.05 per cent and 9.16 per cent respectively and for southern Odisha these figures were 9.10 per cent and 5.78 per cent respectively.

(d) In case of casual job, in 1983, the share in coastal Odisha was 32.31 per cent, which has declined to 26.97 per cent in 2004–05. Similar figures for north Odisha during the same time span were 40.91 per cent and 37.79 per cent respectively and for southern Odisha, these figures were 35.79 per cent and 42.40 per cent respectively.

Table 3.23(A) *Percentages distribution of usual status workers in broad group (UPSS) by their gender in Odisha from 1983 to 2004–05*

		Total			*Male*			*Female*		
		C	*N*	*S*	*C*	*N*	*S*	*C*	*N*	*S*
Own account workers, Employers and unpaid family labour	1983	55.98	46.04	55.12	54.95	46.54	54.72	59.38	45.06	55.63
	1993–94	57.98	52.56	50.41	57.98	51.27	49.11	57.98	54.62	52.21
	1999–00	54.03	45.13	42.51	52.47	43.77	43.16	58.73	47.24	41.53
	2004–05	63.34	53.05	51.83	61.47	50.00	50.49	68.16	57.69	54.15
Regular salaried/wage employee	1983	11.71	13.05	9.10	14.17	18.31	14.53	3.63	2.76	2.09
	1993–94	10.11	8.56	5.68	12.04	11.94	8.41	4.27	3.19	1.92
	1999–00	11.03	8.57	2.93	13.27	11.96	4.18	4.30	3.30	1.03
	2004–05	9.69	9.16	5.78	11.52	12.21	7.59	4.99	4.50	2.62
Casual wage labour in public work and other types of Work	1983	32.31	40.91	35.79	30.88	35.16	30.75	37.00	52.18	42.28
	1993–94	31.91	38.88	43.91	29.98	36.79	42.48	37.75	42.20	45.87
	1999–00	34.94	46.30	54.55	34.26	44.26	52.66	36.97	49.46	57.44
	2004–05	26.97	37.79	42.40	27.02	37.79	41.92	26.85	37.80	43.23

Source: NSSO, Unit level data various quinquennial rounds.

Summing Up

Overall, though there is no clear picture of transformation in the nature of employment, yet it has shown a kind of transformation. In all the regions, the share of regular workers have decreased, though rate of decline is much higher in case of southern Odisha. Self-employment has increased in coastal and northern Odisha while the casual labour has decreased in these regions. Hence, in northern and coastal Odisha, the decline in the share of both regular job and casual job might have appropriated in the self-employment category. However, in case of southern Odisha, the share of casual labour shows an increasing trend over 1983 to 2004–05 (even after a 12 percentage point decline in casual labour in 2004–05 over 1999–00). Compared to north and coastal Odisha, southern Odisha is disadvantaged not only in terms of a low level of regular job and higher reduction in regular jobs over time, but also in terms of a higher level of casual labour and increasing trend in casual labour over time. These factors partly explain why the poverty situation in Southern Odisha continues to remain high.

Growth of Employment at the Regional Level

After examining the structure of employment across different regions of Odisha, we have further tried to map out the growth of employment across the major sectors and across gender in all the regions in Table 3.24 and Appendix A3.16. The table depicts:

Overall, the secondary sector has grown at a better and consistent rate than the service sector and the primary sector. However, there is a common feature of fluctuations in the growth rate of all the sectors.

(a) As far as the employment growth of primary sector is concerned, during 3rd to 5th, 5th to 6th and 6th to 7th quinquennial rounds, the annual compound growth rate in coastal Odisha remained 2.54 per cent, –1.30 per cent and 0.69 per cent respectively, which is a persistent decline; (ii) in northern Odisha during the same time period the growth rates are 1.92 per cent, 0.60 per cent and –0.55 per cent respectively, which also shows a persistent decline from a low level and finally became negative employment growth; similarly, figures for the same time period in southern Odisha are 2.02 per cent, 2.35 per cent and –1.00 per cent respectively. This also shows a persistent decline and the negative employment growth rate in 1990s is higher than north Odisha.

(b) As far as the employment growth of secondary sector is concerned during 3rd to 5th, 5th to 6th and 6th to 7th quinquennial rounds, the annual compound growth rate (i) in coastal Odisha remained –1.89 per cent, 14.99 per cent and 5.97 per cent respectively; (ii) in northern Odisha during the same time period the growth rates are 0.70 per cent, 7.11 per cent and 6.26 per cent respectively; similarly, figures for the same time period in southern Odisha are 2.25 per cent, 4.21 per cent and 7.36 per cent respectively. Over all there is a sign of employment growth in the secondary sector.

(c) As far as the employment growth of service sector is concerned during 3rd to 5th, 5th to 6th and 6th to 7th quinquennial rounds, the annual compound growth rate (i) in coastal Odisha remained 1.52 per cent, 6.05 per cent and 1.12 per cent respectively; (ii) in northern Odisha during the same time period the growth rates are 1.69 per cent, 2.07 per cent and 4.83 per cent respectively; similarly, figures for the same time period in southern Odisha are 0.65 per cent, –5.61 per cent and 19.23 per cent respectively. Overall, there is a sign of marginal improvement.

Across the rural urban sector, with certain limitations (due to fluctuations in the growth rate) it can be inferred that, the urban growth rate remained better than their rural counterpart.

(a) During 3rd to 5th, 5th to 6th and 6th to 7th quinquennial rounds, the annual compound growth rate of employment in the primary sector (i) in rural coastal Odisha remained 2.36 per cent, –0.61 per cent and –0.94 per cent respectively, whereas the same in the urban coastal Odisha are 4.97 per cent, –10.40 per cent and 20.18 per cent respectively; (ii) in rural northern Odisha during the same time period the growth rates are 1.70 per cent, 0.69 per cent and –0.70 per cent respectively, whereas the same in the urban northern Odisha are –5.21 per cent, 13.62 per cent and –3.87 per cent respectively; similarly, figures for the same time period in rural southern Odisha are 1.96 per cent, 2.64 per cent and –1.26 per cent respectively, where as the same in the urban southern Odisha are 9.45 per cent, –5.78 per cent and –9.80 per cent respectively.

(b) The figures in the secondary sector are: (i) rural coastal Odisha remained –2.79 per cent, 13.52 per cent and 9.78 per cent respectively, whereas the same in the urban coastal Odisha are 1.31 per cent, 15.82 per cent and –2.31 per cent respectively; (ii) in rural northern Odisha during the same time period the growth rates are 1.10 per cent, 7.34 per cent and 8.78 per cent, whereas the same in the urban northern Odisha are 2.26 per cent, 3.35 per cent and –2.06 per cent respectively; similarly, figures for the same time period in rural southern Odisha are 0.58 per cent, 4.73 per cent and 8.68 per cent respectively, whereas the same in the urban southern Odisha are 10.71 per cent, 1.84 per cent and 6.48 per cent respectively.

(c) These figures in the service sector are: (i) rural coastal Odisha remained 0.19 per cent, 6.34 per cent and 3.71 per cent respectively, where as the same in the urban coastal Odisha are 4.18 per cent, 2.69 per cent and 2.59 per cent respectively; (ii) in rural northern Odisha in the same time period the growth rates are 0.81 per cent, 1.52 per cent and 4.70 per cent respectively, whereas the same in the urban northern Odisha are 6.22 per cent, –0.97 per cent and 6.74 per cent respectively; similarly, figures for the same time period in rural southern Odisha are 2.57 per cent, –16.74 per cent and 36.21 per cent respectively, whereas the same in the urban southern Odisha are –5.32 per cent, 10.42 per cent and 4.78 per cent respectively.

Given this trend and structural analysis of employment in the following section we have examined the characteristics of labour and its association with poverty. Linking regional employment growth with poverty reduction reveals that the growth of employment in primary sector is more sensitive to poverty reduction.

Table 3.24 *Trend in the employment growth among different regions of Odisha from 1983 to 2004–05*

Year	*Primary*	*Secondary*	*Service*
	Coastal		
3rd to 5th	2.54	–1.89	1.52
5th to 6th	–1.30	14.99	6.05
6th to 7th	0.69	5.97	1.12
5th to 7th	–0.31	10.39	3.56
	Northern		
3rd to 5th	1.92	0.70	1.69
5th to 6th	0.60	7.11	2.07
6th to 7th	–0.55	6.26	4.83
5th to 7th	0.02	6.68	3.44
	Southern		
3rd to 5th	2.02	2.25	0.65
5th to 6th	2.35	4.21	–5.61
6th to 7th	–1.00	7.36	19.23
5th to 7th	0.66	5.77	6.09

Source: NSSO, Unit level data various quinquennial rounds.

Section IV

Household's Labour Market Characteristics and Poverty

The incidence of poverty should have been either lower in Odisha compared to all-India or both should have been in equal magnitude, if work participation rate or the types of job alone were the sole determinants of poverty. In order to explore some of the household characteristics in general, and labour market characteristics of the household in particular, and the likelihood of the households in poverty, the recent quinquennial round (61st round) employment unemployment data has been analysed. The focus of the analysis is on the pattern and not the trend.[17] There are ten variables we have found (and data set allowed) that are likely to contribute to the probability of a household being in poverty. The dependent variable is the poverty (calculated from monthly per capita consumption expenditure) and the independent variables are occupation, household type, regions, social group, landholding, education, worker-population ratio, nature of employment, child-women ratio, wage rate and seasonality. We have examined the relationship of poverty[18] with some of these independent variables below.

Table 3.25 presents the percentage of below poverty line households by their occupation across the regions of rural Odisha. The broad inference from the Table is that irrespective of the different regions of Odisha as we move from self-employed in non-agricultural household to self-employed in agricultural households to manual labour households, the incidence of poverty increases. Similarly broadly, irrespective of the type of household as we move from coastal

Odisha to Northern Odisha to Southern Odisha, the proportion of households living below the poverty line has increased.

Table 3.25 *Association of poverty with types of household and regions in rural Odisha*

Household type	*Coastal*	*Northern*	*Southern*	*Odisha*
SENA	15.82	39.24	34.24	26.49
SEA	23.86	45.89	58.98	38.98
Manual	22.29	50.62	62.51	42.33
Total	21.32	47.33	57.43	38.34

Source: NSSO unit level data, 2004–05.

Table 3.26 *Association of poverty with social groups and regions in rural Odisha*

Social group	*Coastal*	*Northern*	*Southern*	*Odisha*
ST	45.01	57.32	65.91	59.31
SC	25.47	51.12	59.42	41.41
Others	17.58	37.60	42.29	26.58
Total	21.32	47.33	57.43	38.34

Source: NSSO unit level data, 2004–05.

Table 3.26 presents the percentage of below poverty line households by their social groups and across different regions of rural Odisha. Poverty decreases as we move from ST to SC to others category of social groups of household. This is true irrespective of the different regions of Odisha. Similarly, irrespective of the social identity of the household, the percentage of households living below the poverty line has increased from Coastal to Northern to Southern Odisha.

Table 3.27 presents the percentage of below poverty line households by their per capita land possession in the household and across different regions of rural Odisha. The results show that broadly as we move from landless households to households with larger size of landholdings, the percentage of households below poverty line decreases. However, there are some exceptions. In case of coastal Odisha, there are no below poverty line households among the landless households, which might be due to regular and salaried people in such households in coastal area. But the exception in the case of Northern rural Odisha is a different case, where some households are poor with more than 4 hectares of land with them. Across the region broadly, as we move from Coastal Odisha to Northern Odisha to Southern Odisha, the percentages of people living below the poverty line has increases for almost all categories of landownership.

Table 3.28 presents the percentage of secondary plus persons in the households and probability of households being poor in different regions of Odisha. It clearly shows that presence of secondary or secondary plus educated persons in the household is negatively associated with poverty in all the regions of rural Odisha.

In Table 3.29, we have examined the associations of poverty with household's worker population ratio. However, it has not shown any systematic or clear association with poverty.

In Table 3.30, the association of poverty with child-women ratio has been examined, which depicts a clear relationship irrespective of region. As the CWR increases, the proportion of households being in poverty also increases.

Table 3.27 *Association of poverty with per capita land possession of household and regions in rural Odisha*

Land holding	*Coastal*	*Northern*	*Southern*	*Odisha*
Land less	0.00	55.86	27.22	39.29
0.001 to 0.4.	21.88	44.96	47.36	34.82
0.41 to 1.0	19.43	49.84	77.30	40.91
1.001 to 2.0	23.04	49.90	59.60	41.84
2.001 to 4.0	15.68	36.64	62.79	36.91
> 4.001	14.46	57.12	57.99	43.03
Total	20.94	46.72	57.04	37.66

Notes: No data has been recorded on landholding for 116 rural sample households' cases. Hence the total figure in this table is different than other tables.

Source: NSSO unit level data, 2004–05.

Table 3.28 *Association of poverty with the education of household and regions in rural Odisha*

Education	*Coastal*	*Northern*	*Southern*	*Odisha*
Nil Secondary educated	24.93	53.34	62.07	44.26
25% Secondary educated	18.29	43.69	32.83	27.88
50% Secondary educated	12.73	21.40	14.67	16.38
75% Secondary educated	5.98	8.96	22.89	8.92
100% Secondary educated	3.69	0.00	0.00	1.83
Total	21.32	47.33	57.43	38.34

Source: NSSO unit level data, 2004–05.

Table 3.29 *Association of poverty with the WPR of household and regions in rural Odisha*

HH WPR	*Coastal*	*Northern*	*Southern*	*Odisha*
Nil HH WPR	7.16	24.43	27.59	15.58
25% HH WPR	21.73	55.13	52.76	33.23
50% HH WPR	23.34	51.74	65.80	39.52
75% HH WPR	24.55	55.75	59.40	45.44
100% HH WPR	14.18	30.10	53.20	33.23
Total	21.32	47.33	57.43	38.34

Source: NSSO unit level data, 2004–05.

Table 3.30 *Association of poverty with the CWR of household and regions in rural Odisha*

CWR	*Coastal*	*Northern*	*Southern*	*Odisha*
Nil CWR	8.65	24.70	38.09	20.99
25% CWR	9.82	43.03	63.64	31.70
50% CWR	28.13	60.56	64.62	44.01
75% CWR	25.47	53.94	72.48	43.34
100 CWR	28.41	58.52	65.86	48.99
Total	21.32	47.33	57.43	38.34

Source: NSSO unit level data, 2004–05.

Conclusion

The objective of this chapter was to analyse the growth experience (both output and employment) of Odisha and its effects on overall poverty reduction. The analysis shows a high degree of product transformation from primary to service sector, but the secondary sector remained passive in this process. The growth rate since 1970s, made the structural transformation (in output) to happen, however, that growth has failed to trickle down and reduce hunger and deprivation. In other words, the rate of change in output has not much affect on the rate of change in poverty as there is less employment growth in the primary sector.

About 63 per cent of the total workers are found in primary sector in Odisha. The rate of growth in output is higher than the rate of growth in employment in the previous 27 years in Odisha. Further, the rate of annual compound growth rate of both output and employment is found to be higher in service sector than in primary sector during 1990s and post-90s period. Since most of the workers are dependent on primary sector and the rate of growth is lower in the primary sector, poverty situation has worsened. Further, the study of the employment transformation reveals that there is negligible shift of workforce from primary to secondary to service sectors. The overdependency on primary sector might be due to the less impressive growth and employment in the secondary sector.

The low and negative employment elasticity exhibits the lack of absorptive capacity or displacement of employee from the particular sector. Hence, the rate of change in employment did not bring about reductions in poverty. However, the growth in employment during the late 1970s to early 1990s, perhaps accounts for some reduction in poverty in the state. But the momentum in growth in employment in subsequent period did not keep pace. However, the analysis of output and employment shows that the poverty reduction is more sensitive to the primary sector growth.

There is almost no transfer of labour from rural to urban sector. There is transfer of some labour force from agriculture to non-agriculture, yet the percentage of workers in the agricultural sector remained high. One would expect theoretically that labour productivity in the agricultural sector or in the primary sector should increase, but this has not happened in Odisha. Therefore, the Odisha economy needs a big push, especially in the secondary sector for overall development of the state and for poverty reduction. The analysis shows hunger and deprivation in Odisha can be related more to the growth (both employment and output) of primary sector rather than overall growth rate.

The regional analysis within Odisha reveals interesting results. For most of the indicators analysed, Southern region of Odisha remained at a disadvantaged position and Coastal region remained the best performer. The Northern region remained in the middle of the other two regions in terms of overall performance. In Southern Odisha, poverty level is the highest, casual labour is much higher, share of ST population is much higher, literacy level is much lower. Due to these interrelated factors, hunger and deprivation are likely to be much higher in Southern region of Odisha.

Among all the major factors, social group is the most disadvantaged. The ST population has the highest level of poverty regardless of regions. Poverty level among ST population in the Southern region is much higher compared to their counterparts in the other two regions. Next in order is the SC population who are also disadvantaged, especially in Southern region followed by Northern region.

On the whole, Southern region of Odisha and ST and SC communities play a dominant role in the overall deteriorating situation of hunger and deprivation in Odisha.

Within Odisha, the southern part of Odisha showed a similar image (in terms of WPR, over dependency on primary sector etc.) as Odisha shown compared to all-India. The labour market characteristics and its relationship with poverty reveal – education, land possession, child women ratio, social group and regions have shown close association with poverty.

From the above discussion, different relationships between the employment and poverty have been found. The findings suggest that, land and wages are the two important components of poverty. It is however, unclear about other aspects in the labour market, its impact on the employment and earnings and thereby hunger and deprivation, which has further research scope. The next Chapter covers the land distribution aspects in greater detail.

Notes

1. Dantwala, M.L. (1973), *Poverty in India: Then and Now (1870–1970)*, MacMillan, Delhi.
2. The modern analysis of sectoral transformations is originated with the work of Fisher's, 1935, 1939 and Clark, 1940. This has been systematically discussed further by Kuznets, 1973. In India, Chakravarty, 1993 uses these models for empirical validations.
3. To know more on the structural transformations and poverty, also see Ravallion and Datt's 1996 article – how important to India's poor is the sectoral composition of economic growth, 1996.
4. To evaluate economic development in terms of structural changes of an economy, both gross as well as net output have been considered in the literature, e.g. V.K.R.V. Rao (1987) used net output while Nagraj (2006) uses gross output. Here, we have used net domestic product.
5. Papola (2005) has shown structural changes among the Asian countries taking data for a period of 1960 to 2002. The study shows that in 1960, the share of industry to

country GDP, was 49, 25, 19, 28, 18, 20, 16 and 16 per cent for China, Indonesia, Thailand, Philippines, Malaysia, Republic Korea, Pakistan and India respectively. But in 2002, these figures have changed to 51, 45, 43, 33, 47, 41, 23 and 25 per cent. Observing this change in the share of secondary sector, one would have expected the share of secondary sector in case of Odisha to be higher.

6. The formula has been applied in the empirical data by some researchers, such as Hari, 2003; Babu, 2005.
7. The 1950s has shown a negative growth rate; hence we did not interpret it.
8. Some of the studies on the employment structure have considered NSSO data is better, as the scope of the data provided by the NSSO is more for a deeper analysis of employment structure (Visaria, 1978; Chadha and Sahu, 2002; Hari, 2003). However, since the NSSO data is based on sample survey and the large sample is available since 1972–73, the Census data would have been much better with regards to time series analysis, as it provides all kinds of information. Though the Census provides data for a long period of time, it faces the problem of definitional change over time, which requires some adjustment of the data. And even after such adjustment, as Visaria has pointed out, in case of women work participation rate, the information provided by the NSSO is more reliable than the Census.
9. Necessary adjustments have been done to compare different rounds (as the initial three rounds used +5 age groups to calculate WFPR) and to get the absolute numbers (by adjusting to the Census population) of workers to calculate growth of employment. We have taken usual status workers (both principal and subsidiary status).
10. Some of the subsectors have been clubbed together as at the disaggregated level the percentages are becoming very negligible to reports in the NSSO report.
11. The year 1987–88 being as a drought year, we have also calculated the annual compound growth rate during 1983 to 1993–94, where the overall employment growth rate remains at 2.04 per cent and primary sector's growth rate remains at 2.13 per cent.
12. Here we have considered data since 38th round, as the unit level of data are available since that point of time. In the earlier rounds it was difficult to get the data for the regional level.
13. Employer as a separate category was not captured during 1983 round of NSSO.
14. Here we have dealt only with the post-1990s data as it is during that period when both the output and employment are showing the pace.
15. We have taken population here from 1981 as the unit level employment data is available since 1983, which will be discussed further.
16. The data considered are since 38th round as the unit level of data are available since that point of time.
17. Conventionally poverty ratio has been estimated from the monthly per capita consumption expenditure data but to understand the labour market characteristics of the poor, the abridged monthly per capita consumption expenditure data provided in the employment-unemployment survey has been used by some authors (Sundaram and Tendulkar, 2003). We have also used a similar procedure here. However, some of the labour market components like wage rates will be further discussed in Chapter 4 to study its time series behaviour.
18. The poverty line is taken from Planning Commission's documents. For rural Odisha the cut-off line is ₹ 325.79 (Uniform reference period).

4

Absence of Distributive Justice

Access to Land and Related Resources in a Comparative Perspective[1]

Land is one of the principal factors of production of agriculture, which helps to produce and supply food to human beings. Though, there are other productive forces, such as agricultural implements, irrigation, energy, fertilisers, and agricultural credit, which influences production and production relations, land plays a pivotal role for improving livelihoods of the people in an agrarian economy. Therefore possession of land, which also establishes the class relations and mode of production and establishes the production relations and social relations, plays a vital role in an economy in reducing conditions of hunger and deprivation in a household. Landlessness, therefore, may push households into hunger and deprivation.

It is argued that, land reforms also associated with better human development. Bagchi (2004) argued that efficient land reforms would lead to higher capability enhancing facility and hence to a better human development. As access to land increases their purchasing power, they can utilise the facilities provided through public provision. It could be argued that inefficient land reform is responsible for the failure of effective demand with people. Further, this will lead to failure of social provisioning for human development, as the public provisioning also requires minimum implicit cost. In the absence of land reforms, such minimum purchasing power may not be created and hence people would be more vulnerable. This further leads to reduce the capability to face 'exogenous shocks' such as displacement, natural calamities and so on.

This chapter attempts to examine land distribution and agricultural wages system in Odisha and explore their relationship with hunger and deprivation. To study the relationship between land and poverty in both static and dynamic context, both time series (quinquinneal rounds data) and cross-sectional data have been explored. This chapter comprises of four sections including the present section. Section two presents theoretical perspective of impact of distribution on poverty. Section three presents empirical aspects of the process of land distribution and poverty. Data from National Sample Survey (unit level data, 2002) and Census of India are used for the purpose. Section four examines the relationship between the land distributions with hunger and deprivation, where the response of other forces of production has been discussed.

Land and Poverty

Theoretical Arguments (Land as an Instrument of Poverty Reduction)

Land is an important factor to consider for different reasons: (i) it makes production possible; (ii) the production process generates employment, (iii) generated income may facilitate exchange, which may lead to increased consumption; (iv) increased consumption may reduce poverty.

In the case of technology-intensive production, however, this logic may lead to reduced employment opportunities by economies of scale through land concentration. Raj (1990) in his article 'Mechanisation of Agriculture in India and Sri Lanka', points out that there is evidence of large landholders undertaking mechanisation in order to reduce dependency on tenants and casual labour. Therefore, while it may lead to higher productivity, land concentration can also lead to unemployment and poverty.

Where production takes place at the household level, farming may ensure that the household is spared from food insecurity and related problems. In India, Kerala and West Bengal have exemplified poverty reduction through improving agrarian structure. It is interesting to examine the processes that enabled the land reforms in Kerala and West Bengal and juxtapose the same against the agrarian structure in Odisha. The sources of data include Census of India, the National Sample Survey Organisation and Planning Commission of India for the period 1951 to 2002.

It is observed that if egalitarian distribution does not happen, poverty reduction due to high agricultural output may not hold true (Kannan et al., 2003; Kumbhar, 2001). There is evidence to show that a growth-led poverty reduction strategy has not been successful in many Indian states. Examining the relationship between the incidence of rural poverty and agricultural performances over the period of 1956–57 to 1973–74, Ahluwalia (1978) found clear evidence of an inverse relationship for India but not at the level of individual states. His estimates pointed that although Kerala, Odisha, West Bengal, Punjab and Haryana and Uttar Pradesh showed a significant trend of increasing agricultural output per head of the rural population, rural poverty had not shown a significant decline. Krishna et al. (2005) stated that agricultural growth alone is hardly sufficient to achieve poverty reduction on any significant scale. It is also argued that the unequal distribution of outputs can be caused either by unequal distribution of wealth and assets among the people (i.e. a particular kind of agrarian structure), or due to the inefficiency of different social institutions like welfare state[2] (to distribute the produced output equally). Assuming that unequal distribution could be the main reason for poverty, then the question is which type of redistribution should be carried out to eliminate poverty?

(a) Should there be redistribution of wealth and assets among the poor to relieve them from poverty? For example, land reforms can reduce poverty. Sen (1981) reported that providing land entitlement has its own merit in protecting people from poverty. He considered land a production-based entitlement as it produces food for the household and also as an exchange entitlement for the household as it serves as collateral at time of need. This logic is valid for productive land. Unproductive land may not yield enough to get out of poverty. Hence redistribution of essential commodities through public provisioning may be desirable. (b) Should the redistribution exercise be confined to the redistribution of agricultural output alone? For example, different institutions distribute different basic necessity commodities to citizens through different forms of public provision[3] (Dreze and Sen, 1989).

Kerala has undertaken strategies of output distribution as well as asset distribution (Kannan, 1995 and UN/CDS, (1975); Sebastian, 2002). The output distribution strategy in West Bengal has been recognised as the reason behind poverty reduction in the state (Lieten, 1996; Sen, 2001; Raychaudhury, 2004). In Odisha, although both strategies have been adopted immediately after India's independence as part of abolition of intermediary system and ceiling; as a part social welfare state, neither has provided adequate results, despite the fact that initial asset distribution was better than the other two states. This brings us to the question of relationship between land ownership and poverty reduction. Can land ownership bring about food security and reduction in poverty? The next section addresses the types of agrarian structure (and by implication which type of land reforms) that are appropriate to support a poverty reduction strategy.

Agrarian Structures and Rural Poverty

In the absence of more direct information, the pattern of land distribution can be considered as an appropriate indicator to understand the agrarian structure. Some of the hypothetical possible agrarian structures are discussed below:

Scenario 1

An agrarian structure where land concentration may generate growth but not necessarily lead to poverty reduction. In this kind of structure, since a few households hold a large amount of land, there may be demand for labour (in the absence of technology) and possibly a rise in wage rates, which may help to reduce poverty.[4] According to Raj (1974), in the growth-led poverty reduction 'a presumption is that higher income for those below poverty line (and more employment for the unemployed) can be provided only by achieving higher rates of growth of output in agriculture as well as in the economy as a whole and redistributing the increased output so realised'. Or an emphasis on introducing productivity-enhancing technology could lead to trickle-down effects and poverty reduction (Bardhan, 1984; Alhuwalia, 1978). But in many states, in rural economy neither higher growth nor trickle-down effects have occurred. Both approaches (growth expansion and strategic employment expansion) have failed, either partially or wholly, to reduce poverty. Therefore, it could be argued that an agrarian structure aiming to increase agricultural output may not necessarily reduce poverty.

Scenario 2

An agrarian structure where land is fragmented may offer better growth prospects, if the required infrastructure is in place, otherwise it can hamper growth. This type of an agrarian structure may have led to growth and poverty reduction in

Kerala and West Bengal, but it has not happened in Odisha. Rather, over time the agrarian structure became less egalitarian and created a huge number of landless households.

Scenario 3

An agrarian structure, which comprises of both the approaches mentioned above, may reduce poverty but it also facilitates a surplus leading to higher growth.

Depending on his/her value judgements, one may prefer the growth-inducing agrarian structure or the poverty reducing agrarian structure – or the combination of both.[5] The above discussion provides the analytical linking of agrarian structure and poverty; in reality agrarian structures are continuously changing, depending on the circumstances facing the economy. Therefore, one type of agrarian structure may not necessarily lead to poverty reduction in a similar way across region and under different circumstances. The following section draws from empirical studies to understand different strategies of poverty reduction.

Trends in Land Distribution and Poverty

Patterns in the Agrarian Structure in Odisha, Kerala and West Bengal

The agrarian structure has an important impact on different groups' access to land, credit and other resources. While the agrarian structure in a society is heavily influenced by external changes in that society over a period of time, agrarian reform can also influence and modify this structure.[6] Kerala and West Bengal are two prominent examples, and in Odisha the land tenure system implemented a mixed agrarian structure which was improved through constitutional provisions, i.e. abolishing the Zamindari system.

Agrarian structures here have been examined by analysing two datasets: the Census of India, which provides information on the distribution of workers/self-supporting earners by occupation, and the NSSO, which provides information on the distribution on landholdings across different size classes. Table 4.1 presents the distribution of main workers/self-supporting earners by occupation from 1951 to 2001. The following can be inferred from the Table: (i) the proportion of cultivators (workers cultivating their own land) always remained higher in Odisha than in Kerala and West Bengal, and the level also remained high. In 1951, the proportion of the self-supporting cultivator class in Odisha was as high as 55.84 per cent (excluding 5.5% of tenant workers and around one per cent of rental class). In Kerala and West Bengal, the rates were 15.14 per cent and 27 per cent respectively. (ii) The share of agricultural labourers was much less in Odisha and West Bengal (14% and 13% respectively) than in Kerala (24.5%) in 1951. (iii) Tenant occupation was around 9 per cent in Kerala and West Bengal during the initial period (1951), but in Odisha it was around 5.5 per cent. In Odisha, therefore, the issue of poverty among non-landowning peasants is less relevant because there were few non-landowning cultivators in Odisha. (iv) The renting class, which has not been shown in the Table 4.1 is less than 2 per cent in all the four categories considered here.

Table 4.1 *Distribution of main workers/self-supporting earners in percentages by their occupations, 1951 to 2001*

State	*Cultivator*	*AL*	*HI**	*OW*	*Total population*	*WPR*	*MW/TR*
Odisha							
1951	55.84	14.13	5.53	22.9	1,46,45,946 (95.94)	38.2	25.3
1961	56.82	17.01	6.93	19.24	1,75,48,846 (93.68)	43.66	—
1971	49.16	28.28	3.63	18.92	2,19,44,615 (91.59)	31.22	—
1981	46.94	27.76	3.3	22	2,63,70,271 (88.21)	38.01	13.84
1991	44.31	28.68	3.13	23.88	3,16,59,736 (86.62)	37.53	12.67
2001	35.82	21.88	4.22	38.08	3,68,04,660 (85)	38.8	32.83
Kerala							
1951	15.14	24.5	9.53	49.26	1,35,49,118 (86.86)	33.5	17.3
1961	20.92	17.38	8.68	53.02	1,69,03,715 (84.89)	33.31	—
1971	17.8	30.69	4.28	47.23	2,13,47,375 (83.76)	29.12	—
1981	13.06	28.23	3.69	55.01	2,54,53,680 (81.26)	30.53	12.61
1991	12.24	25.54	2.58	59.64	2,90,98,518 (73.61)	31.43	9.24
2001	7.12	12.4	3.35	77.13	3,18,41,374 (74.0)	32.3	19.9
West Bengal							
1951	27.01	13.17	9.33	49.98	2,63,02,386 (76.12)	35.1	9.3
1961	38.5	15.3	4.21	41.99	3,49,26,279 (75.55)	33.16	—
1971	31.97	26.45	2.7	38.87	4,43,12,011 (75.25)	27.91	—
1981	29.76	25.23	3.52	41.49	5,45,80,647 (73.53)	30.17	6.32
1991	28.4	24.56	3.9	43.13	6,80,77,965 (72.52)	32.19	6.09
2001	19.79	19.64	6.24	54.33	8,01,76,197 (72.02)	36.8	21.91
India							
1951	43.83	14.25	8.4	31.94	35,66,28,312 (82.67)	39.9	26.7
1961	52.8	16.71	6.38	24.12	43,89,36,918 (82.02	42.98	
1971	43.34	26.33	3.52	26.81	54,79,49,809 (80.09)	32.92	
1981	41.58	24.94	3.47	30.01	66,52,87,849 (76.30)	36.77	9.03
1991	38.72	26.09	2.38	32.81	83,85,67,936 (74.27)	37.46	8.98
2001	33.11	20.29	3.9	42.71	1,02,86,10,328 (72.1)	39.1	22.18

Source: Census of India, Various Issues; WPR – Work Participation Rate, MW – Marginal Worker, TW – Total Worker, AL – Agricultural labour, HI – Household industry, OW – other workers, Bracketed figure is the share of rural populations. * This column represents the share of tenants and not HI for 1951.

Notes: The calculation is based on main workers as it is not possible to find out the occupational distribution of the marginal workers for all the time point considered here. But the marginal workers' component has been increasing in all the states of India. When we include this component, in 2001 agricultural labour occupation increases to 35%, 16%, 33% and 26.7% in Odisha, Kerala, West Bengal and India respectively.

Therefore, it would appear that as far as access to land is concerned, during the initial period, distribution was better in Odisha compared to Kerala and West Bengal. If we assume that the agrarian structure is linked to poverty, the level of poverty should, therefore, have been lower in Odisha than in the other states – which is not, however, the case. In 2001, the picture is very different: (i) The percentage of the cultivator class decreased in Odisha but still remained higher than in Kerala, West Bengal and India's average. (ii) The proportion of agricultural labourers increased in Odisha and became the highest, compared to the other two states and India's average. (iii) The share of household industry workers showed a decline in all categories but the rate of decline was lower in Odisha. (iv) The share of other workers also increased significantly in all cases, although the rate was more remarkable in the case of Kerala.

With the limitations of census data, any study of the links between agrarian structure and poverty has to be confined to occupational groups. However, the data collected by the NSSO provides exact patterns of landholding by households. Therefore, although it is based on a sample survey, it has advantage over the census data, in that the relationship can be explored more precisely. Table 4.2 uses the NSSO data.

Table 4.2 *Percentage distribution of operational landholdings in rural areas, 1953–54 to 2003*

	Years	*Landless*		*Marginal*		*Medium holdings*		*Large holdings*	
		HH1	*HH2*	*Household*	*Area*	*Household*	*Area*	*Household*	*Area*
Kerala	1953–54	36.27	18.07	46.91	9.18	33.61	69.05	1.41	21.77
	1961–62	30.9	16.73	52.73	12.39	29.82	73.06	0.72	14.62
	1971–72	15.74	11.71	59.91	15.76	28.06	68.97	0.32	15.07
	1981–82	12.76	12.99	62.57	20.27	24.2	74.44	0.24	5.32
	1991–92		5.94	86.08	52.47	7.95	45.99	0.03	1.55
	2003	4.8	38.6	56.2	57.2	5.1	42.8	0	0
Odisha	1953–54	12.29	4.53	35.97	2.12	54.88	66.58	4.62	31.3
	1961–62	7.84	32.59	10.21	1.52	54.12	73.18	3.08	25.32
	1971–72	10.57	25.12	16.57	2.81	56.64	82.48	1.67	14.71
	1981–82	7.66	27.77	16.05	2.29	54.7	60.1	1.48	14.59
	1991–92		26.94	43.71	22	29.12	74.34	0.22	3.68
	2003	5.6	26.8	57.2	42.83	15.9	56.24	0.1	0.91
West Bengal	1953–54	20.54	0.89	47.76	3.9	49.12	78.19	2.23	18.01
	1961–62	12.56	33.88	12.74	2.31	52	86.26	1.38	11.43
	1971–72	9.78	30.94	19.79	4.32	48.62	89.23	0.65	6.42
	1981–82	17.21	22.14	38.51	8.65	38.65	83.81	0.7	7.54
	1991–92		13.9	69.49	40.04	16.62	59.96	0	0
	2003	5	30.7	61.4	57.93	7.9	42.08	0	0
India	1953–54	23.09	10.87	30.67	1.25	49.86	51.28	8.6	51.1
	1961–62	11.69	26.86	15.1	1.46	50.55	56.07	7.49	42.47
	1971–72	9.64	27.42	14.93	1.69	52.04	59.3	5.61	39.01
	1981–82	11.33	26.06	23.96	2.45	45.83	64.56	3.85	32.99
	1991–92		19.78	50.31	15.5	28.84	69.25	1.06	15.24
	2003	6.6	31.2	48	22.4	20.3	65.7	0.6	11.88

Note: The total area of 1981–82 for Odisha is not adding to 100 in the NSSO report also. HH1 = share of land less based on land ownership; HH2 = share of landless based on operational holdings. Size classes: Marginal – 0.002 to 1.00, Medium – 1.00 to 10.0, large > 10 hectares.

Source: Various NSSO rounds on Landholding; Sharma, 1994.

The Table indicates the following points: (i) During the initial period (1953–54) landlessness was not as much a problem for Odisha as it was for Kerala or West Bengal. The percentage of landless households was consistently lower in Odisha than in Kerala, West Bengal or the Indian average for ownership holdings and for operational holdings. (ii) In 1953–54, landless households (based on operational holdings) in Odisha, Kerala and West Bengal constituted 4.53 per cent, 18.07 per cent and 0.89 per cent respectively. But by 1991–92, these figures had changed to 26.94 per cent, 5.94 per cent and 13.9 per cent for Odisha, Kerala and West Bengal respectively. This implies a significant increase of landless labour in Odisha as compared to other states. But during 2003, the relative position has changed not because of any change in the operational holdings of the state but the same has increased for Kerala and West Bengal. During 2003, the percentage of landless households (based on operational holdings) in Odisha, Kerala and West Bengal constituted 26.8 per cent, 38.6 per cent and 30.7 per cent respectively.

Trends in Poverty

To examine the poverty in Odisha, this study has used headcount ratio of absolute poverty. Until 1972–73 (period I), the headcount ratio of poverty was calculated from the NSSO data on consumption expenditure. Since data was collected from small sample size, the results obtained were relatively less reliable. However, since 1973/74 (period II) in each quinquennial round, the NSSO collects similar data from a large sample size. So, estimation of poverty from this data is more reliable (even at the state level). The estimated figures of both have been put separately in Table 4.3. Poverty trends in each period (1957/58 to 1973/74 and 1973/74 to 2004–05) can be seen and analysed independently; combining both periods together cannot be done. For period I, we refer to Ahluwalia's figures (1978), which shows only rural poverty and for period II, we refer to the Planning Commission's estimated figures.

Table 4.3 *Trends of poverty in Kerala, West Bengal and Odisha*

Period I	*Headcount ratio*			
	Kerala	*West Bengal*	*Odisha*	*India*
1957–58	59.6	62.3	66.6	53.4
1959–60	62.3	61.4	63.4	48.7
1960–61	57.8	40.4	62.4	42
1961–62	50.3	58.3	49.3	42.3
1963–64	52.8	63.3	60	49.1
1964–65	60.7	64	61.9	50.4
1965–66	70.7	56.5	62.1	51.1
1966–67	67.1	64.3	64.2	57.4
1967–68	63.4	80.3	64.7	57.9
1968–69	64.6	74.9	71.2	53.5
1970–71	62	70.1	65	49.1
1973–74	49.3	66	58	47.6

Source: Alhuwalia, 1978.

As depicted in Table 4.3, an important feature for all the states and for the nation as a whole is a marked fluctuation in rural poverty over time. As far as the levels are concerned, in 1957–58 the headcount ratio was highest in rural Odisha (66.6%), while the levels for India as a whole, Kerala and West Bengal respectively were 53.4 per cent, 59.6 per cent and 63.2 per cent. Further looking into change, a markedly different picture emerges during 1967–68 when the percentage of the population below the poverty line (BPL) in West Bengal was highest and rose up to 80.3 per cent. Similarly, Kerala's data during the 1965–66 period show an increase of BPL population to 70.7 per cent. In both cases, Odisha remains in the second highest with 64.7 per cent in 1967–68 and 62.1 per cent in 1965–66. At the end of period I (1973–74), West Bengal had the highest percentage of rural poor with 66 per cent followed by Odisha (58%) and Kerala (49.3%).

	Period – II											
	Headcount ratio											
	Kerala			*West Bengal*			*Odisha*			*India*		
	R	*U*	*T*	*R*	*U*	*T*	*R*	*U*	*T*	*R*	*U*	*T*
1973–74	59.2	62.2	59.79	73.2	34.5	63.43	67.3	56.3	66.18	56.4	49.2	54.88
1977–78	51.5	59.5	52.22	68.3	38.7	60.52	72.4	53.6	70.07	53.1	47.4	51.32
1983	39	48.7	40.42	63.1	32.2	54.85	67.5	50.6	65.29	45.6	42.2	44.48
1987–88	29.1	43.4	31.79	48.3	32.8	44.72	57.6	44.1	55.58	39.1	40.1	38.36
1993–94	25.8	24.6	25.4	40.8	22.4	35.6	49.7	41.6	48.6	37.3	32.4	35.9
1999–2000	9.4	20.3	12.7	31.9	14.9	27	48	42.83	47.2	27.1	23.6	26.1
2004–05	13.2	20.2	15.0	28.6	14.8	24.7	46.8	44.3	46.4	28.3	25.7	27.5

Source: Planning Commission of India (various rounds).

During period II (1973–74 to 2004–05), the relative position of the states remained the same as at the end of period I. However, an overall declining trend is noticeable, although the rate of decline differs across states. It is interesting to note that there is change in period II. In Kerala, there is a more or less a gradual decline until 1993–94, when a sharper decline begins, which is again followed by a marginal increase. In West Bengal, the rate of decline was relatively slower prior to 1983, but after that it was relatively faster than in India as a whole and also for Odisha. In Odisha, the picture is quite different. In the early 1970s, there was an increase in poverty from 66.18 per cent in 1973–74 to 70.07 per cent in 1977–78, which declined to 65.29 per cent in 1983. However, during the decade (1983 to 1993–94) there is a decline of 17 per cent, but levels remained stationary at around 47 per cent during 1993–94 to 2004–2005. Despite this decline, the level of poverty still remains higher in Odisha than in the other two states and India's average. To relate these trends with land distributions, the study has examined the agrarian structures of these states and all-India, which is discussed in following section.

Linking Agrarian Structure with Rural Poverty

The case study shows that there are several problems in assuming a positive link between better agrarian structures and reduced poverty.[7]

Scenario in Odisha

Level

During the 1950s, the proportion of landless households in Odisha was lower than that of West Bengal and Kerala. The level of poverty should, therefore, have been lower in Odisha. The high proportion of cultivators in Odisha could have generated demand for the agricultural labourers, but this did not happen. Instead, the level of poverty remained higher than in Kerala and West Bengal.

Change

Rural poverty in Odisha declined between 1983 and 1993–94 when there were a high percentage of landless households and the share of agricultural labourers has marginally increased.[8] Since then (from 1993–94 to 2004–05), the rural poverty has declined by just 2 percentage points, whereas the landlessness remained stagnant at around 26 per cent during 1991–92 to 2003.

West Bengal Case Study

Level

During the initial period (1953–54), the pattern of land distribution in West Bengal was even less equitable than in Odisha. About 20.54 per cent of households did not own any land in the state. The share of tenants was more in West Bengal than Odisha (9.53% in West Bengal and 5.53% in Odisha during 1951). However, it may be assumed that landless households were taking leases and cultivating land because the share of landless households in terms of operational holding was just 0.89 per cent. However, in the 'marginal' group, 47 per cent of the households operate just 3.9 per cent of the total operated land.

Change

During the late 1960s and early 1970s, there were consistent improvements in the agrarian structure of West Bengal, but poverty has increased in the state. While rural poverty started declining in 1973–74, a marked reduction was observed from 1983 onwards. This coincides with the time at which the positive effects of earlier land reform efforts could have been reaped. However, despite these efforts, the landlessness problem is again reappearing in West Bengal (Government of West Bengal, 2004).

The West Bengal case study questions the argument that agricultural-induced physical infrastructure can reduce poverty.

Kerala Case Study

Level

In 1953–54, the share of landless households in Kerala was higher compared to Odisha and West Bengal, yet the level of poverty was lower. However, the 'marginal' group of workers constituted 46.91 per cent of the total households and were operating 9.18 per cent of total operated area. The share of self-supporting landowning cultivators was lower in the state, but the share of agricultural labourer was higher.

Change

The agrarian structure in the state has gradually improved and rural poverty also declined until 1963–64. However, despite the improved agrarian structure, from 1963–64 onwards the level of poverty increased and remained at a high level until 1970–71. However, a marked decline in rural poverty was observed at two points in time (rather than over a period of time). During 1970–71 to 1973–74, poverty decreased from 62 per cent to 49 per cent. Between 1993–94 and 1999–2000, poverty reduced from 25.8 per cent to 9.4 per cent. These declines occurred despite the stagnation in agricultural output during the mid-1970s to mid 1980s (Kannan and Pushpangadan, 1988).

To sum up, the three case studies pose the following problems: (i) Why is land reform not always an effective instrument in reducing poverty? (ii) Why and how does land distribution become worse over time, as it did in Odisha?

Structures of Land Distributions and Poverty

Table 4.4 presents the incidence of poverty across landholding size by social groups. The Table reflects (i) Across the social groups, the incidence of poverty remained highest among the Schedule Tribe (ST) population (74.03%), followed by Schedule Caste (SC) (52.27%) and then followed by OBC (37.69%) and 'other' category (25.50%). (ii) Across the landholdings class, though in general, the incidence of poverty decreases as we move from a lower landholding class to higher landholding class, there is a significant change in households with more than 2 hectares of land.

Table 4.4 *Percentages below poverty line population by their household's operational holding across social groups*

Landholding (hectare)	*Social Groups*				
	ST	*SC*	*OBC*	*Other*	*Total*
Land less	42.99	17.71	59.29	16.33	31.85
0.001 to 0.4	67.81	55.53	39.29	27.39	44.52
0.41 to 1.0	77.34	45.67	36.20	25.60	47.80
1.001 to 2.0	79.80	44.03	42.61	25.33	51.13
2.001 to 4.0	83.67	65.18	19.05	2.45	42.14
> 4.001	63.30	39.33	4.39	18.27	29.65
Total	74.03	52.27	37.69	25.50	45.83

Source: NSSO, unit level, 61st round, 2004–05.

Examination of Land Distribution and Poverty

One of the major assumptions of the Fifth Five-Year plan in India was to increase agricultural output to 475 per cent during the period 1973–74 to 1978–79. However, it is difficult to realise such a high level of growth in agricultural output. But even if it had been realised, it would have required many redistributive measures to ensure minimum levels of consumption for the bottom 30 per cent of the population. The point made here is that, growth-led poverty reduction may not always be feasible or possible. During the same period, Professor Minhas suggested a redistribution proposal which was built around two main themes; one favouring the reallocation of large landholdings into smaller ones, and the other one was to integrate public work such as with consolidation of landholding and complementary land and water development work (Minhas, 1970).

Irrigated land can play a significant role in poverty reduction and this has been widely discussed in literature. Raj noted that the agricultural labour households, those who have more than 0.4 hectare of land to cultivate in the deltaic region of West Godavari district of Andhra Pradesh, do not find it necessary to enter the rural market, i.e. they can procure the necessary food requirement from their land (1992). Visaria (1978) observed from rural Gujarat and Maharashtra that where there are no irrigation facilities, it showed that the proportion of all households below the poverty line does not decline as own landholding rises even up to 7.5 acres in rural Gujarat and above 10 acres in rural Maharashtra.

Mahendra Dev (1988) has found that households which have less than one acres land constituted 90 per cent of the total population in states like West Bengal, Odisha, Tamil Nadu, Bihar, Assam and Punjab. In these states, rural poverty and small landholdings are closely related, but it is different in other states like Andhra Pradesh, Gujarat, Haryana, Karnataka, Madhya Pradesh, Maharashtra and Rajasthan. It is observed that the incidence of poverty is seen among households with more than 1 acre, and sometimes more than 2.5 acres of land. These results reflect that, not only the size, but the quality of land is important for poverty reduction.

In this respect, Lipton (1985) points out that,

> in developing Asia and Africa, private land forms an increasingly dominant share of rural productive assets, yet in un-irrigated and unreliably rain fed places, there is no association between land owned and operated (between zero and five to 10 acres) and poverty risk. (Lipton, 1985: 13)

However he notes that even a small amount of reliably watered land reduces poverty risk and concludes that,

> Access to even a little bit of land, despite being associated with larger household size, is associated with lower probability of poverty in an average year – if, and only if, the land offers prospects of a reasonably high net return per acre. (ibid.)

Irrigated land can also be a positive externality in labour markets as it may facilitate double cropping – and hence can help in poverty reduction by demanding more agricultural labour and consequently increase in wage rates. In the book edited by Chadha and Sharma (1997), some articles hint at this externality logic. For example, Vidya Sagar's study on agricultural growth and poverty in Rajasthan shows that poverty reduction was linked with irrigation development, a shift in cropping patterns (induced by irrigation) and an increase in real agricultural wages. This occurred because of expansion of rural employment. Gill's study on agricultural growth, employment and poverty in Punjab also credited the poverty reduction achieved by the state to the expansion of agricultural employment and the increase in agricultural wages. However, the author also attributed part of the poverty reduction because of expansion of employment outside the agriculture and even outside the rural areas in small towns and cities. Kashyap's study on irrigation-induced agricultural growth and poverty reduction in the Kheda district of Gujarat shows that irrigation-induced agricultural prosperity is the most important reason for reduced rural poverty and inequality.

However, in the same work, two studies on Uttar Pradesh and Bihar, where poverty reduction has not progressed as well but is still at a high level, finds that inadequate agricultural credit (for cultivation) and inadequate investment (for irrigation) are the reasons for continued rural poverty (ibid.). The role of agricultural investment for poverty reduction is also highlighted in the study by Roy and Pal (2002). They found that private investment in agriculture, rather than public investment, has played a very significant role in reducing poverty in India. They suggest enhancing both public and private investment in agricultural research and rural infrastructure (roads, markets, storage and irrigation). Their study, which covers the period from 1965 to 1999, shows that in Kerala both public and private investment in agriculture remained much higher than in both Odisha and West Bengal. In West Bengal, public investment was at a similar level to Odisha, but private investment was much higher than that of Odisha.

Based on our review, no study carried out so far provides evidence to support the proposition that concentration of land reduces poverty. While one may argue that large landholdings may favour higher productivity derived from economies of scale, there is no evidence that such growth helps in poverty reduction, at least in the Indian context (Ahluwalia, 1978). Rather, increased agricultural outputs can help in reducing poverty via the labour market.

Our observations from the available data and literature do not point to a clear link between the pattern of land redistribution and rural poverty reduction. Therefore, we need to look at this problem from the perspective of broader agrarian reforms. Along with land, reforms also include promoting the viability of small and marginal peasants' holdings by providing cheap credit, inputs, irrigation, collectivising farming, etc. (Byres, 1974). While the impact of land reforms on poverty reduction have been discussed at length, the processes through which land reform reduces poverty have rarely been studied. There is evidence that physical infrastructure played an important role in poverty reduction in West Bengal, as social infrastructure did in Kerala. This justifies looking at the issue of poverty reduction from a broader perspective rather than land redistribution alone.

While economic policies oriented towards agricultural development are important, the social welfare model also plays an important role in poverty reduction. Raj (1992) and Kannan (1995) pointed out that rural transformation, including poverty reduction in Kerala was not due to land reforms alone but because other social overhead capital. Raj noted that over the last two decades the Kerala experience shows that, 'land reforms made a high degree of social and economic transformation possible in its countryside' (Raj, 1992). His study did not elaborate on investment in agricultural land, but emphasised on the importance of rural credit and social infrastructure. Kannan (1995) gave emphasis on importance of social welfare model of the state – especially the nutritional support to the people through the Public Distribution System, and the provision of free education and health care.

Parthasarathy and Murty argue for land reforms from a different angle. In their article 'Land Reform, New Economic Policy and Poverty' (in Chadha and Sharma, 1997) they note that,

> it is well known that land reform failed to contribute directly for improvement of access to land for both the landless agricultural labour and marginal owner. Very little surplus was acquired from the ceiling, much less was distributed and what distributed was mostly dry and land requiring huge investments beyond the capacity of the poor. (Parthasarathy and Murty, 1997: 338)

In spite of that, they formulated a hypothesis,

> even a limited but successful implementation of land reforms has the potential to contribute to human development and the reduction in poverty. (ibid.)

They found this hypothesis to be true in their study of Kerala and West Bengal,

> in West Bengal too, land reforms brought about profound changes in the rural society. (ibid.)

The above review suggests that sustainability and impact of land reforms on sustainable poverty reduction depends on the state's economic and social policies.

Historical experiences, political interest and socio-economic policy played a significant role causing land reforms, sustaining and making them a successful instrument for poverty reduction in Kerala and West Bengal. The absence of such process and policies are the key factors behind the continued higher incidence of poverty in Odisha. Therefore, below we have explored the 59th NSSO data, which is on the situation assessment of farmers. The survey also provides information about the other productive forces in agricultural development.

Other Agrarian Productive Forces

Given the above discussion, next section provides a review of the historical experiences leading to land reforms in Odisha and analyses these state's economic and social policies (considering suitable indicators) for the sustainable poverty reduction through land reforms.

This section first tries to understand how historical conditions in these three case studies may have contributed to the different levels of poverty incidence in each state. Initially, the discussion is limited only to land possession and landlessness. It then proceeds to observe how state economic and social policies helped poverty reduction in Kerala and West Bengal via land reforms.

Case 1: Kerala

Modern Kerala is composed of three parts with different histories and different land tenure systems.[9] Travancore and Cochin in the south, where land tenure was less feudal with many owner-cultivators and progressive agrarian policies, benefited tenants. In north (Malabar), tenancy was dominant. Perhaps this tenancy system was the main reason for the agrarian movement in Kerala, which was born in this region. It must be noted here that the agrarian movement was also intertwined with the national independence movement as well as with caste and religious struggles.

The Moplah community, composed of low caste Muslims, was the first to oppose both the British rule and indigenous rulers and exploiters in Kerala. The Moplah tenant opposition was the prelude to organised tenant movements in Malabar (the north Kerala). While the dissatisfaction of the Moplah was very clear, the suppression of the Khilafat movement[10] catalysed the outbreak of the Moplah rebellion in 1921. Though the British were able to suppress the rebellion, Congress recognised that 'the plight of the Moplah tenants was the root cause for the rebellion' (Bergmann, 1984). The Malabar Tenancy Act of 1929 was intended to relieve the distress of the Moplah and the Congress party's agitation in Kerala to bridge the gap between Muslim and Hindus.

By 1934, the leftwing had organised – for the first time openly – as a faction inside the Congress and they played an important role in the emerging agrarian movements. In April 1936, an all-India Kisan Sabha (Meeting of the Kisan) was formed, where socialist peasants' leaders from Kerala participated actively and made the following demands:

(a) Rent to landlords should not exceed 25% of the total value produced.
(b) Prohibition of all other levies.
(c) Permanent security of tenure through amendments in the Malabar tenancy act.
(d) Abolition of all rent arrears.
(e) Fixed wages for agricultural labourers.

In 1937, a Congress-dominated government came to power under the Madras Presidency. They announced an amendment to the Malabar tenancy act in 1938 but it did not satisfy the peasants. A Malabar tenancy committee was appointed in 1939 to suggest a better solution and finally in 1946 the 'madras tenants and ryots' protection act was enforced to protect the tenants from eviction. Nevertheless, the agrarian movements slowly spread towards Travancore-Cochin as well.

After independence, agrarian movements continued to take place but peacefully. In such a peaceful manner, during the election of 1951–52, the Communist Party of India (CPI) joined with other left parties and formed a United Front, demanding the abolition of landlordism. The United Front won 38 out of 108 seats and became the second largest force in the assembly. Hereafter, agrarian issues were at the forefront of election issues. While the united front called for the distribution of land to the landless and small peasants, and further security of tenure for the tenants, the Congress promised introduction of ceilings on holdings and distribution of excess and crown land. The Praja socialist party also demanded land reforms.

Finally, the government introduced seven land reforms bills dealing with tenancy, land tax, prevention of eviction of hutment dwellers, and compensation for improvements made by tenants, abolition of special rights over land, restrictions of possession and ownership of land. During the same period in Malabar, many new demands were also raised and many tenancy legislation acts were passed, which helped the tenant to a greater extent than before.

Therefore, it appears that the class struggle which originated from the community laid the foundation stone for land reforms in Kerala. After independence, and especially after state formation, these issues were addressed by political parties and by the early 1970s, land reform was successfully carried out (although there were some biases in the distribution of land).

Case 2: West Bengal

West Bengal is another unique case, where land reform took place within the context of a continuous class struggle and strong political support.[11] The Zamindari system then prevailed in West Bengal, under which the Government let villages or small districts to the Zamindars (landlords). While the amount of revenue that a Zamindar needed to deliver to the administration was fixed permanently, they were collecting revenue from the cultivators without any public control. In fact, the system offered freedom that gave the Zamindars the freedom to exploit. When a richer Zamindar could not manage the whole area assigned to him, he would appoint a Sub-Zamindar, which further aggravated the extortion and exploitation, leading to increased indebtedness among the cultivators. Gradually, these cultivators lost their land titles to the Zamindars, who became de facto landlords.

Counters to these developments began in 1885 under Act X, which recognised the occupancy rights of the real cultivators. But landlords opposed this occupancy right (particularly in East Bengal) and fought for the doctrine of high landlordism – whereby the landlord class assume their right as the sole proprietor of the land – and wanted the law to be in their favour; to limit or eliminate occupancy rights and increase the provisions to allow rents to be raised.

However, over the time, the size of the struggling class swelled and the first agrarian movement was the jute boycott campaign which took place in February 1921. This was followed by another movement of refusing to pay tax for irrigation water from the Damodar canal. Due to such rural unrest, in 1923 the provincial government planned to amend the law of 1885 and confer occupancy rights on the sharecroppers, who were then tenants. But the lack of a strong movement among sharecroppers, and strong opposition by the upper strata, forced the government to withdraw this amendment.

In the 1930s, the peasant movement rose up again, demanding the abolition of the Zamindari system. The peasants organised local movements to refuse to pay the rents, which induced the government to pass the Bengal agricultural debtors Act which set a ceiling on interest. The debt settlement board was also established to reduce agricultural debt. Since then, a number of Acts have been passed in the interest of the peasant class. In August 1938, the Bengal tenancy Act was passed to increase the awareness of tenants about their land rights and title.

The peasant movement of 1939 which later became the forerunner of the Tebhaga movement in 1946 demanded that the share of the sharecroppers be increased from 50 to 66 per cent. In September 1946, the Bengal Provincial Kisan Sabha took up the issue of rent reduction and demanded two-thirds of the crop for the sharecroppers. The cultivators simply took the entire crop to their own barns instead of carrying it to the landowner. This incident along with the Tebhaga movement led to the West Bengal Bargadari Act of 1950 which provided for sharing two-thirds of the output with the sharecroppers if the land owners do not supply draft animals, agricultural implements and manure; if the landowner provides all these, then the share would be 50:50.

In 1955, a more comprehensive land reform act came into effect, which mainly provided for the abolition for intermediaries and the imposition of ceilings on landholdings. Further amendments to this act have made the Bargadari right of cultivation heritable (1970). However, since there were no complete official recording of Bargadars, many of these rights were not realised in practice. In the post-1977 era, when the left front government came into power; a major change was brought to the land reforms giving a more visible shape to the long period of struggle by the peasantry in the countryside.

Sen (2001) points out, despite the considerable amount of undistributed surplus land held with the government, benefits of the land reforms could have been greater had the undistributed land been distributed or had the ceiling limit been higher than 17.29 acres of irrigated and 24.2 acres of unirrigated land. In brief, in West Bengal too, it can be seen that the persistent class struggle along with political interest are key factors resulting in land reforms and agrarian reforms as the key factor behind land reforms as instrument of poverty reduction.

Case 3: Odisha

A few small class struggles have been cited in the literature.[12] The available literature focuses on the higher intensity of exploitation through rent and rates of interest in the land revenue system. Hence, landlessness, lower productivity and poverty in the later period may be traced to this lack of a mass class struggle and the apparent acceptance of the exploitative land revenue system.

During the pre-British period there was no established code of rules giving the right of ownership over land to anyone.

> It was generally accepted that state had as much right on the share of the produce as revenue as the cultivator had the right to cultivate. (Jena, 1968: 30)

During this time, the revenue collectors were either originally elected by the people (rather than holding hereditary office) or nominated by the king. They were never given the right of ownership over land unlike the Zamindar (landlord) claiming exclusive ownership over land during the Moguls and the British period (Jena, 1968).

Until the Akbar period there were no significant changes regarding ownership right of land but after Akbar, when the Mogul rulers found it difficult to pay their military troops, a practice of granting the military chief a loan to cover the rent was developed and in the course of time the military chief claimed ownership rights over land. As the Mogul rulers became weaker, private ownership rights became stronger and finally, the collectors of land revenue formed themselves into a group as distinct from state and the cultivators. The collection of discriminatory rates of land revenue can be traced to this time.

After the Moghuls, the Maratha ruled Odisha, and since their aim was to collect revenue they were unconcerned about the condition of the cultivators once the revenue had been collected. In fact, production was concentrated on two groups – the rulers and the collectors. The Marathas created a period of acute suffering and extreme poverty for the cultivating class of Odisha (Jena, 1968: 35). The British government brought landlords from other provinces and made them the Zamindars of Odisha which aggravated the situation. In many cases the Zamindars enjoyed the status of owners of the land. This was not the system in all parts of Odisha. There were two other types of land revenue systems prevailing in Odisha at the time: the Mahalwari and the Ryotwari.[13] The state was divided[14] into 24 different princely states according to the land revenue system.

According to Samal (2000),

> there was little isomorphism in the land revenue system within the Odisha context. Odisha was fragmented into three divisions, each being attached to a centre of control not located within the state for administrative convenience of the rulers and each division experienced a different mode of Land Revenue administration. (ibid.)

The Zamindari system was highly exploitative, unfair and profiteering which forced the tenants into abject poverty and ignominious subjection. The Ryotwari system was a little better, as it contained some enlightened aspects and was benevolent in design, but also failed to deliver the goods. The Mahalwari system was in between the two, which did not ensure benevolent attitudes to all households.

The reformulation of the state of Odisha was driven by a mass movement (Jena, 1968), therefore the existence of mass consciousness in the state cannot be denied. But why didn't this lead to class struggle? Sahu (1997) suggests three reasons for the limited success of tenancy legislation in Odisha: (a) the stranglehold of colonialism; (b) the contradicting class interests; and (c) the complex agrarian structure.

From this discussion, it is observed that in Odisha, the cultivator class was disorganised and less radical, perhaps due to different agrarian structures and land revenue systems, and the political leaders were (and continue to be) from the landlord and moneylender class. Therefore, the political leaders faced conflicting class interests, one of the crucial reasons for many peasants' demands failing to achieve any success.

Economic and Social Policies

Table 4.5 shows the agricultural investment, which may be linked to sustainability of land reforms. Whether it is public investment or private investment, Odisha remained at much lower level compared to other two sates. In the case of private investment, Odisha never invested more than ₹ 50 per hectare during the last 35 years (1965 to 1999), whereas Kerala invested 3 to 11 times more during the same period. This has been shown in Table 4.6 which shows land productivity.

Table 4.5 *Public and private investment in agriculture, 1980–81 prices (₹/ha)*

	Public Investment			*Private Investment*		
	1965–73	*1974–86*	*1987–99*	*1965–73*	*1974–86*	*1987–99*
Odisha	92 (2.91)*	235 (6.15)	207 (–1.68)	48 (–3.34)	45 (–0.74)*	49 (2.77)*
WB	143 (3.94)**	202 (0.78)	206 (0.76)**	120 (6.69)**	144 (–0.29)	147 (7.06)**
India	162 (3.62)**	315 (1.64)*	265 (0.43)	154 (2.98)*	202 (0.98)*	288 (7.44)***
Kerala	188 (–4.59)	387 (1.89)*	340 (2.36)	172 (–1.74)*	530 (5.18)**	591 (5.85)**

Source: Compiled from Roy and Pal (2002). *, **, *** implies significant at 1%, 5% and 10% level respectively.

Table 4.6 *Level of agricultural productivity, 1980–81 prices (Ag SDP, ₹/Ha)*

	1965–68	*1980–82*	*1996–98*
Odisha	1,773	2,982	2,685
India	2,353	3,431	5,727
WB	4,915	5,407	12,165
Kerala	5,576	6,806	10,965

Source: Compiled from Roy and Pal (2002).

The highest land productivity achieved in Odisha is still lower than the lowest productivity recorded in the other two states. Lower land productivity has always been a problem for Odisha, as the following humorous lines by a royal poet of Odisha notes:

> Chhamu jaha dele harase, pau pau gala barase; Bachhi bachhi kari dele padia, srabana masare bhange nadia; Emanta bhuin, chakhunda bunile uthai nahin.
> *Jadumani Rahasya Katha*

Whatever king has kindly granted took years to get; a hard field where a coconut cannot be broken even after a month of rain; such a barren land where even the Cassiatora plant will not grow.[15]

The literature reviewed above notes that rural credit was extended to cultivators after land reforms. The impacts of agricultural investment and rural credit on poverty reduction can be examined by looking at the increase in agricultural earning. Table 4.7 shows the changes in agricultural earnings of male labourers over the period 1964–65 to 1999–2000 from the Rural Labour Enquiry (RLE) reports on wages and earnings of rural labour households.

The Table 4.7 shows that – (i) Real wages were stagnant during 1964–65 to 1983. (ii) In Odisha; wages were lower compared to Kerala, West Bengal and all-India. (iii) Thirdly, overall social policies may also have strongly supported poverty reduction, especially in Kerala where there is ample evidence that social security has been an instrument for poverty alleviation (Kannan, 1995, 2002;

Sebastian, 2002). While some social security schemes like the public distribution system do exist in Odisha, they function poorly due to both demand and supply-side problems. As Table 4.8 shows, the per capita expenditure of the social sector is much lower in Odisha compared to Kerala, but not so in West Bengal.

Table 4.7 *Agricultural real wage earnings (male ₹/day at constant prices 1964–65)*

States	*1964–65*	*1974–75*	*1977–78*	*1983*	*1987–88*	*1993–94*	*1999–2000*
Kerala	10.44	10.28	14.28	12.36	16.99	21.55	30.52
Odisha	5.50	3.96	5.34	3.64	6.67	8.59	9.07
West Bengal	7.58	5.93	7.68	5.29	10.46	12.24	14.28
All-India	5.77	5.28	6.93	5.44	8.54	10.96	12.99

Source: Estimated from RLE Report on Wages and Earnings: Various Issues, Consumer price index of agricultural and rural labourers annual reports.

Table 4.8 *Per capita and percentages of social sector expenditure to total expenditure*

	Per capita expenditure		*Per cent of total expenditure*	
	1990–91	*2000–01*	*1990–91*	*2000–01*
Kerala	508	1646	43.7	39.9
Odisha	352	1104	36.5	36.8
West Bengal	415	1200	46.9	36
All states	419	1214	38.6	36.8

Sources: Social sector expenditure from RBI, Report on State Finance and Population from Census of India.

(i) During 1991 while the per capita expenditure on social sector for Kerala was ₹ 508, the same for Odisha was ₹ 352, which was much lower than Kerala and also lower than West Bengal (₹ 415) and all-India (₹ 419). During 2000–01 same order has been maintained, Kerala at the top with ₹ 1,646 followed by all-India ₹ 1,214 than West Bengal ₹ 1,200 and at the bottom Odisha with ₹ 1,104.

(ii) In both the period, Odisha was having lower per capital social sector expenditure even compared to all-India. However, in terms of social sector expenditure as per cent of total expenditure do not show much variation during 2000–01, except Kerala which is a little higher than all others.

The following points can be drawn from this discussion:

(a) Mass class-consciousness and persistent class struggle between the landowning class and landless class are crucial factors behind the occurrence of land reforms – both in Kerala as well as West Bengal. In Odisha, such mass class-consciousness has not yet developed but is stemming from liberal development strategies of the Government of India, a land reforms initiative commenced in the state. This, however, failed.

(b) Once land reforms have taken place, it is the social and economic policies of the state which turn land reforms into an effective instrument of poverty reduction – as seen in Kerala and West Bengal. In Odisha such policies were very weak.

Both of the above points may point to the reasons behind higher landlessness and poverty in Odisha. As the HDR points out, declines in institutional credit and agricultural extension services have adversely affected small peasants in rural West Bengal (Government of West Bengal, 2004). This may be the cause of increased landlessness.

Summary

The chapter attempted to examine the extent to which land distribution contributed to the existing hunger and deprivation. It did not find a clear relationship between the operational holdings and income poverty, which might be due to lack of overall agrarian reforms. Attempting to explore the relation between landholding and poverty, it is observed that (from the case studies) that while the occurrence of land reforms may rely on historical process (notably class struggle), political interest, the land tenure system or conscious economic policies of the state, their sustainability and impact on sustainable poverty reduction depend on the state's economic and social policies. It can be concluded that land reforms or landholdings are necessary but not sufficient for poverty reduction. While one would disagree with the concentration of land, the point made is that ensuring equitable landholding alone is not a panacea for poverty reduction, but rather, that poverty reduction through land reforms is conditional on complementary state economic and social policies.

From Chapter 2 it is observed that, there is prevalence of different land revenue system, where the land was indirectly under the disposal of the landlords or the Zamindar. But the improvement over such land relations has been brought in the post-independent period, where attempts have been taken to abolish Zamindary system and hence land reforms were taken up. But the institutionally corrected land relations could not sustain due to lack of other supplementary reforms (such as irrigations, agricultural credit) to enhance other productive forces to help agricultural production. Hence, after a decade or so the institutional efforts to ameliorate the land relations has failed and perhaps led to the existing hunger and deprivation.

The above analysis and discussion also draw the following as concluding remarks. First, the growth-led poverty reduction hypothesis is disproved by the literature, which makes clear that the significant growth in agricultural output in Kerala, Odisha, West Bengal, Punjab and Haryana and Uttar Pradesh has not contributed to poverty reduction during the 1960s. Chapter 3 also has shown – during 1990s, Odisha had healthy growth, but poverty has not reduced much. Secondly, the social justice approach to poverty reduction has been effective in all the three states of Kerala, West Bengal and Odisha. The first way of ensuring social justice (distributing output and services at subsidised prices) became insufficient when the amount of resources available for distribution became smaller, while the number of dependents for such public assistance became larger. For example, in the case of Odisha, although the share allocated to the social sector is comparable with that of Kerala and better than that of West Bengal, the per capita amount of assistance in Odisha remained very low. Further, large dependence on public provisioning together with insufficient growth is a threat to the sustainability of such schemes. This issue will be further discussed in Odisha's context in Chapter 7. Finally, when the intensity of poverty is very high in case of Odisha, it becomes extremely difficult for the concerned households to even access public support. Therefore, adopting the alternative approach to social justice, by providing land entitlements to the landless through land reforms, is important from a poverty reduction point of view. In Kerala and West Bengal, however, many other conditional factors played an important role in land reforms leading to poverty reduction:

In Kerala, the overall rural transformation (particularly social and economic transformation) was made possible not by land reforms alone but rather by combining land reforms with supporting measures in labour and credit market including the agricultural investment (both private and public). In West Bengal, it was not land reforms which directly reduced poverty, but rather, as pointed out by the Economic Review of West Bengal (cited in Government of West Bengal, 2004) and Roy and Pal (2002), agricultural investment (public investment for minor irrigation and private investment for bore wells), non-land inputs such as credit, seed, fertilisers, and irrigation facilities which played a significant role in bring agrarian change and reducing rural poverty (See Lieten, 1996: 121). Although Odisha had an advantage over Kerala and West Bengal in terms of both occupational distribution and land distribution at the initial period, this has not only failed to reduce poverty but the pattern of land distribution has gradually worsened. Such conditions can be attributed to deficiencies in rural credit systems (Sarap, 1991), agricultural investment and irrigated areas. These may have contributed to keeping the agricultural wages at a low level in Odisha, unlike in Kerala and West Bengal. In other words, comprehensive agrarian reform may have been able to save the smallholder from becoming landless. Further, the failure of the institutional system to retain the land with the small holder leads one to infer that the cause of poverty in Odisha lies not only with weak policy but also in its institutional structure.

From these lessons, we can make the following concluding policy points:

(i) As Odisha's experience of land possession/landlessness has shown, landholding alone does not ensure poverty reduction. Rather, irrigated and productive landholding is important as shown by West Bengal and Kerala. Therefore, efforts should be made by both private and public agents to transform the land into productive assets.

(ii) In a developing economy concentration of land is not desirable, despite the fact that it may facilitate economies of scale and increase growth of agricultural output with the help of technology. This is because the introduction of technology can cause unemployment, and therefore, in the absence of strong and adequate redistribution measures, worker households may be driven to a state of poverty. An examination of the NSSO data also shows that concentration of land is most visible in Odisha.[16] Therefore, a minimum amount of landholding should be considered as a basic need for every agricultural household in an economy. This may justify institutions stipulating that, for the welfare of the people, if a household (or person) possesses only the specified minimum amount of land, then legally, neither he/she be allowed to sell nor any other household (or person) allowed to buy the piece of land. This paper argues that the scenario 3 type of agrarian structure is better, at least from a poverty reduction perspective.

Raj (1974: 11) stated 'Even a toe-hold on land means a great deal to those who are seriously handicapped because they have none', but the attempts has failed in case of Odisha due lack of other supportive measures, such as lack of to irrigation, agricultural credit, public or private investment in agriculture and other social security support. Hence, Raj's statement should be qualified by saying that other factors – at least a minimum irrigation (or an adequate amount of water) is necessary to make even a toehold land an effective instrument for poverty reduction.

So far we have discussed the economic growth and distributive justice aspects of hunger and deprivation, which shown-neither the growth of output and employment nor the resources distribution is impressive to reduce hunger and deprivation. This situation further gets accentuated with different existing institutional rigidities and occurrence of natural calamities in the state. Social exclusion is one outcome of such systematic rigidity that can cause hunger and deprivation being discussed in next chapter.

Notes

1. As pointed out earlier, distribution implies at least two kinds, such as distribution of resources, (e.g. land distribution) and distribution of output. The latter can be found in two ways – distribution of output to the factors of production (such as land, labour, capital and organisations gets their shares in terms of rent, wages, interest and profit respectively) and distribution of output through social security schemes. Here we are dealing with the impacts of land distributions and factor payments exclusively to the labour (wages) with hunger and deprivation.
2. Analytically it is clear that an agrarian structure has an impact on poverty. A better agrarian structure can impact on poverty both from the production and distribution perspectives. In contrast, the concentration of land with a few groups of people over long periods of time – as it has been under different land tenure systems (e.g. in Permanent Settlement) in India – obstructs agricultural production, adversely affects income and productive employment for the landless, and creates unequal access to social and political power (Government of West Bengal, 2004) – all of which can generate poverty.
3. For example, in India the Public Distribution System directly distributes essential commodities to the population; other countries distribute food stamps (see Swaminathan, 2000).
4. This may not always hold true however, and the arrival of technology may push unskilled labour into poverty. In such instances, Minhas (1970) suggests '... take away the land from the landlord and large farmers and redistribute it among the small holders and build a floor ceiling as well like an upper ceiling'. Basically, smaller holdings can be encouraged, which would lead to growth and may create demand for labour as well. Further employment can be generated through other employment programmes for the landless poor.
5. West Bengal and Kerala, for example, have agrarian structures which ensured both growth and a minimum level of holdings. In Odisha, however, fragmentation into unsustainable small plots has occurred.

6. According to Daniel Thorner (1980), in India the different agents of agrarian structure consisted of Maliks (a category who had a right to share of the produce without participating in any labour), Kisan (the subject peasants who possessed the land and instruments of production, particularly cattle and gave a share of the produce to the Maliks), Mazdurs (untouchable field labourers, who were ploughmen with no right to the land). However, in a Marxist concept of agrarian structure, the agents may include feudal and capitalist landlords, rich peasants, middle peasants, poor peasants and landless proletariat.
7. A 'better' agrarian structure may not concentrate the wealth (land) among the few households and must not have many resourceless households; in brief and limited sense, larger sections of the populations must get access to land, credit and other resources. But as the discussion here is confined to land only, our observations from Tables 1, 2, and 3 only refer to land.
8. The situation is somewhat different for the marginal group of operational landholders. During 1981/82 the proportion of marginal landholders was 16.05 per cent and at that time they were operating only 2.29 per cent of total operated area. By 1991/92 the proportion of marginal group had increased to 43.71 per cent and their operating area had increased to 22 per cent of the total operated area. This implies that the situation among the marginal group of operational holders is better than among the landless group.
9. Several authors have documented the successful story of land reforms in Kerala. We refer in detail to two such studies, by Oommen (1975) and Bergmann (1984).
10. Khilafat movement was an anti-colonial movement of Muslims of India based on the hostility of the British to the Turkish Sultan after the World War I.
11. Here, we have closely referred to Bergmann (1984) and West Bengal HDR (2004).
12. For this section we have closely referred to Samal (2000), Padhi (1999), Sahu (1997)
13. In the Zamindari system, the Zamindars were the intermediaries between the state and the actual tillers. They collected land revenue, keeping a part of it as commission and paying the remaining to the government. Under Ryotwari system, the cultivator was recognised as the owner of a plot of land subject to the payment of land revenue. Under the Mahalwari system, the revenue settlement was made, village-by-village with the village headman who was the intermediary between the government and the actual cultivators.
14. Such as Odisha division, Ganjam-Koraput tract, Sambalpur tract and Northern Odisha.
15. This is a popular quotation in Odisha, which depicts the grievances of a royal poet to his king, who had granted a piece of fertile land to him for his excellent service and talent. But not only did it take a long time to get the land but it also turned out to be unproductive. Hence, the poet was unhappy and one day when the king enquired about his unhappiness, the poet replied the king with the above lines.
16. During 1953–54 although Odisha was having only 5% landless households, which was much lower than other two, another bottom 36% of were having only 2.1% of total land and bottom 50% around 10% of total land, which may lead to very small holdings.

5

Social Exclusion

A Source of Hunger and Deprivation

This chapter looks into the social exclusion aspects of hunger and deprivation in Odisha. The socio-cultural history of India is such that, caste as an institution divides the society into different groups in a social hierarchy and that hierarchy determines the political economy power of a person. The higher is the order in a social hierarchy, the higher is the wealth, education, occupation, income and well-being and hence lower order would mean hunger and deprivation. In most Indian states including Odisha, the existence of the institutional norms, which shapes the social hierarchy, determines the socio-economic power of a person and whether he will face hunger and deprivation. Therefore, hunger and deprivation among the socially lower order bears a historical reason. In the social history of Odisha, there exists groups, which have been excluded from various mainstream activities and their social exclusion reinforces the hunger and deprivation among them.

In its economic sense, social exclusion occurs out of lack of effective access to land, labour market, educational institutions and health care systems. Such denial of access may occur due to citizen's acts, i.e. in the free market, a group of citizens may accumulate wealth faster than others and hence the others become socially excluded over a period of time. The denial may also occur due to the norms of institutions or system, e.g. in Indian context, the institutions of social hierarchy based on caste had not only restricted the access of a group of people to certain facilities but also treated them as 'untouchables', especially the scheduled caste and indigenous people. Significantly, the macro picture that has been examined in the previous chapters have revealed that irrespective of the place of residence (based on rural and urban location as well as based on regions), the SCs and STs people in general suffer higher incidence of hunger and deprivation (in most respects) in Odisha.[1] However, there have been efforts to uplift the status, especially the social status of these social groups, since the first quarter of 20th century and after independence. Intensive conscious policies have been developed in the form of affirmative actions for their development. However, even after six decades of such efforts, the outcomes are far from satisfactory and these groups fall behind other social groups. Therefore, this chapter investigates into the details of the exclusions of these two groups in the development process.

The chapter has four sections. Section 2 discusses the concept of social exclusion and its development globally, in the Indian context and more specifically in Odisha. Section 3 examines the trend and magnitude of social exclusion from the macro data. Section 4 presents the micro evidences.

Conceptual Discussion on Social Exclusion and Poverty

On the concept, application, scrutiny and relevance of social exclusion, Sen (2000) pointed out that – 'no concept of poverty can be satisfactory if it does not take adequate note of the disadvantages that arise from being excluded from shared opportunities enjoyed by others. To be excluded from common facilities or benefits that others have, can certainly be a significant handicap that impoverishes the lives that individual can enjoy'. Sen has used the capability and deprivation approach to explain the social exclusion and hence, he went back to the Aristotelian ideas of lack of freedom or the Smithian idea of deprivation as social exclusion (ibid., 5–7). Within the capability deprivation, Sen discussed six important fields where the social exclusion can cause deprivation and poverty. Those are: (i) Inequality and relational poverty, (ii) Labour market exclusion, (iii) Credit market exclusion, (iv) Gender-related exclusion and inequality, (v) Exclusion from health care and (vi) Exclusion from the food market.

However, since Sen focused on the macro picture of social exclusion, there is little discussion on the caste-based aspects of social exclusion as it exists in India and widely discussed by many other researchers even prior to Sen, in the Indian context.

Indian Context

Social exclusion in India is visible in many forms but caste-based exclusion is the most visible form. Social exclusion occurred due to exclusion from valuable resources like land, education and work. As Appasamy et al. (1995) put it, 'social exclusion is a process, which restricts the access of certain social groups to valued resources and entitlements, relegating them to the status of social outsiders'. By the early phase of 20th century, when the problem of social exclusion was brought to the notice of intellectuals, the matter of 'untouchable' was at the centre of the debates and hence, the efforts were made for the upliftment of these groups thorough public provisions.

On the matter of persistence of caste-based social exclusion in India, Ghurye (1969) pointed out the outstanding features of Hindu societies. These are: (i) segmental division of society, (ii) hierarchy, (iii) restrictions on feeding and social intercourse, (iv) civil and religions disabilities and privileges of the different sections, (v) lack of unrestricted choice of occupation and (vi) restrictions on marriage, i.e. insistence on the practice of endogamy. Therefore, the structural basis of the Hindu society was based on the idea of exclusion and encouraged the social exclusion in the absence of effective positive description.

De Haan and Nayak (1995) have carried on a study on 'Social Exclusion and South Asia', where it is pointed out that the expectation that development would bring about improvement in the well-being of all individuals is a naïve belief. In other words, the economic development processes bypass or exclude certain groups of individuals, who are recognised as socially excluded group at a later stage. Further, social exclusion may not only arise out of the process of economic development, but it may occur due to the traditions, socio-economic system, cultural practices, etc.

The social exclusion also occurs in many other dimensions, such as exclusion from education, housing, property ownership, democratic participation, and access to health services and public goods. It may also include exclusion of old and infirm, widows, physically handicapped, etc. Following the widespread notion, social exclusion may also include the rights notion, i.e. exclusion from 'rights'. Violations of those rights also may lead to social exclusion. However, the dimension of social exclusion that has taken central importance in India is the exclusion based on caste. Caste-based exclusion is the most potent form of exclusion because this sociological characteristic is 'fixed at birth and hence is completely inflexible'.

The caste-based social exclusion is based on the premise of a rigid social hierarchy. Based on the traditions, Indian population was broadly divided into four groups, (i) the Brahmin, (ii) Kshatriyas, (iii) Vaishyas and (iv) Shudras. There were also thousand of sub-castes but the most important features of that system was that the upper castes comprising mostly, Brahmins (occasionally the other two castes as well) have invariably been the dominant group in social and economic interaction and hence accumulating most of the social produce. The Shudras were to serve these upper castes and within Shudras there was also group of large number of untouchables or the Harijans.[2] Hence, the most discriminated people in the society are the so-called untouchables.

On the determinants of social exclusion, Nayak (1993) stated that the determinants can be based on race and ethnic minorities, immigrant communities, political dissidents, the old, single mother, etc. but in case of developing countries like India, poverty alone can be the significant determinant of social exclusion. Employment and the caste-based exclusions are the crucial factors behind social exclusion. Nayak also pointed out, while the social exclusion based on all other categories denotes an 'end state' of the process of development and it is the end state that we regarded as denoting from exclusion, which is less probably in the case of caste-based exclusion, as the nature of exclusion is fixed or rigid. In other words, there is a likelihood that a person can escape from abject poverty or from child labour or unemployment or bonded labour, but the exclusion based on caste as in the Indian context is fixed at birth itself and hence, from the birth itself a person gets excluded from the society. Therefore, there are aspects of exclusion that cannot be characterised, when a person is described as being simply poor or unemployed. For instance, it is considered that contact with a schedule caste person is polluting in many parts of India. Untouchables are prevented from entry into certain temples and untouchable students have to sit separately in the educational institutes. Untouchables are forbidden use of village wells, that are exclusively preserved for the upper caste members and even they are prevented from using public wells, tube wells etc.

Kannan (1994, 1990, 1992), Papola and Rodgers (1992) and Deshpande (1983) have shown the higher segmentation of labour market in India. The key factors of the process of segmentations are caste, community and gender. The people from lower social group like scheduled caste have limited occupational choices despite the affirmative action policies in the labour market. Similarly, the gender-based discrimination is found across all categories. All these contribute to the process of social exclusion. Kannan (2003) while examining 'India's shining' argues that the SCs and STs have been left behind in the process of development.

Therefore, in India, social exclusion cannot be mapped out only with the economic deprivation, rather the cultural and social deprivation has to be part of it. While the secondary literature does capture some aspects of both economic and socio-cultural deprivation, it does not capture certain other kinds of socio-cultural deprivation (untouchability), which very much prevails in the Indian society.[3]

Thorat (2004) analysed the subject of social exclusion (especially the caste based) from human rights perspectives. The United Nations human rights framework was taken up and the Hindu religious and social order (particularly, the caste system and untouchability) in its original and classical form which comes in direct conflict with the universal human rights framework is discussed. He also presents a societal situation, where the religious and social systems directly conflict with universal rights and become the cause of ubiquitous human rights violations. He also examined, 'the presence and continuations of antagonistic social, economic, and religious cultural elements make the enforcement of human rights (by the various wings of the state) difficult despite the provision of legal measures'.

The concept of human rights dates back to the post-seventeenth century western liberal political thought. The United Nations included a comprehensive list on human rights, which covers civic, political, economic, social and cultural rights. These rights have taken its universal application with the UN Declaration of 1948. It is the first two of the articles that make it clear that freedom, equality and fraternity are the foundation of human rights. Article 1 states, 'all human beings are born free and equal in dignity and rights. They are endowed with reason, conscience and should act towards one another in a spirit of brotherhood'. Article 2 states, 'everyone is entitled to all rights and freedoms, set forth in this declaration, without distinction of any kind, such as race, colour, sex, language, religion, political or other opinion, social origin, property, birth or other status'. These two articles also indicate the principle of justice which is the corner stone of the human rights formulation.

The Indian Constitution drafted during 1947 to 1949, coincided with the UN Declaration of Human Rights in 1948. The Indian Constitution also emulates the UN Declaration in some sections of the Constitution, especially in the 'Fundamental Right' and in the 'Directive Principle' sections, which stated '(it is solemnly resolved) to secure to all citizens Justice: Social, Economic and Political, Liberty of thought, expression, belief, faith and worship. And equality of status and of opportunity, and to promote among them all fraternity, assuring the dignity of individual and unity'. The Constitution also states that 'the state shall not discriminate against any citizen on grounds only of religion, race, caste, place and birth or any form'. The Directive Principle adds, 'the state shall promote with special care the educational and economic interest of the scheduled caste/tribes (SCs and STs) and shall protect them from social injustice and all forms of exploitation'.

As per these constitutional provisions, the government has initiated a number of measures for protecting the interest of the untouchables (scheduled castes or SCs). Based on the 'Protective Sphere' and Article 17 of the Indian Constitution, it is declared 'untouchability' as abolished and forbade its practice in any form and the enforcement of disability arising out of 'untouchability' was to be an offence punishable by law, according to the Anti-Untouchability Act, 1955. The government reviewed this Act in 1976 to make the law more stringent and effective. The enactment of the 'Protection of Civil Rights Act 1955' (PCR Act)[4] took place.

Since the normal provisions of the Indian Penal Code were inadequate to provide safeguards to SC/ST against several crimes, the need for the additional Act was felt in 1989.[5] This implies various conscious policies (through reservations) to improve the access of these social groups into the economic, educational and political spheres. However, such promotional legislations do not cover discrimination and the violation of rights in the private economic activities, such as discrimination in various markets, like in land, capital, consumer goods, wage, employment etc. (Thorat, ibid.) and hence fails to improve the conditions of this social groups.

Such failures relate to the Hindu social order, because the socio-cultural rigidity and regional disparities makes the enforcement and practice of the human rights or constitutional laws difficult. Three central features of the Hindu social system, which may go against the principle of liberty (or freedom), equality and fraternity (which are the base of human rights) are: (a) it involves a division of people in social groups (castes); the social, religious, cultural and economic rights of each member of these castes are predetermined in advance by birth into that caste, (b) the unequal and hierarchical (graded) division of these rights among the castes, and (c) the provision of strong social, religious and economic ostracism supported by social and religious ideology to maintain the Hindu social order. Therefore, as Ambedkar puts it, the doctrine of inequality is the core and heart of the philosophy of Hinduism (see Thorat, ibid.) and hence, he stated, liberty to be real, there must be social equality and economic security; further, to realise freedom, knowledge to be made available to all because deprivation of knowledge is denial of power to use liberty. However, the social system remained such that, even in the first decade of the 21st century, there are numerable evidences on the violations of human rights against the socially excluded groups or against the untouchables.

As Thorat (2004) presents some of the empirical evidences on the violations of civic, political and religious rights of the untouchables' castes, where the violations of civil rights include prohibition to the untouchables to use public water bodies, such as well, taps, temple, tea stall, restaurant, community bath, road and other services. At the all-India level, Thorat found that, the average number of cases of human rights violation of untouchable registered annually was 480 during 1950s; 1903 during 1960s; 3,240 during the 1970s; 3,875 during 1980s and 1,672 during 1990s. Further, he pointed out that on an average, annually 30, 000 cases of general crimes and atrocities were committed on the former untouchables during 1981 to 1997.

Thorat also presented evidences from some of the micro level studies from Karnataka and Andhra Pradesh in the southern part of India, from eastern part of Odisha and Gujarat from the western part of India. Though the author has shown a significant extent of the violations of human rights against the former untouchables, yet further investigations are required as the subject matter enquired in different states were different, hence a comparison cannot be drawn but some absolute picture can be derived.

Given the broad aspects of social exclusion and its forms of existence in different parts of world and different parts of India, in the following section we have examined the forms of existence of the same in Odisha.

Odisha Context

Though the intensity of social exclusion is very high in Odisha, yet there are very few studies carried out on the subject in the state. And those studies are primarily based on the caste based social exclusion. It is the caste based exclusion which also puts them into economic exclusion or exclusion from the labour market. The available studies are presented below.

The caste-based social exclusion in Odisha is a part of the Hindu social order (Behuria, 1990). However, to improve such situations many researches have tried to bring out the realities in their work. Bailey's study at Bisipara (1959) was one of the influential studies in Odisha, which recorded the caste-based aspects of social exclusion. The intensity of the caste-based exclusion is such that, it led to conflicts between SCs and STs (even in 1990), while a person from a former group enters into the village temple, the latter considers that the God in the temple has been polluted due to the entry.

Mohanty (1984) also presented the social root of backwardness (A study of class, caste and power). Mohanty pointed out two major reasons in this regard. First the social formation has happened mostly among the upper castes, mainly among Karans, Brahmins and Kshatriyas and they were the beneficiaries of the temporary settlement in the coastal area and permanent settlement in the inland. While the agrarian base gradually weakened in rural Odisha, these groups initiated the movement for Odia nationalism and led the freedom struggle in Odisha. However, it has offered new opportunities in the modern environment, mostly to them rather than to other social groups. 'Odisha witnessed several mass movements, ranging from Adivasi (tribal) revolts to peasant uprisings and militant anti-feudal struggles in the princely states, besides the cultural movements opposed to Brahmanism. But these movements could not effectively challenge the evolving power structure' (ibid.). The old power structure of the upper castes dominancy also was found in the post-independence period. In the whole process, discrimination against the SC/ST took place historically and culturally. In the post-independence era, the positive discrimination like reservations of seats for SCs/STs in the legislature as well as services benefited some sections of these groups but these sections have not produced an autonomous leadership capable of negotiating with the upper caste leaders.

Barik (1987) also provides a good description of the Brahmanism and the restrictions on temple entry matter.

Praxis (2001) comes out with the social dimensions of poverty in the Balangir district of Odisha. The work describes the socio-cultural aspects of poverty and gender dimensions of livelihoods respectively. Regarding social hierarchy, Praxis pointed out caste, social norms and other forms of discriminations, leading to perpetuation of poverty, especially among the lower castes and tribes, which are treated as inferior. On untouchability, they reported the strict norms related to inter-dining and accepting water from other communities 'Pichhalu Barik's little grand daughter touched a tube well in the village Khairmal. The villagers refused to take water from that tube well. They called a meeting of the villagers and threatened Barik's family. He had to apologise for the act of his granddaughter'. On gender perspective, Praxis has pointed out the discriminations in production sphere, household sphere and public sphere. In the production sphere, there is higher wages for males compared to females. In the household sphere, it reveals that, women engage in all the household activities, which ranges from cleaning the house, cooking and taking care of children to other household management activities. 'There are more than 20 activities that women are engaged in on a daily basis. However, on public sphere, women do not have any participation especially in the public offices and traditional institutions'. Therefore, the intra-household disparity against women also remained higher. This implies that women from a lower caste have to confront a multiple burden of exclusion. Since women from a lower caste are normally engaged in menial and low earning job, their scope further get synchronised from gender disparities' point of view.

Therefore, though the analytical inceptions of the concept of social exclusion can be related to economic problem, over a period, this has been put in such social order that a group of people can be identified based on their social identity than economical status. Therefore, the attempts for ameliorating social exclusion by improving economic conditions of the socially excluded group is less of a success in the country and especially in the state like Odisha. However, Rawls' theory of justice can help the situation, that social and economic inequalities are to be arranged so that they are both (i) to the greatest benefit of the least advantaged, consistent with a just saving principle and (ii) attached to positions and offices open to all under condition of fair equality of opportunities. However, this has not been realised so far in the reservation policy of India. The following section examines the extent of social exclusion from some of the macro indicators of Odisha.

Profiles of SCs/STs Population in Odisha

The social groupwise distributions of population based on some indicators of hunger and deprivation in Odisha are presented in Chapter 2 and 4. Some other additional indicators are presented here, especially related to social exclusion. These are income poverty, perceptions of the households on getting two square meals, nutritional status of children, employment and education. The analysis has been done considering social groups and gender. Tables 5.1 and 5.1(A) present the percentage distribution of population across the region and social groups (see Appendix A5.1 for districtwise distribution of population by social group) of Odisha based on 2001 Census. The composition of social groups across the regions shows the following:

Table 5.1 *Percentage distribution of population of Odisha by regions, 2001*

Regions	*TOT_P*	*TOT_M*	*TOT_F*	*P_SC*	*M_SC*	*F_SC*	*P_ST*	*M_ST*	*F_ST*
Coastal	47.84	48.03	47.64	53.49	53.67	53.32	12.10	12.23	11.97
Northern	34.69	34.76	34.62	29.71	29.73	29.70	52.87	53.09	52.64
Southern	17.47	17.21	17.74	16.79	16.61	16.99	35.03	34.68	35.39
Total	100	100	100	100	100	100	100	100	100

Source: Census, 2001.

Table 5.1(A) *Percentage distribution of population of Odisha by social groups, 2001*

	Rural			*Urban*			*Total*		
	SC	*ST*	*Others*	*SC*	*ST*	*Others*	*SC*	*ST*	*Others*
Coastal	19.82	6.18	74.00	11.72	2.64	85.65	18.48	5.60	75.92
Northern	14.37	37.25	48.39	13.02	14.84	72.14	14.15	33.73	52.12
Southern	15.79	47.71	36.50	16.83	12.03	71.14	15.89	44.38	39.73
Odisha	17.19	24.61	58.20	12.75	8.10	79.16	16.53	22.13	61.34

Source: Census of India, 2001.

(i) Out of the total population of Odisha, 48 per cent live in the coastal region, 35 per cent live in the northern region and 17 per cent live in the southern region. Such low level of population in southern region may be due to out-migration.

(ii) Out of the total SCs population in Odisha, 53 per cent live in the coastal region, 30 per cent live in the northern region and 17 per cent live in the southern region.

(iii) Out of the total STs population in Odisha, only 12 per cent live in the coastal region, 53 per cent live in the northern region and 35 per cent live in the southern region.

Table 5.1A shows the distribution of population by social groups across the three regions in Odisha. The results show the following:

(i) In Odisha as a whole, the proportion of SCs population is 16.53 per cent and the proportion of STs population is 22.13 per cent.

(ii) As regards regional variation of social groups, southern Odisha has the highest concentration of STs (44.38%) as compared to other regions. In addition, the proportion of SCs in southern Odisha is 15.89 per cent. Taking into account SCs and STs population together, southern Odisha has the highest percentage (60.27%) as compared to other regions.

(iii) Out of the three regions, coastal region has the highest concentration of SCs population (18.48%).

(iv) As regards rural-urban difference, a higher proportion of STs population resides in the rural areas of southern Odisha (47.71%), and a higher proportion of SCs population resides in the rural areas of coastal Odisha (19.82%).

Table 5.2 *Percentage of below poverty line population across social groups and by the regions of Odisha, 2004–05*

	Based on social group			
	Coastal	*Northern*	*Southern*	*Total*
ST	66.76	71.99	82.92	75.25
SC	36.91	66.38	67.46	52.63
OBC	27.23	47.09	65.53	38.32
Other	22.45	30.28	37.53	25.37
Total	29.82	56.59	71.27	46.61

Source: NSSO, 2004–05, Unit level data.

Table 5.2 presents the percentage of below poverty line population based on income poverty from 2004–05 NSSO. The table depicts that, overall, the incidence of poverty is the highest among the STs population (75.25%) followed by SCs population (52.63%) and OBCs population (38.32%). The proportion of below poverty line population is lowest among the category of 'other' social group or the higher castes (25.37%). This shows the magnitude of social exclusion among the STs and SCs population in the whole of Odisha and Southern Odisha in particular.

Table 5.3 *Percentage distributions of households by the type of response on the question whether all members of the household get two square meals a day by social groupwise in Odisha, 2004–05*

		As ascertained by a direct question		*As judged by the investigator*		*All*		*Getting inadequate food for some months of the year*		*Getting inadequate food all months*	
		India	*Odisha*	*India*	*Odisha*	*India*	*Odisha*	*India*	*Odisha*	*India*	*Odisha*
Rural	ST	57.4	77.6	38.2	12.6	95.6	90.2	3.7	7.7	0.5	1.2
	SC	53.7	81.1	42.2	11.5	95.9	92.6	3.3	5.1	0.7	2.2
	OBC	44.6	72.2	54	22.6	98.6	94.8	1	3.5	0.2	1.2
	Others	41.8	55.2	55.6	42	97.4	97.2	1.9	2.3	0.6	0.5
	Total	47.2	72.5	50.2	20.9	97.4	93.4	2	4.8	0.4	1.3
Urban	ST	39.7	69.8	59.3	29.8	99	99.6	0.8	0.3	0	0
	SC	34.3	65.2	64.5	33.8	98.8	99	0.8	0.5	0.3	0.2
	OBC	33.5	55.2	65.9	44.2	99.4	99.4	0.4	0.5	0.1	0.1
	Others	24.7	41.6	75	58	99.7	99.6	0.1	0.2	0.1	0.2
	Total	29.7	52.2	69.7	47.2	99.4	99.4	0.4	0.4	0.1	0.2
All	ST	55.7	77.1	40.3	13.6	96	90.7	3.5	7.2	0.4	1.2
	SC	49.6	79.2	46.9	14.1	96.5	93.3	2.7	4.5	0.6	2
	OBC	41.9	70.1	56.9	25.3	98.8	95.4	0.9	3.2	0.2	1
	Others	34.8	50.9	63.5	47	98.3	97.9	1.2	1.6	0.4	0.4
	Total	42.4	69.6	55.5	24.7	97.9	94.3	1.6	4.1	0.4	1.1

Source: NSSO, 2004–05.

Table 5.3 presents the percentage distribution of households by the type of response on the question whether all members of the household get two square meals a day across the social groups. The results suggest that SCs and STs population compared to other social groups report a higher level of deprivation in terms of not being able to get two square meals a day, both during some months of the year as well as throughout the year. This is true for Odisha as well as all-India. However, in this respect, the SCs and STs population in Odisha report relatively more deprivation as compared to SCs and STs population in India as a whole. The SCs and STs group is particularly vulnerable in the rural areas of Odisha.

Table 5.4 *Nutritional status of children (based on Stunting (H/A)) by region and across different social group wise in Odisha, 1998–99*

	Coastal		*Northern*		*Southern*		*Total*	
	–2 SD	*–3 SD*	*–2 SD*	*–3 SD*	*–2 SD*	*–3 SD*	*–2 SD*	*–3 SD*
SC	47.59	27.78	55.13	20.78	52.24	14.71	50.69	22.84
ST	60.00	26.67	44.44	15.28	53.33	23.81	49.46	19.71
OBC	47.31	22.16	40.25	13.84	45.45	18.18	44.28	18.29
Other	34.60	10.38	30.65	12.90	19.35	0.00	32.24	9.84
Total	43.36	19.44	42.89	15.38	47.21	17.41	44.0	17.61

Notes: The Body Mass Index is the ratio of the weight in kilogrammes to the square of the height in metres (kg/m^2). Each index of the children (1 to 47 months) nutritional status is expressed in standard deviation units (SD) from the median of the international reference population.

Source: NFHS, 1998–99.

Table 5.4 depicts the magnitude of stunting across the social groups by regions in Odisha in 1998–99. The table shows that in the case of moderate stunting, SCs children are most disadvantaged with 51 per cent of them moderately stunted, followed by children from STs (49.46%), OBCs (44.28%) and other caste (32.24%). In case of severe stunting, SCs children are most disadvantaged with 22.84 per cent of them severely stunted, followed by children from STs (19.71%), OBCs (18.29%) and other caste (9.84%). On the whole, SCs and STs children's nutritional status are much lower compared to other caste groups, regardless of regions in Odisha.

Table 5.5 *Nutritional status of children (based on wasting (W/A) by region and across different social groups in Odisha, 1998–99*

	Coastal		*Northern*		*Southern*		*Total*	
	–2 SD	*–3 SD*	*–2 SD*	*–3 SD*	*–2 SD*	*–3 SD*	*–2 SD*	*–3 SD*
SC	57.93	20.83	62.82	23.38	58.21	32.84	59.31	24.31
ST	50.00	20.00	61.81	26.90	58.10	27.62	59.14	26.43
OBC	49.73	18.38	62.26	22.01	62.12	24.24	56.59	20.73
Other	42.18	11.32	48.39	14.75	37.50	12.90	42.95	12.17
Total	49.04	16.46	60.27	22.85	56.67	26.39	54.52	20.75

Note and Source: same as Table 6.4.

Table 5.5 presents the magnitude of wasting across the social groups in Odisha. The table shows that, as far as moderate wasting is concerned, SCs and STs children are most disadvantaged with 59 per cent of them being moderately wasted, followed by OBCs (56.59%) and other (42.95%) caste. In case of severe wasting, STs children are most disadvantaged with 26.43 per cent severely wasted, followed by SCs (24.31%), OBCs (20.73%) and other castes (12.17%). As compared to the Coastal Odisha, children from Northern and Southern Odisha are relatively more wasted.

Table 5.6 *Nutritional status of children (based on under weight (W/H) by region and across different social groups in Odisha, 1998–99,*

	Coastal		*Northern*		*Southern*		*Total*	
	–2 SD	*–3 SD*	*–2 SD*	*–3 SD*	*–2 SD*	*–3 SD*	*–2 SD*	*–3 SD*
				Based on Caste				
SC	17.24	2.07	26.92	5.19	31.34	2.99	23.10	3.11
ST	20.00	3.33	36.81	8.28	24.76	2.86	30.47	5.71
OBC	16.67	3.23	32.08	6.29	28.79	4.62	24.57	4.63
Other	18.48	1.42	19.67	4.84	29.03	6.25	19.80	2.61
Total	17.66	2.27	31.00	6.55	27.88	3.72	24.40	4.05

Note and Source: same as Table 6.4.

Table 5.6 presents the magnitude of under weight children across the social groups in Odisha. As regards moderate under weight, STs children are most disadvantaged with 30.47 per cent moderately underweight, followed by OBCs (24.57%), SCs (23.10%) and other castes (19.8%). In severe forms of under weight, STs children are most disadvantaged with 5.71 per cent severely under weight, followed by OBCs (4.63%) SCs (3.11%) and other castes (2.61%). Like wasting, for underweight as well, children from Northern and Southern Odisha are relatively more vulnerable.

Table 5.7 *Percentage distribution of work participation rates, Main workers and agricultural labourers across the Social groups by different districts of Odisha, 2001*

	Total				*Total*				*Total*			
	Work Participation Rate				*Main Workers*				*Agricultural Labour to total workers*			
	Total	*GEN*	*SC*	*ST*	*Total*	*GEN*	*SC*	*ST*	*Total*	*General*	*SC*	*ST*
Coastal	33.8	31.9	36.7	45.7	75.0	77.6	69.5	66.5	29.4	22.6	45.7	47.0
Northern	41.6	38.6	42.0	47.1	64.6	70.5	62.4	57.7	35.7	24.8	44.2	48.4
Southern	47.5	43.3	45.1	51.8	59.2	64.8	58.4	55.7	42.1	34.5	45.2	46.1
Odisha	38.8	35.0	39.3	49.0	67.2	72.9	64.8	57.4	35.0	25.8	45.7	46.9

Source: Census of India, 2001.

Table 5.7 and Appendix A5.2, presents the percentage distribution of worker participation rate (WPR), main worker and agricultural labour across the social groups by different regions and districts of Odisha respectively. The table depicts the following: (i) The WPR remained higher among the STs (49%), followed by SCs (39%) and others (35%) for Odisha as a whole. Further, as the WPR of STs and other social groups are concerned, districts located in the southern part of Odisha show a higher WPR. However, the WPR alone does not show any significant hope for the STs or SCs social group as many of them are marginal workers. (ii) In general, the proportion of main workers is higher among the general category (72.9%) followed by SCs category (64.8%) and then STs (57.4%). (iii) The proportion of agricultural labour remains much higher among the STs (46.9%) and SCs (45.7%) compared to other castes (25.8%). This shows that hunger and deprivation are likely to be much higher among the SCs and STs agricultural labourers.

If we compare WPR across the regions, southern Odisha exhibit a higher WPR across all social groups. However, within southern region, STs population has the highest WPR (45%), and there is no difference in WPR between SCs population and other higher caste group. As regards main workers, the proportion is lower for all social groups in southern Odisha as compared to their counterparts in other regions. Here again, the STs and SCs groups in southern Odisha have the lowest proportion of main workers. As regards agricultural workers to total workers, this proportion shows similar pattern as the state average, regardless of regions.

Table 5.8 *Percentage distribution of literacy rate across the social groups*

	Total	*General*	*SC*	*ST*
Coastal	70.79	77.04	57.25	39.46
Northern	63.19	73.63	57.90	47.07
Southern	41.81	55.96	45.04	29.50
Odisha	63.10	73.90	55.50	37.40

Source: Census of India, 2001.

Table 5.8 and Appendix A5.3 present the educational deprivation across the social groups. For Odisha as a whole, the literacy rate is strikingly lower among the STs (37.40%) as compared to even the SCs (55.50%) and other social group (73.90%).

Table 5.9 *Housing conditions across the districts of Odisha by different social groups, 2001*

	All			*SC*			*ST*		
	Good	*Livable*	*Dilapidated*	*Good*	*Livable*	*Dilapidated*	*Good*	*Livable*	*Dilapidated*
Coastal	26.0	60.4	13.6	17.4	61.7	20.8	13.8	65.8	20.5
Northern	24.1	68.0	7.9	19.1	68.9	12.0	15.3	74.7	10.0
Southern	22.9	72.2	4.9	22.5	70.8	6.7	19.1	76.6	4.3
Odisha	25.6	64.4	9.9	19.1	64.8	16.1	17.1	74.6	8.3

Source: Census of India, 2001.

ST population has the lowest literacy rate regardless of region. For Odisha as a whole, only 37 per cent of the STs population are literate, followed by SCs population (56%) and other higher castes (74%). Here again, STs population residing in southern region has the lowest literacy rate (30%) as compared to any other groups across the regions.

Districtwise literacy rate by social group can be seen in the Appendix A5.3. Table 5.9 and Appendix A5.4 present the housing conditions of SC, STs and all caste categories. The highlights of the table are that: (i) In general, most of the depleted houses are found among the SCs with 16.12 per cent followed by all other caste (9.95%) and then by STs (8.25%). However, with respect to incidence of depleted household across the districts and social groups is concerned, maximum percentages of depleted household are found among the scheduled castes in the developed region, i.e. costal Odisha (Jagatsinghpur with 29.68 per cent, Jajpur with 28.11 per cent and Cuttack and Puri with 26 per cent each).[6] (ii) As regards 'good' housing condition, for Odisha as a whole, STs groups are relatively more disadvantaged (17.11%) followed by SCs groups (19.06) against the all-Odisha figure of 25.64 per cent. Here again a relatively lower proportion of SCs and STs population residing in southern region have 'good' housing condition as compared to their counterparts in other regions. As regards dilapidated houses, the highest proportion (one-fifth) is for SCs and STs group residing in coastal region.

Table 5.10 *Infant mortality rate across the social groups of Odisha*

Social groups	*Infant mortality rate*			
	1992–93		*1998–99*	
	India	*Odisha*	*India*	*Odisha*
ST	107.3	160.8	83.0	98.7
SC	90.5	113.4	84.2	83.9
OBC	—	—	76.0	95.6
Other	82.2	115.3	61.8	79.1
Total	78.5	112.1	—	89.5

Source: National Family on Health Survey.

Table 5.10 presents the Infant Mortality Rate across the social groups in Odisha. The table shows that: IMR though declining for all social groups in the last decade, it is the highest among STs (98.7 for Odisha and 83.0 for all-India in 1998–99) followed by SCs (83.9 for Odisha and 84.2 for all-India in 1998–99). Thus, while there is no difference in IMR between Odisha and all-India for the SCs group, the differential is large for the STs group.

Table 5.11 presents the average daily earnings of males and females of rural labour household in agricultural occupations by their social groups. The table depicts,

Table 5.11 *Average earnings (₹) of men and women in rural labour household in agricultural occupations by their social groups, 1999–2000*

	All classes		*Scheduled Caste*		*Scheduled Tribes*		*M/F*100*		
	M	*W*	*M*	*W*	*M*	*W*	*All*	*SC*	*ST*
Odisha	28.69	22.31	28.5	23.38	26.37	22.72	77.76	82.04	86.16
India	40.58	28.57	41.89	29.6	33.19	26.44	70.40	70.66	79.66
O/I*100	70.70	78.09	68.04	78.99	79.45	85.93			
Kerala	94.52	62.22	97.25	62.37	76.28	56.34	65.83	64.13	73.86
Chhattisgarh	26.31	22.92	25.56	21.83	27.16	23.21	87.12	85.41	85.46

Source: Rural Labour Enquiry Report, 1999–2000

(i) There is a substantial difference between the earnings of rural labour living in Odisha compared to all-India.
(ii) However, across social groups, there is hardly any difference between the average wage rate between all classes and scheduled caste but the gap is visible in case of STs when compared with other two groups.
(iii) Given that higher proportion of workers among SCs/STs are agricultural labour (shown in Table 5.7, around 45 and 47% respectively), the total earnings of SC/ST may be lower compared to 'other' category among whom the share of agricultural labour is 25 per cent. Hence, there is a higher probability of incidence of hunger and deprivation among SCs/STs.

The secondary data analyses thus far shows that social exclusion is prevalent in Odisha and the magnitude is higher in Odisha than all-India. In order to understand social exclusion in its entirety, qualitative analysis will be more helpful. An attempt has been made in the following section to understand the phenomena at the micro level.

Micro Evidences on Social Exclusion

A few case studies were carried out among Scheduled Caste community *(Ghansi)* in the Kundamal village of Palasagora gram panchayat of Boudh district in Odisha. Boudh district is chosen because the share of SCs population in this district is higher compared to other districts of southern Odisha. In addition, to our observations, the prevalence of caste-based exclusion persists in this part of the state. The village is situated around 6 kilometres from the panchayat headquarter and around 18 kilometres from the block headquarters. The caste composition according to 2001 census is around 41 per cent (SCs), 6 per cent (STs) and 52 per cent (other castes).

The case studies were carried out among 19 households of the Ghansipara stratum. In our enquiry, the respondent is the head of the household. From the 19 households,[7] there are mainly three different occupational categories. About 17 are agricultural labourers and are also engaged in leather work, one is driver and one is peon in a bank. The findings from the case studies are presented below:

Landlessness

Land is the major source of livelihood in Odisha and land provides protection against risk and calamity and is a source of social power in rural communities.

Out of the 19 households, only three possess agricultural land. Of them, two households have half acre of land each, that they received from the government as ceiling land. The household, where the respondent is the peon possesses six acres of less fertile and single crop land. The remaining 16 households are landless and depend on other's land for their livelihood. Since most of the households do not possess land, they lack the power and access to social and economic resources. The landless households work for wages and when there is no work, they borrow from the landed people or from moneylenders. This renders them weak and therefore have no or less power. This perpetuates the poverty situation even more. Landlessness, therefore, in most of the households of the community (labour is the only asset) excludes them from the mainstream society. This exclusion is further perpetuated in other spheres of their lives.

Illiteracy

Out of the 19 head of the households, 11 are illiterate, while others have some primary and middle level of education. This low level of education could be due to the low landholdings. The households are unable to decide on education, since basic needs are not met and there is a need for children to contribute towards the family income pool at a very young age. There is some awareness among the households of the importance of education and the government scheme for free and compulsory primary education for all. However, the households feel that providing free books and free mid-day meals are not enough, as they are unable to meet other associated costs, such as clothing for instance. Moreover, if the student is not going to school, then he/she also marginally contributes to family income directly or indirectly.

Household's Assets and Livestock

There is hardly any asset or livestock with the households. There is only one two-wheeler, which the bank peon possesses and around 12 households possess a bicycle each, purchased through a bank loan or obtained as dowry. No household possesses any four wheelers, tractor, water pump, thresher, bullock cart, telephone, fan, refrigerator or mobile phone. However, the bank peon possesses a radio and 7 households possess a pair of oxen and few sheep and goats with them. This lack of assets and livestock creates a situation where they have nothing to depend on in times of economic distress. They cannot even start a business due to low access to credit from financial institutions as they have no collaterals. Hence, they continue to remain within the poverty cycle and there is intergenerational transfer of poverty. In addition, when assets are not available there is less of interdependency with other households, and results in exclusion.

Exclusion in Public Spheres

The most common form of exclusion of the community is ban of temple entry and entry even to the peripheral area of the home of other caste households. The person from this community (Ghansi community) is considered as polluted person. The second form of exclusion found is in the seating arrangement in the classroom or at the time of marriage function of the other caste. In the classroom, students from this community sit separately. In certain cases, parents of superior caste students ask the teacher to make the seating arrangement of their children slightly away from the lower caste (Ghansi) students. In some other cases, untouchability is experienced by the students outside the school, i.e. inside the village and hence in the classroom too, the students from lower (Ghansi) stratum themselves sit separately.

People from this community have to sit in a distance place, away from the place where the superior caste people sit and take their food. These caste groups are often given a place for eating, which is unclean. The place is filled with dust and cow dung like a village road, where one may find every kind of waste. Yet these people take their food in that place due to several reasons.

The other spheres, where social exclusion exists is in the work place, especially in the agricultural fields, where the upper caste workers do not like to work together with the workers from this community. The workers from this community have to forgo the work/livelihood. There are occasions where workers of this community exclusively work together, but that is not always the situation. These sorts of exclusion forced them to remain in their traditional menial jobs, which are less remunerative and less competitive.

As awareness about Constitutional provisions for Scheduled Castes and Schedules Tribes has increased, there is a resistance among the communities especially towards social exclusion. But there is also fear among SCs and STs communities about their loss of job and livelihood and current status quo where social customs are not too discriminatory.

Summary

This chapter attempted to understand caste-based social exclusion and its association with hunger and deprivation in Odisha. From the review of literature and available data and their analysis, it may be inferred that, social exclusion (caste based) does prevail in Odisha to a great extent. The social exclusion is so deep that even after 60 years of constitutional efforts a segment of people remained backward in terms of almost all basic necessities. This could be the reason for the higher incidence of hunger and deprivation among the SCs and STs population. It is also inferred from the micro level case studies that there is prevalence of exclusion (in explicit or implicit form) of Dalits and landless from various socio-economic spheres, such as exclusion in the socio-cultural institutions like educational institutions, temple, workplace and financial institutions, etc. In addition to SCs and STs communities, Dalits are also vulnerable to social exclusion as revealed by the case studies.

So far the study has inferred that neither the growth (output and employment) nor the distributive justice is enough to reduce mass hunger and deprivation in the state. Further, analysis based on both employment and possession of land has shown a pattern where the SCs/STs are more vulnerable to hunger and deprivation. This chapter further confers that within SCs/STs, there are socially excluded groups such as Ganda, Ghansi, Chamar, Hadi, etc. whose access to capability enhancing facilities are restricted by the structural factor of caste system, which is a barrier for the SCs/STs to access over social infrastructure like education. While the SCs/STs remain at the lower rung of both economic power and social ladder, the women from these communities are further held down by the cultural norms, which restrict the access of women to education, health and to other capability-enhancing programmes. Hence, the role of public provisioning is crucial here to reduce the impact of hunger and deprivation. The next chapter looks at the public provisioning and hunger and deprivation.

Notes

1. The exclusion of SCs, is related more towards the Indian social system and the creation of social barriers for these groups of people. The exclusion of STs is related to their place of dwellings as located far from the mainstream society, which might be the reason why they were not having adequate access to the modern social institutions and bounded with their own culture and remained backward. Further, in the subsequent period with the implementations of forest laws and construction of dams and industries, some of the STs populations were displaced and they migrated from remote to the plain area but they could not progress at par with the other social groups in the plain area.
2. Caste people who were also known as avarna (as they fall beyond the above four caste groups) or panchmas (the fifth caste group).
3. See, Sainath, 1996; Praxis, 2001; Currie, 2000.
4. It provides penalties for refusing admission to hospitals, educational institutions, hotels, temples etc.; for refusing to sell goods or vendor services, for preventing the exercise of any rights under Article 17 (which includes access to shops, public restaurants and place of public entertainment or wells, tanks, public bathing places, roads and place of public resort); for demanding unlawful compulsory labour from the untouchables, like scavenging, sweeping, removal of carcasses, flaying animals, or removing the umbilical cord or any job of similar nature (see Thorat, 2004).
5. The 1989 Act specifies 18 types of atrocities for penalty: forcing the eating of obnoxious substances; dumping waste matter on land; intimidations during voting; mischievous litigation; false information; public humiliation; denudation; wrongful occupation on land; bonded labour; outrage of modesty; sexual exploitation; fouling of water resources; obstruction of entry to a place of public resort; eviction of habitation; mischief with explosive; destruction of building; and suppression of evidence (Commission for Scheduled Caste/Scheduled Tribes, 1998).
6. The higher percentages of depleted house in Jagatsinghpur and Jajpur may be due the impact of Super Cyclone of 1999, which affected these districts.
7. There are 22 households in that stratum but 3 households were absent during the time of survey.

6

Public Provisioning to Reduce Hunger and Deprivation

In the absence of distributive justice (discussed in Chapter 4) and prevalence of social exclusion (discussed in Chapter 5), a second best approach to tackle hunger and deprivation is public provisioning for basic needs that would also help in enhancing capabilities of the poor, socially excluded and the marginalised. If distributive justice and public provisioning are combined, as in Kerala and China, then capability of the poor will be strengthened further. This chapter therefore, discusses 'Public Provisioning' in Odisha in the absence of distributive justice to see how far inclusive and effective it is. The results show that even in the access to public provisioning, the upper social groups have better access over the public services compared to the lower social groups. The chapter has four sections. Section one clarifies the concept and relevance of the growth-mediated and support-led security; section two reviews the secondary literature on different schemes in India and in Odisha in the post-independence period, it also discusses the nature of support-led security, such as, promotional security and protectional security and then it examines the functioning and performances of different welfare schemes in Odisha; section three presents the evidences from a micro-level survey in selected villages in Odisha (how different schemes are working in the grassroots level to achieve the desired welfare of the people) and section four concludes the chapter.

Growth-mediated Security and Support-led Security

Most of the social security theories are based on the experience of developed countries. There was a linear relationship between social security measures and economic growth in that social security schemes were associated with the level of development. Hence, there are debates about the validity of the social security schemes in the case of developing and underdeveloped countries, characterised with low product growth and high population growth, illiteracy and poverty.

However, countries like Chile, Costa Rica, Cuba, China and Jamaica have shown their success in achieving development goals through public provisioning with low level of economic growth (Dreze and Sen, 1989). Sen (1981) and Dreze and Sen (1989) stated that, it is the failure of command over food that may have caused hunger and not necessarily by the output failure. Therefore, amelioration of purchasing power (income) is more important to reduce hunger but it is a poor focal point, as income itself depends on ownership and exchange of many other commodities. Moreover, there are other essential commodities such as educational services, health care, clean water, or protection from infectious epidemics, which are not sold and bought in the market in a usual way; hence the real income may not provide a good idea of the command over food.

The authors have brought two principles to explain the associations: (a) it is not the aggregate opulence, but the translations of this into patterns of individual prosperity depends on the distributions of income across the population, and (b) the capabilities enjoyed by the people that can add to their individual prosperity, not only depends on income as many of the commodities are not available in usual market (such as health, education and sanitations), but are provided through public provisioning. Therefore, it was argued that 'a high level of GNP per head provides an opportunity for improving nutritional and other basic capabilities, but that opportunity may or may not be seized'. In such situation, public provisioning often plays a crucial role in transforming the opportunities into tangible achievements.

Sen and Dreze distinguish between two major approaches: (1) the 'growth-mediated security' and (2) the 'support-led security'. The former approach puts forth, 'promote the economic growth and take the best possible advantage of the potentialities released by greater general affluences, including not only an expansion of private incomes, but also an improved basis of public support' (pp. 183). The examples can be South Korea, Hong Kong and Kuwait. The latter approach describes, 'to resort directly to wide-ranging public support in domains such as employment provision, income distribution, health care, education and social assistance in order to remove destitution without waiting for a transformation in the level of general affluences' (ibid.).[1] The examples can be China, Sri Lanka and Kerala etc.

Though the debates on the growth-mediated or support-led strategies remained mostly unsettled due to the merit of the respective strategies (and hence one would like to have both the strategies in an economy), Sen and Dreze have shown that 'public support programmes can work even when a country is quite poor in terms of GNP per head. This makes it possible to something immediately about conquering deprivation and raising the quality of life without having to wait quite some time before ploughing back the fruits of economic growth into improved health and longevity' (ibid., 253).

However, the significance of either of these strategies cannot be overlooked – while economic growth is required to enhance the public resource-based needs, the public support (or the social security mechanism) is required for equitable growth. Therefore, both economic growth and public provisioning are essential in an economy. However, while economic growth can be considered as a necessary element in an economy, it may not be sufficient for extending public support. Even in the support-led security, the minimum amount of growth cannot be ignored. Therefore, the developing economy like India or the backward state like Odisha may simultaneously go for support-led security to reduce hunger and deprivation at the earliest, rather than wait for growth to happen and then transform it into development. The following sections look into the functioning of some of the public provisioning schemes in Odisha.

Evolution, Implementation and Functioning of Public Provisioning in India and Odisha

Public provisioning in the independent India started with the Constitutional commitments by the Government of India to ensure and promote the livelihoods of Indian citizens. Since then, various welfare schemes have been introduced especially to help the poor. The areas covered in those schemes include food and nutrition, education, health, housing and other contingent aspects. These schemes have been working since independence though in different names. The details of the schemes are presented in Annexure E. However, to reduce the impacts of hunger and deprivation, what is more important is the design, implementation and functioning of these schemes at macro and micro level.

The salient features drawn from the secondary data analysis are the following (the details of each scheme are presented in Annexure F): (1) There are almost all types of public provisioning schemes existing in the state, though their implementation and functioning are far from satisfactory, (2) Most of the public provisioning schemes cover at least food and nutrition, education, health, housing and other protectional measures such as old age pension. However, almost all these schemes face the common problems of (a) inadequacy (allocation and coverage), (b) irregularity, (c) targeting errors and (d) being implemented through inefficient institutions. Below, these issues have been further explored from the primary survey conducted in Odisha.

Evidences from Micro Study

A micro-level study is carried out in two phases, at the institutional level and at the household level. The study on the functioning of different institutions is carried out in Sason Panchayat of Sambalpur and the study of household beneficiary from a specific public provisioning scheme (i.e. National Rural Employment Guarantee Act) is carried out at household level in the Thuamul block of Kalahandi district.

Evaluation of Various Schemes at the Institutional Level in Sason Panchayat

In order to assess the functioning of various institutions in Sason Panchayat of Sambalpur district in Odisha, the present researcher collected information from the service providers or officials in these institutions through a set of relevant questions pertaining to the design and implementation issues. The results and analyses of various schemes at the institutional level are presented below.

The major social securities that are provided in the Panchayat are food, housing, health, education, and women and children's welfare etc. Table 6.1 provides the list of different institutions that exist in Sason Panchayat (detail discussion on Sason Panchayat is given in Annexure G). Table 6.1 shows that based on number of institutions and population size, the Sason gram panchayat is not poor, but what is more important is whether good services have been provided to the people. Below, we have discussed the functioning of each of these institutions one by one in detail.

Table 6.1 *List of different institutions in Sason GP*

Name of the institution		*Numbers*
Educational Institutions	Education Guarantee School (EGS)	3
	Upper Primary schools	7
	Middle English Schools (two are private aided)	3
	High School	1
	College	1
	Nearest University Centre	30 kilometers
	Swami Vivekanada coaching centre	1
	Day-care centre	1
Library		None
Health	Subcentre	1
Ration shop		3
Police station		1
Main post office		1
Banks		2
Self-Help Group		20
Anganawadi centre		6

Source: Institutional survey at Sason.

Functioning of the Educational Institutions

Two points are captured here: (1) how adequately and smoothly the basic education has been provided to the students of the panchayat? In order to understand this, we have analysed various issues including the teacher student ratio, teacher classroom ratio, type of the school buildings and other infrastructural facilities at the school, roles of School Inspector and the Parent Teacher Association. (2) As providing supplementary nutrition through mid-day meal (MDM) has become a regular feature in India, we have also tried to understand how the MDM is functioning in different schools of Sason gram panchayat.

Basic Education

As shown in Table 6.2, there are 3 EGS Schools, 7 UP schools, 3 M.E. schools, one high school and a college in the Panchayat. Apart from that, there is also a Vivekananda Vidya Pitha, and a missionary led day-care centre in the Panchayat. In nine out of the 10 schools, the number of teachers is inadequate based on one teacher per class.

Sason is the only school where there is full strength of teachers. The worse condition in this regard is the Phalsa Pali UP school. The school was established in 1972 but still there is only one teacher for five classes. On an average, for all the schools together, one teacher has to take care of at least two classes.

Table 6.2 *Establishment of different school in Sason Gram Panchayat till 31.07.2004*

School's name	Village name	Years of establish ment	Classes offered	Number of teachers	Number of Students	Average number of students per class	Average number of students per teacher	Average number of teachers per class
EGS – 1	Dharuapali	2002	1 to 3	1	32	10.67	32	0.33
EGS – 2	Mahulmunda	2002	1 to 3	1	26	8.67	26	0.33
EGS – 3	Talpadar	2002	1 to 3	1	43	14.33	43	0.33
UP – 1	Nua Khuri Gaon	1959	1 to 5	2	70	14	35	0.4
UP – 2	Phalsa Pali	1972	1 to 5	1	31	6.2	31	0.2
UP – 3	Rani Khenda	1961	1 to 5	4	197	39.4	49.25	0.8
UP – 4	S. Kanta Pali	1972	1 to 5	2	86	17.2	43	0.4
UP – 5	A. Sason	1902	1 to 5	5	112	22.4	22.4	1
UP – 6	Sdhasingha	1948	1 to 5	4	65	13	16.25	0.8
UP – 7	Chatargada	1986	1 to 5	2	43	8.6	21.5	0.4
Total	10 schools		44	23	705	16.02	30.65	0.52

Source: Institutional survey at Sason.

Further, as the teacher-student ratio shows, it is lowest in Chattargada (21) and highest in Rani Khenda (49). But as the table shows, the average number of students per class (not classroom) is lowest in Mahulmunda (9) and highest in Rani Khenda (39). Table 6.3 shows the distribution of students and teachers across schools by their gender. The table shows that the highest proportion of students are studying in Rani Khenda (27%) and the minimum proportion in Mahulmunda. The number of teachers by their gender shows that the number of female teachers is marginally higher than male teachers. The distribution of students by their gender shows that the number of male students is somewhat higher than female students.

Table 6.3 *Distribution of number of students and teachers across their gender in different schools of Sason Gram Panchayat*

Name of the school	No. of students			Percentage of students in each school	No. of teachers		
	Girls	Boys	Total		Female	Male	Total
EGS – 1	19	13	32	4.54	0	1	1
EGS – 2	9	17	26	3.69	0	1	1
EGS – 3	18	25	43	6.10	0	1	1
UP – 1	28	42	70	9.93	1	1	2
UP – 2	15	16	31	4.40	0	1	1
UP – 3	106	91	197	27.94	3	1	4
UP – 4	39	47	86	12.20	1	1	2
UP – 5	53	59	112	15.89	4	1	5
UP – 6	30	35	65	9.22	2	2	4
UP – 7	22	21	43	6.10	1	1	2
Total	339	366	705	100	12	11	23

Source: Institutional survey at Sason.

All these statistics, inadequate teacher per class, low teacher student ratio and less number of students per class depict that the system as a whole may require rethinking and call for reorganisations, keeping the economies of scale and modernisation facility for education in mind.

School Infrastructure

Another important feature to achieve satisfactory result in education is the infrastructure available in the school. Table 6.5 shows that there is very little facility available for the EGS schools. In fact, these schools are even not having their own buildings. The classes are conducted in the youth clubhouse or community house of the respective villages. As regards to the modern facilities like computer education or mike set or own vehicle of the schools, or furniture (bench and desk for students), none of these schools has these facilities although some of these schools are very old. There are toilet facilities in three schools out of 10, but usually those toilets are not in a condition to use.

Table 6.4 *Number of classes offered in different schools and classroom available in Sason*

School's name	*Classes offered*	*Numbers of room available for class*
EGS –1	1 to 3	1
EGS –2	1 to 3	1
EGS –3	1 to 3	1
UP –1	1 to 5	4
UP –2	1 to 5	2
UP –3	1 to 5	4
UP –4	1 to 5	4
UP –5	1 to 5	4
UP –6	1 to 5	4
UP –7	1 to 5	3
Total	44	28

Source: Sason GP, Institutional survey

Further, as seen from Table 6.4, the number of rooms available per class is very poor. (around 0.59 to 0.63 of a room is available for a class). Due to shortage of room on the one hand, and shortage of teacher on the other, the teacher has to do the adjustment and accommodate students from two classes in one room. More often, since there is no clerk or peon in any of these schools, the head-master has to attain many official works.

Table 6.5 *Infrastructural facility in the schools of Sason Gram Panchayat*

Name of facility	*Name of the schools in the study area*										*No: of schools having each facility*
	EGS 1	*EGS 2*	*EGS 3*	*UP 1*	*UP 2*	*UP 3*	*UP 4*	*UP 5*	*UP 6*	*UP 7*	
Play ground	×	×	×	%	×	%	%	%	%	×	5
Building in condition	×	×	×	%	×	×	%	%	%	%	5

Cont'd...

...Cont'd

Toilet facility	×	×	×	×	×	%	%	%	×	×	3
Proper water source	×	×	×	%	%	%	×	%	%	×	5
Adequate furniture in the class rooms	×	×	×	×	×	×	×	×	×	×	0
Adequate ventilation in the class rooms	×	×	×	%	%	%	%	%	%	%	7
Mike set	×	×	×	×	×	×	×	×	×	×	0
Physical education	×	×	×	×	×	%	%	%	×	×	3
Computer education	×	×	×	×	×	×	×	×	×	×	0
Own vehicle	×	×	×	×	×	×	×	×	×	×	0
Number of facilities available in each school	0	0	0	4	2	5	5	6	4	2	

Source: Institutional survey at Sason.

As revealed from Table 6.5, overall it appears that the school infrastructure is very poor.

Parent Teacher Associations

For the smooth functioning of the educational system and for better results, various meetings have been conducted among the teachers, student's parents and other villagers. Mainly, there are three different meetings that have been conducted in this regard are by the parent teacher association (PTA), the mother teacher association (MTA) and the village educational committee (VEC).

In many schools, the PTA meets thrice in a year (at the beginning of the year, after six months and before the announcement of final result). In some schools, PTA meeting is conducted each month (e.g. Rani Khenda school). Similarly, the MTA are called every two months, but as it was reported in many cases, mothers do not come for the meeting. Sometimes, as revealed by the teachers, the contractor has prevented some of the parents to attend the meeting, so that the contractor can influence the PTA president and secretary to get the contract for the schoolwork. The VEC meets every month but it hardly brings any effective result because of the supremacy of the village mukhia (village president) and the contractor. Apart from the above discussion, another activity that is going on in all the primary schools is the Mid-day meal scheme, which has been discussed below.

Functioning of Mid-Day Meals Schemes

Mid-day meal (MDM) has been provided in all the schools of Sason gram panchayat and all the students take meals from the schools. Apart from that, there is a charitable institute, called Mission of Mercy (MoM), headquartered in Bangalore), which provides breakfast, MDM and afternoon tea and snack, where more than 300 people (including children) take their food daily. However, although the MoM supplements to a great extent, it has been learned from the school that to a great extent students are eating in schools.

In all the schools the mid-day meal system runs more or less in a similar pattern. It was learned that for one student the provision is 100 gram of rice, 15 grams of pulses, 1 gram of salt, one gram of oil and 10 paise for vegetable expenditure.

There is no separate kitchen to cook the meal in many of these schools. Meal is cooked in the school premises. There is a cook and a helper to prepare and distribute the food. The cook gets ₹ 200 per month, whereas the helper gets 100 rupees per month.

The general problems that have been found are:

(i) The allocated quota is not enough for a student, it was mentioned that 150 to 200 grams of rice per student might be enough. The problem is many students, especially SC and ST students, come with an empty stomach to the school. Hence, they require more quantity of food than what actually is distributed. Some of the schools (at least in one school surveyed) used non-iodized salt for cooking.

(ii) Access to fuel is a big problem.,

(iii) There is no separate room for kitchen.

Table 6.6 *Facilities available for MDM in different schools of Sason Gram Panchayat*

Features of midday meal scheme	*Name of the schools in the study area*										*Number of schools having each facility*
	EGS 1	*EGS 2*	*EGS 3*	*UP 1*	*UP 2*	*UP 3*	*UP 4*	*UP 5*	*UP 6*	*UP 7*	
Separate kitchen	×	×	×	×	×	×	×	%	×	×	1
Kitchen room in condition	×	×	×	×	×	×	×	×	×	×	0
Separate women to cook	×	%	%	%	%	%	%	%	%	%	9
Regular supply of materials for cooking	%	%	%	%	%	%	%	%	%	%	10
Regular grants for meeting other expenses	×	×	×	×	×	×	×	×	×	×	0
Good quality of the food items provided as perceived by the teachers	×	×	×	×	×	×	×	×	×	×	0
Adequate utensils	×	×	×	×	×	×	×	×	×	×	0
Non-discrimination while serving	%	%	%	%	%	%	%	%	%	%	10
Number of facilities available in each school	2	3	3	3	3	3	3	4	3	3	

Source: Primary Survey.

Table 6.6 provides the availability of some of the basic facilities to run MDM schemes in Sason Gram Panchayat. As the table depicts, there is no separate kitchen in any of the schools except one (A. Sason), but there (where kitchen room is available) too the kitchen is not in a good condition. In almost all the schools there are two women who have been appointed by the village committee for cooking. Since the payments are not made regularly, these women are not interested to cook. For example, we observed that in EGS-1, due to non-payments for 4–5 months, two women left their jobs and now there is no cook in that school. However, the food gets cooked on a rotation basis by one of the family members of each student.

While the MDM is provided in all the schools regularly, the quality of food is not good. As one of the headmaster complained, she asked the cook not to use the pulses since there were so many insects in it. She also showed the unused pulses to us that she kept to be returned to the authority. Many a times same is the case for rice as well. As shown in Table 6.6, we observed some other difficulties in the implementation of MDM schemes (e.g. inadequacy of utensils, non-availability of additional grants etc.)

Integrated Child Development Scheme (ICDS)

During 1975, the ICDS has been introduced in India for the betterment of women and children. The target groups under this scheme are children in the age group of 0–6 years and expectant and nursing mothers in the age group of 15 to 45 years who belong to BPL families. Now the state has covered all the 314 blocks under ICDS programme apart from 12 projects in urban areas. As of March 2007 there were 326 projects with 36,219 Anganwadi centres operating in Odisha for the efficient functioning of the ICDS. The schemes provide a package of services covering supplementary nutrition, immunisation, pre-school education, health check up, referral services and nutrition and health education for women (Economy Survey of Odisha, 2008). Our survey on social security also studies the functioning of ICDS Institutions. Since Anganawadi centres are the main organisations through which ICDS works, we took the interviews of the Anganawadi workers and Beneficiaries of Sason Panchayat to understand the functioning of ICDS schemes. The results are presented below.

Functioning of the Anganawadi Centres

There are five Anganwadi centres, located within the Sason Panchayat. All these centres are located in the school premises except in Kantapali, where it is a little far from school that runs in a community house. There are two rooms in each of these centres. One is the classroom usually and roughly 10* 14 feet in size, and the other is the store cum kitchen room (roughly 5*10 feet size). But based on our observations in a few cases the food is cooked just near the premises in an open area rather than in the kitchen room, whereas in some other cases it is cooked in the home and brought into the school to distribute among the students.

In each of the centres there are two staff members. One is the Anganawadi worker and the other is the helper. The workers get a salary of ₹ 1,000 a month and the helper gets ₹ 500 per month. But as some of the workers complained, they have not yet received their previous two month's salary. It was also observed that in all the centres both the worker and the cook are from forward clan or caste.

In all the centres there is a separate pucca building (including Kanthapali, although they do not have their own building). There is enough ventilation facility in the classroom although there is no electricity in any of these centres except in Kanthapali. There is no bench or desk in any of the centres. However, there is a pair of table and chair for the worker. While we visited, in all the centres, the children were sitting on the cement floor, even without mat. Again some of these floors are broken. Again, as it is also rightly pointed out by one of the worker that without fresh whitewashing of the wall, the room is looking dirty. Further this has been general complaint that there is no place to keep their materials. None of the centres is having any cooler or Aquaguard. Children drink water from the tube well in the school premises.

The timing of Anganwadi centres ranges from 7.30 am to 12 noon. The school activities start with prayer. Then the class/lesson starts. There is time for break, play etc. The food items cooked are mainly wheat and dal. They also mix Gur (khandasheri) in it. Based on the calculation, they cooked 80 grams of wheat and 14 grams of dal per person, which is enough for the children but it is inadequate for the nursing mothers and pregnant women. However, many workers complained that there is no regular supply of food to the centre and when the supplies of ration reach the centre there is at least 15 kilograms less in every quintal of wheat or dal. Hence, when there is no supply, not only the centre stops providing food, but even when there is supply the centre provides less food to the beneficiaries because of the leakage in the process. The villagers and the officials at the block level do not act on the complaints of the workers about the leakages or irregularities in supplies of ration.

The quality of these wheat and dal are badly commented upon by one of the workers. She said – '*Ita kana gute? Sarkar Chhua aar maa mankar Bhalar lagi Karuchhe, aar kharap jinish deuchhe*'. The rough English translation would be – 'What is this? The government is trying to do some thing good for the children and women but providing bad food'. However, as the general comments have been received both from the parents of the children and the Anganawadi workers, the system is improving over a period of time but leakages and low quality still persist.

Apart from supplementary nutrition to the children, the ICDS also provides nutritional support to the pregnant and nursing mothers. Although less quantity of food has been a general complaint, it was also learnt from the Anganawadi worker of Nuakhurigaon and could be seen from the table below that, the coverage is also inadequate. In fact, the target of 40 persons in a centre makes the job much harder for the Anganawadi worker to exclude a pregnant/nursing female, just because she is slightly and relatively better off but still lives below the absolute requirements. In such cases, the schemes itself needs universalisation rather than targeting. Above all, this is an investment for a healthy future generation.

The Anganawadi worker also provides iron tablets to the adolescent girls. In all the centres the tablets have been distributed. But it was reported that some of the girls do not take these tablets. In all the centres the beneficiaries have a positive response on the Anganawadi workers. The workers do visit the villages regularly. They create the awareness about pulse polio, iodized salt and other minimum necessary knowledge. However, it is reported by the Dalit settlement that the Anganwadi workers visited less number of times to their settlement, which might be either due to cultural stigma of untouchability or may be due the uneducated mass of the settlement, who do not have much idea on the rules and regulations of the schemes and they cannot exercise their rights when knowledge exists.

The number of beneficiaries at the Anganwadi centres is presented in Table 6.7. On the whole, our field observations suggest that the ICDS schemes are running well in spite of many difficulties faced by the workers. The system needs to be strengthened in its coverage and regularity.

Table 6.7 *Details of the Anganwadi beneficiaries in the study area*

Name of Anganwadi	*Different categories of beneficiaries in the Anganwadi*						*Total number of beneficiaries*
	Pregnant women	*Nursing mother*	*0 to 6 month*	*6 month to one year*	*1 to 3 years*	*3 to 6 years*	
Nuan Khurigaon	2	2	0	2	17	17	40
Rani Khenda	6	—	20	38	44	56	164
A. Sason	0	0	11	14	20	59	104
K. Pali	6	6	6	7	16	62	103
S. Singh	5	9	9	17	13	40	93

Source: Institutional survey.

Functioning of the Self-Help Groups in the Study Villages

Self-Help Groups (SHGs) are mostly informal groups whose members pool their savings, and realigned within the group, on rotation and need basis. The interaction with some SHGs during the filed study revealed that apart from their pooled savings, they also got loan from different banks that they use for different purposes. According to the information gathered, there are around 18 SHGs that are functioning in the different villages of Sason GP. Most of them are initiated with the help of ICDS instructors and a few by different NGOs. Members of some of these groups were interviewed.

Although there are 18 SHGs, we could meet 6 groups and discussed with them to elicit information about the functioning of such SHGs. The details about the *establishment, strength, membership fee, numbers of times received loan, repayment status of the loan and the difficulties faced by the groups* are reported in Table 6.8 below. As the table shows, some of these SHGs were formed recently, while other few are quite old. The strength of the group varies from 10 to 15 members and the membership fee of the group ranges from ₹ 20 to ₹ 100. All these SHGs have received loans from the bank and have been repaying the same. Different groups use the loan for different purposes, such as for production, food processing, marketing and business purposes and so on. Some of them generated profit out of those loans and some incurred losses as well. A few illustrative examples are narrated below.

Information collected from the interviews of members of a successful SHG is quite pertinent in this context. We interacted with the Jai Bajrang Mahila Gosthi, Gandapali of Nuan Khurigaon. The group was started during September 2000. There are 14 members in this group. Initially the member's fee was ₹ 50 but now it has been reduced to ₹ 20. This group is well known in that locality for their success. Initially they took a loan of ₹ 25,000. They took half an acre of land on rent and started vegetable cultivation. As vegetables are essential commodities for almost all households, and as there is easy transport facility to town from the village, the group could involve themselves in marketing as well. They divide the work – some had to be involved in the production and some other had to do the marketing, some had to look after account etc. In one way, Smithian division of labour have been discovered in their activity. That was the first time the group enjoyed the taste of the profit from their own efforts. The activity was continued for a couple of years. Many media persons from television channels came to the village and took the interviews of these members. Many things have been flashed in the news on their success. But unfortunately, the process got halted because of inadequate water, which caused the reduction of output from the vegetable cultivation. After some time, they received another loan from Balangir Anchalik Bank. This time they invested on paddy cultivation and rice business. In fact, the capital has been shared into two purposes to share the risk of failure – one into paddy production and the other into rice business. Perhaps, they rightly expected – there was rain failure and hence the paddy production got affected but since they were in the business of rice as well, they could reap the profit from rice business. At present, although the group is having a bank balance of ₹ 70,000 with them, there was a huge desire for adequate water facility to carry on their further activities like agricultural production (where they are efficient). It was also recorded that, there is an NGO called Atma, who gives training to the members of these groups.

All other findings that have been observed in the field are reported in Table 6.8. It is found that all the SHGs are reaping at least some benefits. But the quantum of benefits received is very meagre, which of course can be increased if adequate training could be given and marketing would be facilitated to the existing SHG members. Further, if the SHG groups themselves can open a co-operative shop for marketing purposes or if any other co-operative society would help the SHG for marketing purpose, then the profit margin of the SHG members will increase. Again, when some groups are trying to go for large-scale agricultural production, suitable seed is not available, nor there is assurance of water facility for them. Empowerment through SHGs appears insufficient at present. It is also observed that in almost all the cases, members are not very much clear about the financial intricacies of the matters with the bank, but they do it as instructed by their leader. We also have a word of caution for the bank: just providing loan to the SHGs alone may not help them enough, but regular motivation, training and supervision are equally important for the proper use of those loans. Loans though necessary are not a sufficient condition for success.

The activity process of the SHGs definitely enhances the self-esteem of the members, but their economic gain from the process is constrained by inadequate infrastructure and market conditions available to them. However, as the development of the SC/ST through SHGs are concerned, it is slightly tardy because of lack of capability with SC/ST and lack of close involvement of the official in the process of development. As there is provision that the SHGs constituted by SC/ST would get higher subsidy, the main motto becomes how the subsidised

Table 6.8 *Details of the SHG at Nuakhurigaon*

Name of the SHG	*Date of forming*	*Total members*	*Member's fee (₹)*	*No. of times loan taken*	*Purpose*	*Loan repaid*	*Main problem faced*
Biswa Sawn Sahayak Gosthi	10–12–2002	10	₹ 10 weekly	₹ 7,000 12–07–03	Cultivation	Yes	It was a loss, due to lack of water output come down, and further there was marketing problem. Inadequate training for other activities to do.
Ma Binapani SHG	2001	11	₹ 20 to ₹ 50 monthly	₹ 15,000	Divided the loan and each one uses for different purpose	Continuing	Looking for training
Jai Bajrang Mahila Gosthi, Gondpali	09–09–2000	14	₹ 50 Initially, Now ₹ 20	₹ 25,000, ₹ 2,62,000	Production and business	Paid	Water problem is the main problem, Marketing is another problem, Not getting support for good seed for ground nut cultivation
Ma Samaleswari	28.02.04	10	₹ 100	₹ 20,000	Masala business, Thinking for fish cultivation	Paid	Local demand is not enough. Looking for training
Jagruti	2000	11	₹ 20	₹ 25,000 ₹ 3 lakhs ₹ 1 lakh	Fish cultivation, poultry	Paid	Local demand is not enough.
Jagatmayee	21.09.97	15	₹ 20	One lakh eighty five thousand	Tent house business	Paid	Cultural constraint for women to go and do the tent house arrangement. Hence, used labourer.

Source: Field survey.

money will be shared between the SHG and other officials rather than seriously thinking about the development of the SHG. Such act of both the SHG members and officials fails to bring a sustainable livelihood for the intended group.

Health Security

Health is another basic security that individuals require, and the government of India is providing the same. We made an attempt to understand the constraints and opportunities provided by the public health system in Sason Panchayat.

Access to health care was found to be poor in Sason Panchayat. There is no Primary Health Care (PHC) or subcentre health care facility located in the Panchayat. However, there is a subcentre under Debeipali PHC, which covers most of the population of Sason GP along with covering many from other GPs. This subcentre has a male worker and a female worker. They try to provide health care in the Panchayat with the help of country Dhain (traditional birth attendant) and Anganawadi workers. We could take the interview of a female health worker and the insights drawn are presented below.

Nirmala Mishra is the female health worker who is officially supposed to take care of the health care in the Sason GP as well. She stays in Sason GP in a rented quarter as government quarter is not available. From the discussion it was known that, the basic functions that they do are – immunisation, pulse polio, school visit, visit the pregnant and lactating women and the children, provide free medicine to the TB patients, arrange family planning camp etc. However, given the constraints of limited infrastructure, they try to extend the services to the extent possible.

She also mentioned the problem of inadequate supply of medical instruments, bed, room, staffs, different furniture, storage space etc. Due to this inadequate infrastructure, many patients also do not prefer to come to the PHC (Debeipali). As per our discussion with the health worker, patients generally come to the health centre only for minor ailments. But in our earlier visit to Debeipali PHC, we came to know a child delivery was going on. Further, we came to know that the people who cannot afford to go to near by Burla medical college, they go to the health centre as a second best option.

In our first institutional survey, we had visited Debeipali PHC itself, which is the nearest PHC to Sason GP. This enabled us to know a little more about the health facility in that locality. Sason GP belongs to Dhankauda block. There is 90,000 population living in the block. For this 90,000 populations there is one PHC, which is having 18 subcentres. Each subcentre is expected to have one male and a female health worker but till now there are 15 (excluding two vacant) female workers and 5 male workers (excluding 1 vacant). There is one doctor in charge, one contractual doctor and a pharmacist. Recently, an Additional PHC under Debeipali has opened up but there is only one doctor, one pharmacist and an attendant. So altogether there are 3 doctors, two pharmacists and 22 health workers to look after the 90,000 population in Dhankauda block. Given these limited supplies of health manpower, the health centre maintains the regularity and it is functioning for 24 hours. The outdoor remains open from 8 am to 12 noon in the morning and 3 to 5 pm in the afternoon.

We have also collected some important statistics in the institutional survey on the PHC. Since there is no PHC located in Sason GP, we collected the same from the nearest PHC, i.e. Debeipali PHC (around 7 to 8 kilometres from Sason GP). Table 6.9 provides the details of these statistics. The overall conclusion is that the greatest constraints are the deficit in health manpower as well as supplies of medicine, drugs etc. Further there are also structural problems like non-availability of staff quarters and good educational institutions for the education of the staff's children. This also created disincentives for the staff to join or continue there.

Table 6.9 *Health facilities in Sason Panchayat*

Number of available doctors and supporting staffs	*Doctor – 2, Nurses – 0, Pharmacist – 1, Peon – 2*
Whether there is in-patient treatment available	Yes
Number of beds	6
Whether pharmacy is available	Yes
Diseasewise particulars of last month	Could not obtained
Toilet facility	Yes
Availability of water supply	Yes but irregular and limited
Number of rooms	8 to 9
Ventilation in rooms	Adequate
Mode of waste disposal	Well
Labour room/facility availability	Yes
Clinical service availability	Yes
ICU in the hospital	No
Emergency care unit	No
24 hour service availability	Yes
Ambulance service in the hospital	Yes
Any special service/consultancy available in the hospital	No

Source: Institutional survey in Sason GP.

Apart from this public provisioning of health care, there are a few doctors who do private practice but it was very limited. In the sense, they practice at home and do not have any equipment or separate office/institute to practice.

Food Security

Food is the basic need of people, which must be ensured than anything else in human life. In Sason GP, food assistance has been provided through three institutions: (1) to the preschool students and pregnant mothers by ICDS, (2) to the primary school students though mid-day meal schemes by schools, and (3) to other people through Public Distributions System (PDS). Apart from that, there is a charitable institute, which also provides food to some people (more than 300 people per day) of the Panchayat across the age groups. However, the major source among all these is the food assistance provided through the PDS. Therefore, we examined the functioning of the PDS outlets in order to understand the design and implementation issues.

There are two ration shops in Sason GP that cater to the population of Sason GP as well as some other villages. For this study, one of these two ration shops is covered. In addition, one more ration shop is covered from two of the ration shops located in adjacent villages outside Sason GP. The distribution of ration cards for the BPL households in the selected two villages compiled from the records of Sason Panchayat is presented in Table 6.10.

Table 6.10 *Distribution of below poverty line ration cards across villages in Sason GP*

Village Name	*BPL*
Gurupali	12
Sadhasingha	22
Ajitpur Sason	161
Nuan Khurigaon	61
Rani Khenda	69
Paab Pali	20
Kanta Pali	83
Total	428

Source: Field survey.

Out of 919 households in Sason GP, 428 (46.5%) are possessing BPL ration cards, 112 (12.18%) are possessing Antyodaya ration card and 10 are possessing Annapurna cards.

Out of the two ration shops, one is located in Ajitpur Sason. The shop runs all the days in the week, except Sunday. The working time of the shop is from 9.30 am to 12.30 pm in the morning and 3.30 pm to 6.30 pm in the afternoon. Different commodities provided through the shop are sugar, wheat, rice and kerosene. The difference between the market price and prices charged at the ration shop for the specified commodities are given below. As shown in the Table 6.11, the price difference is less for rice between ration shop and open market, especially for the BPL households. As shown earlier, the majority of the households come under BPL and very few under Antyodaya or Annapurna. Therefore, from food security point of view, the objectives are not met adequately. However, this does not mean that the cardholder does not enjoy the facility. As the ration shop owner said, after the arrival of the rice, within a day or two most of the people offtake their quota. However, this was not the case in case of sugar. Many households do not offtake sugar. Therefore, the shop owner earns some profits by selling the sugar elsewhere and adjusts the running costs ₹ 4,000 per month on the operational aspects (payment of ₹ 3,000 to two people who are employed and a monthly room rent of ₹ 1,000).

The other ration shop that we visited is the one located at Kumbhar Kahnta. The shop owner has been running the shop since 1993. The shop and the store house are located in his own house. He could not procure the list of the cards. But our investigation with the beneficiaries from ration shop shows that he is a very honest person. Whenever the households reach there, they get ration even after the official time. And kerosene being a very essential product, the shop keeper tries to distribute it to all the households including even the households without any cards.

Table 6.11 *Ration shop price and market price of different commodities*

Commodities	*Ration shop price*	*Market Price*
Sugar/kg	₹ 13.50	₹ 18.00
Rice/kg	₹ 6.40 for BPL ₹ 3. 00 for Antyodaya Free for Annapurna	₹ 7.00
Kerosene/lit	₹ 9.35	No supply of kerosene in the open market

Source: Field survey

According to the shopkeeper, initially it was very difficult for him to run the shop. All the costs he had to bear himself (for example, the transport cost, the room rent, and the wage for a person). But over the time period, he himself improved his level of living. He is having a big house where he can run the shop, in the process he saves the room rent and since it is his own house he can use family labour as well and now he can save the labour cost. However, now he also bears the transport cost.

Regarding the offtake of PDS commodities by the households, the shopkeeper has different views (1) Rice being the basic necessity, people mostly buy it immediately, although some people cannot offtake immediately due to financial constraints with them. However, they do offtake their rice quota later on. But the point he has made is the low and fixed entitlement of rice per card. It helps the cardholder to a limited extent, especially when the size of the family is large. Hence, he was suggesting if the quota can be increased on the basis of family size.

(2) Second point that he made is on sugar, the sugar offtake by the BPL households is low. Hence, if same subsidy would be transferred to rice then the BPL household's food security can be ensured in a better way.

(3) About our question about how he is able to adjust kerosene when he distributes to the non-card holders as well, his response was that the rich households (who have electricity) and the extreme poor do not require all of their entitlements. Hence, it is that surplus which enables him to distribute to others.

However, from the food security point of view the main point that emerges is the transfer of subsidy from sugar to rice and to increase the rice entitlement and normalise it according to the family size.

Evaluation of the Functioning of National Rural Employment Guarantee Act

This section presents results of an evaluation study conducted by the present researcher on the functioning of NREGA in Thuamul Rampur block of Kalahandi district. Odisha Rural Employment Guarantee Scheme (OREGS) has been formulated under Section 4 of the National Rural Employment Guarantee Act, 2005 (NREGA) with the objectives of (1) providing not less than 100 days of guaranteed employment in a financial year to every household in the rural areas as notified by the Central Government under Section 3 (i) of the NREGA and whose adult members, by application, volunteer to do unskilled manual works; (2) create durable community assets in the rural areas, and (3) strengthening the livelihood resource base of the rural poor. Hence, the scheme aims to target the poor through occupation. In Odisha, 24 districts have implemented the scheme in two successive phases. The districts who have implemented this in the first phase are Balangir, Boudh, Deogarh, Dhenkanal, Gajapati, Ganjam, Jharsuguda, Kalahandi, Kandhamal, Kendujhar, Koraput, Malkanagiri, Mayurbhanj, Nabarangapur, Nuapada, Rayagada, Sambalpur, Sonepur and Sundargarh and the districts who have implemented in the second phase are Bargarh, Anugul, Balasore, Bhadrak and Jajpur.

As per the stated objectives, the scheme should work for the poor and ensure their livelihood. However, at the micro level the performance of the scheme is far from satisfactory. A case study on the functioning of Odisha Rural Employment Guarantee Scheme has been conducted in the Thuamul Rampur block of Kalahandi district. Through interviews of 15 households and a group discussion conducted in the Balangi village of the Badachatrang Gram Panchayat of Thuamul Rampur block in the Kalahandi district of southern Odisha, the findings are presented below.

Financial Devolution of the Scheme

Financial support for the scheme comes from two sources – one, the central government bears 100 per cent for wage work but if there is materials need, then the central government bears 60 per cent and the state government has to bear another 40 per cent. This fund directly goes to the project director of the District Rural Development Agency. Of the total fund, 20 per cent is under the control of Zilla Parishad, 30 per cent under Block Samiti and rest 50 per cent under gram panchayat. At the district level, the collector is the principal person to supervise the functioning of the scheme, the Zilla parishad's share is supervised by the project director of the DRDA, the Block Samiti's share is supervised by the BDO and the Gram Panchayat's share is managed by the Sarapanch and executive officers. Given this financial distribution, there are also certain rules and regulations on the nature and types of work that have to be done and payments to be made, e.g. first class mason would be paid ₹ 90 per day and second class mason would be paid ₹ 75 per day. If it is earth work then the workers may be paid ₹ 100 for general type of soil, ₹ 135 for hard soil and ₹ 210 for rock stone. There are also several other features of the scheme.

Findings at the Village Level

a. *Possession of job card:* The issuing of job card was made according to the BPL survey of 2002. In other words, during 2002 there was a survey for identifying the BPL households and those households having the BPL cards were given the job cards. Rest of the households, who were left out from the BPL survey or the newly formed households during the last five years did not get their job card. However, during 2007 May–June, a survey was conducted for these non-ID households to provide job card but till 2008 February, the job cards have not yet been issued.
b. *A 100 day employment:* Out of the 15 households, none of the households had completed 100 days of employment in the last financial year. In fact, the last work through OREGS for the villagers was done during June 2006. Since then, the villagers had not been provided with any work in spite of the

fact that none of the households have 100 days of employment and they are looking for employment.

c. *Wage rate of 100 rupees:* It was also reported that, the workers gets ₹ 55 instead of ₹ 100 per day per worker for earthwork. The reason provided by the programme officer is that the workers are not fully aware about the rules and regulations and hence they are confused. For example, a small piece of work is done by a group of workers, even then the workers demand ₹ 100 for each worker. But the problem the programme officer faces is that, the work done by these workers is not worth ₹ 100 per worker per day. Hence, the programme officer measures the volume of work and then finds out the average and pays the workers accordingly, and in the process the average wage rate comes down. However, the workers stated that they finished the piecework in a place and they have surplus time in which they can do some additional work but there is no additional work nearby. Therefore, it seems coordination, arrangement and management problems persist. However, people fail to get ₹ 100. Delayed payment is another constraint the workers face. The payment gets delayed by 2–3 months in almost all cases.

d. *Missing of job card:* Many of the job cards are missing since the cards were taken to the Panchayat office (by the contractor) for entering the number of days of work into the card. Some of the workers have also already lost their cards.

e. *No photo in the job card:* We learned that though there was a photo session in which photos of all the eligible persons were taken, yet none of the job cards were having a photo.

f. *Misreporting and zero day work:* Some of the job cards were checked and rechecked with the workers and the numbers of days were not matching. In the cards the number of days listed is more than the actual days of employment they received. In certain case, the card has been untouched, i.e. there is no entry about the days of work done by the workers.

g. *General problem pointed out by the programme officer:* People are neither aware nor conscious about the OREGS and hence also do not put demand for work in a written format, i.e. no official demand has been put by the people for work. In such case the official takes the interest and initiates the process but then all the workers irrespective of job card come for the work and hence the programme officer finds it difficult to manage. Above all, the workers demand payments according to their days of work rather than the volume of work they do, e.g. for the removal of 10 cm length, 10 cm width and 1 feet depth normal soil earn ₹ 100 for a worker. But in such case, two workers finish that job and each of them demand ₹ 100. So the programme officer faces the problem.

h. *Less number of officials for regular supervision:* Kalahandi is the most backward district and Thuamul Rampur is the most backward block of the district, where there is always shortages of staff. There are four junior engineers for 16 GPs. It is therefore difficult for the engineers to supervise the work regularly.

i. *Unutilised Fund:* As can be seen from the table below (Table 6.12), in the block, another 40 per cent of fund is yet to be utilised.

Table 6.12 *Physical and financial performance on NREGS, Thuamul, Rampur block of Kalahandi district, March, 2007 for the year 2006–07*

		In Nos.								Rs In Lakhs						
Sl. No.	Name of the GP	Households as per BPL survey 2002	Households/ Families Registered	Job-Card Issued	Households Applied for Employment	Households Provided Employment	Persons Applied for Employment	Persons provided Employment	Person-days Generated	OB as on 1.4.2006	Receipt During 2006–07	Misc. Receipt	Total Availability	Total Exp.	Balance Left	% of Exp to TFA
1	2	3	4	5	6	7	8	9	10	11	12	13	14	15	16	17
1	Badchhatrang	1,738	1,527	1,527	1,261	1,261	2,444	2,444	48,280	Nil	46.21	Nil	46.21	36.2737	9.936	78.49
2	Dumerpadar	1,376	1,092	1,092	743	743	1,516	1,516	22,664		32.83		32.83	17.78259	15.047	54.16
3	Gopalpur	1,309	1,075	1,075	830	830	1,646	1,646	33,117		30.08		30.08	21.96814	8.112	73.03
4	Gopinathpur	1,034	792	792	474	474	797	797	14,957		22.06		22.06	11.54619	10.514	52.33
5	Gunpur	1,424	947	947	448	448	835	835	15,589		27.36		27.36	15.2735	12.087	55.82
6	Kaniguma	1,597	1,216	1,157	889	889	1,814	1,814	38,722		36.19		36.19	27.44814	8.742	75.84
7	Karlapat	1,249	967	967	431	431	853	853	13,010		30.87		30.87	9.70916	21.161	31.45
8	Kerpai	980	860	860	508	508	1,022	1,022	18,922		21.38		21.38	10.85979	10.520	50.79
9	Mahulpatna	988	962	962	764	764	1,269	1,269	18,975		20.04		20.04	13.73242	6.308	68.52
10	Malligaon	684	679	679	333	333	700	700	14,803		13.12		13.12	10.57893	2.541	80.63
11	Nakrundi	897	904	904	578	578	1,002	1,002	15,917		24.70		24.70	13.29658	11.403	53.83

Cont'd...

...Cont'd

12	Odri	872	753	753	600	600	1,281	1,281	22,077	22.10	22.10	14.4341	7.666	65.31
13	Padepadar	861	855	855	574	574	1,164	1,164	16,544	18.59	18.59	19.22702	–0.637	103.4 2
14	Sindhipadar	878	881	881	203	203	311	311	6,611	14.38	14.38	4.52098	9.859	31.43
15	Talnagi	785	660	637	362	362	679	679	11,774	15.62	15.62	5.76596	9.854	36.91
16	Th. Rampur	1,834	1,830	1,830	916	916	1,509	1,509	32,525	43.78	43.78	20.54578	23.234	46.92
	Total	18,506	16,000	15,918	9,914	9,914	18,842	18,842	3,44,487	419.31	419.31	252.96298	166.347	60.32

Source: Block office, Thuamul Rampur, Kalahandi.

Summary

The chapter examined the role of the various public provisioning schemes in reducing the intensity of hunger and deprivation in Odisha. It found that a large number of schemes have been introduced in the state covering the area of nutrition, education, health, housing and old age pension. Most of these schemes are central schemes and some are state assisted schemes. Recently, some schemes are also introduced by some voluntary and non-government organisations. However, despite the existence of these schemes, the incidence of hunger and deprivation has remained at a high level in Odisha. Most of the schemes have programmatic and operational deficiencies. At the macro level some of the factors pointed out are the inadequate entitlement and inadequate coverage of these public provisioning which are the reasons behind the slow pace of ameliorating hunger and deprivation.

The micro survey reveals that at the institutional level, various institutions do exist though less in numbers. But their functioning to achieve the desired goals is very poor. Such poor performances are due to several factors – inadequacy of the institution (such as inadequate infrastructure and manpower) to manage the task; influence of the few elite groups for the functioning of the institution and lack of regular government attention etc. From the household perspective, the analysis reveals the following deficiencies in the implementation of schemes – inadequate coverage, irregular and insufficient entitlement, wrong targeting of household, wrong targeting of commodities and meagre assistance (subsidy). Therefore, when the support-led development has been considered, an alternative to achieve the development quicker must have adequate coverage with adequate amount. At the household level, our observation is that many households fail to access and utilise the services extended through public provisioning due to higher implicit cost.

The failure to access public provisioning is partly determined by social identity. The socially lower status people have minimal access to land and other basic capabilities, which reduces their awareness and ability to contest their own right assured by the state through public provisioning. Along with these failures, the external shocks intensify the problem of hunger and deprivation in the state, which have been discussed in Chapters 7 and 8.

Note

1. The authors have presented, (during 1985) with a low level of per capita GNP (310 and 380 dollar respectively), China and Sri Lanka could achieve better level of IMR (35 and 36 respectively) and life expectancy at birth (69 and 70 respectively), whereas with high level of per capita GNP (2,010 and 6,730 dollar) South Africa and Oman have high level of IMR (78 and 109) and low level of life expectancy (55 and 54).

7

External Sources of Hunger and Deprivation

The empirical evidence shown in earlier chapters clearly establishes the proposition that hunger and deprivation in Odisha is largely due to the absence of distributional justice, which makes the socially excluded groups more vulnerable. Both these factors together, make a socially excluded person incapable of having access even to public provisioning. This further gets intensified with external shocks like natural calamities and displacement. Therefore, a greater incidence of hunger and deprivation is found in this group, thereby leading to a vicious circle for the vulnerable populations even if the economy may be growing. This chapter examines the natural calamities' aspects of hunger and deprivation in Odisha.

Occurrence of natural calamity is one of the major reasons for hunger and deprivation in a society. Various types of natural calamities may occur in a society, particularly in case of Odisha they are famine, drought, flood and cyclone. Among the various types of natural calamities, famine, drought and flood are the principal causes in producing hunger and deprivation in Odisha. This chapter attempts to understand the external sources of hunger and deprivation that have taken place in Odisha. Despite the fact that an economy or society is having a sound internal social system (for ensuring both growth and distribution) and capable of controlling the problem of hunger and deprivation efficiently, the problem may still appear and may persist in the society, if there is regular occurrence of external shocks or natural calamities. The two major sources of hunger and deprivation are: (a) internal social system and (b) external shock (natural calamities). The disaster-induced displacement, which also causes hunger and deprivation, will also be discussed here. This chapter is divided into four sections. Section one narrates a theoretical framework of the occurrences of different types of natural calamities and their effects in producing hunger and deprivation. This section also provides an exposition to the theories of famine and its effects on hunger and deprivation. Section two narrates different kinds of natural calamities that have occurred over time and their impacts on Odisha's economy. Section three reviews different preventive steps taken against such natural calamities and examines the effectiveness of such steps. Section four concludes the chapter.

Natural Calamities, Hunger and Deprivation

The linkages between natural calamities and hunger and deprivation could be established in many ways. In an agrarian economy, the frequent occurrence of natural calamities may reduce the operations of productive forces, especially the use of land and labour power. From the demand and supply perspectives, frequent occurrence of natural calamities may lead to hunger and deprivation in two ways: it dampens both the aggregate supply and aggregate demand forces in an economy. For example, occurrence of drought (inadequate rainfall) may reduce the scope of production in the absence of irrigation facilities. This may lead to two further consequences: first, immediately reduces employment opportunities, which again reduces income among the people and hence dampens the volume of effective demand in the economy; second, the contraction in the production base creates a supply shock in the economy. As a result, price of food grains increases and ultimately (both the situations together), push a section of population into the net of hunger and deprivation, especially those who are not having adequate income or assets with them. Although the supply shock and dampening demand together affect the economy in general,[1] the impact of this problem gets reflected more among the landless labourers or among the small farmers, due to lack of regular earning and absence of adequate wealth (assets) with them (Bhatia, 1967). Similar situation may also occur in case of flood and cyclone.

Researchers have provided a number of explanations and theoretical framework. But most of those frameworks have been developed while addressing the problem of famines (Sen, 1981; Devereux, 1993). However, due to the broader base of those frameworks, with some modifications, they may allow us to study other forms of natural calamities as well.[2] The different frameworks that have been used to study famine are discussed below.

A Theoretical Framework to Study Famines

Famine is like insanity, hard to define, but glaring enough when recognised (Devereux, 1993). Famine generally means mass death by starvation. But, it is difficult to provide a robust distinction between famine and other forms of deprivation, such as, chronic malnutrition, seasonal hunger and isolated starvation. All of these happen especially (but not only) among the poor in a developing economy. Dictionary definition of famine indicates, it is a 'crises of mass starvation'. This definition essentially implies three elements: (i) food shortage (ii) severe hunger (important) and (iii) excess mortality. These three elements together constitute an implicit theory, which has been proved wrong by researchers using empirical evidence. The Bangladesh famine (1974) occurred with no food shortage; in case of famines in Darfur (1984), excess mortality was caused by disease and in case of Sahel (1972–74) famine occurred with no excess mortality rate. Sen (1981: 39–40) argues that 'while there is quite a literature on how to "define" famine, one can very often diagnose it like a flood or a fire, even without being armed with a precise definition'. However, in recent times scope for defining famine has been broadened. As Walker (1989) defines, '*famine is a socio-economic process which causes the accelerated destitution of the most vulnerable, marginal and least powerful groups in a community, to a point where they can no longer, as a group maintain a sustainable livelihood'.* This broad canvas of the definition would be quite helpful for the present study to explain the causes of famines in general and hunger and deprivation in particular.

Broadly two categories of views were found in literature as the cause of famines (Devereux, 1993).

1. Accept famine, as an 'act of God' or freak of nature.
2. Believe famine as an 'act of Man'.

Since the first approach accepts famine as an act of God, they isolate the phenomenon of famine from other social, political and economic issues behind it. They place famine in the category of natural disaster such as earthquakes, floods, hurricanes or volcanic eruption. This approach implies vulnerability to famine seen as the consequences of living in an area prone to drought or flood or natural calamities, rather than being the result of prolonged poverty, inefficient government policies or underdeveloped markets. Hence, the best solution this group provides is to move people elsewhere from the drought-prone area (Devereux, 1993: 21–22).

Contrasts to this view, the other group places famine on a continuum with hunger, poverty and social deprivation. Hence, famine is a consequence of human activity (or inactivity) i.e. famine is act of man and could be prevented through socio-economic and political intervention. Therefore, it can be said that famine as a process rather than an event, which grows in the society due to the socio-economic and political carelessness. Proponents of this approach provided three main reasons to infer that famine could be 'act of man' rather than 'act of God.'

(i) There are very few famines in which rich have starved, whereas the poor suffered disproportionately. This clearly shows the reason of famine must be lying on the socio-economic and political factors, which affects a particular group more intensively. Hence, the researchers of this group argue that modifying socio-economic behaviour and bringing political change could prevent famine.

(ii) No recent famine occurred out of Malthusian view (food inadequacy in the economy as a whole). Hence, failure of food production in a place should not be the reason for famine; rather it should be man's responsibility to tackle such scarcity.

(iii) 'Correlation between climate and famine is weaker than that between poverty or war and famine' (ibid.).

Hence, this group holds the view that, in general, famines are 'act of man', and could be prevented by bringing reforms into the socio-economic and political process of the society. However, this does not accept famines are maliciously 'man made'.

Considering both types of explanations, the taxonomy of famine theories has been developed in many forms (ibid.). Table 7.1 provides a particular kind of taxonomy as provided by Leftwich and Harvie (1986: 29–35, cited in Devereux, 1993). The virtue of this framework is that it allows a variety of ways to explain all famines, at all times and in all places, without any disciplinary ideological bias or insisting the explanation on a single dominant theory. It focuses on the social system rather than social sciences. Hence, may be it is worthy to mention the framework here, although slight modifications have been accepted in the application stage of this framework.

Table 7.1 *Taxonomy of famine theories*

1	Food Availability Decline (FAD)
	1.1 Population increase (Malthus)
	1.2 War
	1.3 Climatic factors
2	Ecological Mismanagement
3	Socio-economic and Political Dislocation in the course of 'change' or development
4	Economic Theories
	4.1 Market failure
	4.2 Exchange entitlements (Sen, 1981)
5	Government Mismanagement/political or institutional failure
6	Anthropological or sociological explanations
7	Multi-causal or Eclectic Approaches

Source: Leftwich and Harvie (1986: 29–35) Cited Devereux, 1993, pp. 31.

The detailed explanations about the above taxonomy of famine theories have been provided below. The table reveals that different causes of famine could be broadly: (a) FAD (population growth, climate), (b) market failure and entitlement failure, and (c) political economy explanations (development, government policy, international relation and war).

FAD and Famines: Food Availability Decline (FAD) has been one of the dominant explanations of famines, although it has been loosing its relevance since the mid of 20th century. Not only the world has surplus food availability but many countries also have surplus food availability. However, as shown in the table, two major factors of FAD are population and climate. The following sections briefly discuss these two factors.

Population

In his *An Essay on the Principle of Population* Malthus (1798) pointed out that increasing population is one of the basic reasons of poverty and famine in a society. He argued that food is necessary for the existence of man, and for passion between sexes as well. As a consequence of the latter proposition, population would increase. Malthus predicted population would grow at a geometrical progression and food at arithmetic progression. Hence, there would be overpopulation in relation to food, which would cause hunger or deprivation in the society.

But, this theory of population and famine has been criticised mainly on two grounds:[3]

(a) It failed to foresee the technological improvements in agriculture, which would raise agricultural productivity by offsetting the diminishing return to labour and land. In other words, in reality, the production function is (total product curve) neither uniquely determined nor permanently fixed; hence no reason to believe that the production possibilities available to the society will remain unchanged as population grows. Similarly, population may not grow steadily.

(b) It failed to foresee revolutions in transport and communications, which could increase the possibilities for hitherto unexplored land available for food production and trade, thereby facilitating movement of food from surplus countries to deficit countries (or areas).[4]

Finally, the classical Malthusian theory is challenged by the contra Malthusian, the 'Boserup effect' – population concentration encourages investment in rural infrastructure (roads, irrigation), which reduces vulnerability because of economies of scale and may stimulate technological change (Boserup, 1983). Physiocrats have given the same argument as well. Opposed to this the neo-Malthusians stated, 'in parts of Africa, there is a real population problem (areas with low population density and high population density) but, there is no beneficial Boserup effect'.

However, although the Neo-Malthusians accept the potential for increasing return (technical innovation) in agriculture, still they stick to the view that – famine still is explained in terms of 'too many people, too little food'. They argued that, in rural areas the ratio between *people to arable land* will get reduced. Hence, there would be environmental degradation (deficit in the regional carrying capacity) and famine may follow this. Hence, they suggest birth control measures to check the possibility of famine. The classic example they cite is China's 'one child family' policy.

But as modern demographers have found (using the World Development Report data), there is no significant association between population risk and famines.[5] Instead, as Borlaug pointed out, *person: economic prospects* ratio may be more pertinent than *person: land* relation to explain famines. Hence, he suggests, population density must be expressed in relation to economic opportunities such as arable land, average incomes, mode of production, areas of natural endowments, transport facilities, marketing and infrastructure etc. He also pointed out that 'it is crucial to observe the distribution of resources'. Mathusian crisis may occur if few people control most of the productive land or if some people have been pushed onto unproductive land. Hence population per se matters far less than geographical and economic factors for the occurrence of famines (ibid.).[6]

As far as the World Bank's view on the demographic pressures and risk of hunger is concerned (1981: 112–114), it argued that rapid population growth might retard food production per capita. But subsequently in 1986, it stressed the importance of purchasing power in preference to population growth as a causal mechanism of hunger and poverty.

From the above discussion, it is clear that, although the Malthusian view is not of so much relevance in the modern era, one cannot ignore it unless there would be zero population growth. Similarly, it is difficult to reject the neo-Malthusian view as well. Therefore, as pointed out by Devereux (1993) instead of finding out the link between population growth and famine, it is better to study their common causes, poverty and vulnerability. And as he specifies, any attempt to reduce community's vulnerability to famine requires a more fundamental attack on the social, economic and political mechanisms, which generate poverty and inequality in the society. This implies that famine is an internal problem, i.e. famine occurs due to man's activities. Next we move on to examine climate as one of the reasons behind famine.

Climate

Despite technological changes, the mercy of weather remains as one of the crucial determinants of food harvest. There are large areas of rice/wheat cultivation belt in India, which fall under regions that receive tropical or subtropical monsoons. Many districts (Kalahandi, Nuapada, Balangir) of southern Odisha fall under the semiarid category. These areas suffer from drought, even in a decade of overall excellent agricultural weather. Hence, the lack of sufficient monsoons or seasonal delays may also be a reason. As Timberlake, 1985; cited in Devereux, 1993) wrote, 'A drought on a desert island is no disaster, nor is a flood in an uninhabited mountain valley. It is the relationships between humans and these trigger events which determine whether a disaster will occur, and if so, how big it will be'.

The link from drought or flood to crop failure and to starvation is so simple that, it can be imagined 'drought equals famine'. Although drought causes reduction in food production, the secondary effects of drought include reduction in rural employment, reduction in income for buying, and higher food prices – lesser purchasing power. But it is universally not accepted that climate (or drought) directly causes famine. (Climate or drought is not universally accepted as a cause of famine). As Dando (1980) points out, 'although natural factors cause crop failures, famine are man made'. Hence, it could be their activity (or inactivity) or carelessness, which led to famine. In literature, there are four kinds of criticisms, which reveal the inadequacy of FAD to explain the cause of famine. First, drought or flood has impacts on food production, but not fully on distribution. A robust theory must explain why food does not come in through trade or aid? Second, FAD implies that the impact of food shortage is equal for all people but it cannot explain why some groups of people have better access to food. Third, even in times of drought, if there are sufficient coping strategies or if intervention to prevent famine could be materialised, then, starvation may not exist. Lastly, drought does not always lead to famine (e.g. in the United States, Devereux, 1993), the FAD theory fails to distinguish between situations where people are vulnerable to drought and those where they are not. Hence, it can be concluded that drought or flood causes crop failure but vulnerability to drought causes starvation death and famine. Vulnerability to drought could be checked with proper socio-economic interventions. Other theories that address hunger and deprivation link drought or flood or cyclone with market failure

Market Failure and Entitlement Failure

In relation to food crisis and famine, the other two theories used in the literature are market failure and Sen's 'entitlement approach'. Market failure in this regard has been addressed in two ways namely, pull failure and response failure. The 'Pull failure', indicates collapse of individual exchange entitlements, due to extreme poverty and are pulled into famine. Second, there is response failure, which occurs when market fails to meet effective demands. Response failure occurs when there is hoarding of food, whether for precautionary or speculative motive. While the former is a situation of demand failure the latter is market failure and both the situations together can cause famine – or hunger and deprivation.

Based on the demand side failure, Sen's entitlement approach attempts to explain that famine occurs as cause of entitlement failure. In *Poverty and Famines* (1981), Sen demonstrated that famines can and does occur with plenty of food in a region or country. People have differential access to food, hence, its distribution can shift unfavourably even if aggregate food availability is adequate and constantly rising. In Sen's framework, the entitlement has four components: they are trade-based entitlements ('exchange entitlement'), production-based entitlement ('direct entitlement'), own labour entitlement (from selling labour power) and inheritance and transfer entitlement (gift, bequests). As he states, famine occurs when these entitlements collapse into the starvation set. For example, when farmers face crop failurc or crash in crop's prices, landless labourers loose their employment, or traders hoard grains until its price rise too high for market dependent consumers; in such cases different agents involved in this process may face hunger. Apart from that, the other groups, which depend on the incomes of those affected by famine, will suffer 'derived destitution'.[7]

The entitlement approach is better than the FAD approach in the sense that the former systematically identifies which groups of people will be affected by various threats to food. For example, the FAD approach focuses on the aggregate people: food ratio and does not get into the disaggregated level and finds out the particular group who suffer most in famine situations. However, entitlement approach is also criticised. One of the distinct criticisms against the entitlement approach is in connection with intra-household distribution of food. For example, in entitlement approach, children have no legal claim over food, hence as the critique suggests the notion of 'dependency entitlement' should be introduced. Similarly, the 'relief entitlement' or 'latent entitlement' could be suggested in different context. Secondly, the entitlement approach concentrates on proximate causes of famine, such as market prices, incomes etc. It does not address the underlying causes such as generation of entitlements. Why are some individuals more vulnerable to entitlements collapse? As Devereux, (1993: 185) suggests, a possible solution for such questions is to develop a dynamic version of the theory, which can account for the historical determination and evolution of different types of entitlements over time. The recent criticism against entitlement approach is interference of destitution into the process. 'Starvation is a long and painful process, not a sudden event like heart attack. In this process, famine coping strategies include choosing hunger in order to avoid destitution' (ibid.). Considering the limitation, some researchers work on causes of famine through a political economy approach.

Political Economy Explanations

There are different ways in which the term political economy can be interpreted. Here three indicators have been included, while discussing political economy aspects of famine. These are roles of the development process in controlling famine, government policy and international relations in controlling famine.

Development Process and Famine

Debate about the effects of socio-economic changes on vulnerability remains on speculation, because it is difficult to say what would have happened under alternative conditions. Broadly, two views have been found in the literature. First, some believe that capitalist-style development is the best way to conquer famine, but that takes a longer time in many societies because of the general poverty and lack of political will in those societies. The other commentators taking Africa as case study, argue that inappropriate development strategies may fail to control famine. For example, the pursuit of economic growth through industrialisation and the neglect of agriculture may be the reason for inability to control famine. This kind of development strategy creates two groups, the winners by generating wealth and the looser by making some impoverished. 'While the winners accumulate land assets and power, the loosers (such as peasants and pastoralists) are pushed onto less productive land (Devereux, 1993). And at the time of crises, the impoverished groups have to transfer this unproductive land, along with their labour power and other assets to the wealthy group. And over a period of time polarisation increases. So, based on the views of this group, development process also reinforces inequality and could also be a source to produce a marginalised group. Another criticism against the capitalist development is that it entails a disintegration of reciprocity and cooperative insurance networks; and leads to integration into the market system. As it has been said in the literature, the capitalist development had broken the moral 'moral economy', in the sense, 'where people used to help each other during subsistence crises, now they try to maximise profit from the distress of their neighbour (Devereux, 1993). A solution for reducing hunger and deprivation among the marginalised communities is social security programme but 'safety net' is yet to be introduced in most part of the famine prone countries. This time gap from 'moral economy' to 'market economy' and introduction of social security policy is a transitional period. Sen has coined the term PEST – Pure Exchange System Transition' to describe this period of heightened vulnerability, which occurs after the disruption of the moral economy and before the introduction of the effective social security programmes. Hence, a proper government policy is necessary to remove the problem of hunger and deprivation from the society in general and especially from the marginalised communities in particular.

Government Policy and Famine

Government policy contributes to famine, both directly as well as indirectly. It can directly lead to a famine society through inappropriate and harmful policies towards the vulnerable groups. Such as urban bias policy is squeezing agriculture to support industry, setting up inefficient or corrupt marketing, prohibiting free trade etc. Failing to collect adequate information (famine early warning system), inadequate resources and infrastructure in the famine prone areas also comes in this category. It can indirectly lead to famine situation by failing to intervene in preventing famines.

International Relations and Famine

The humanitarian response to famine by the international community has improved dramatically over the years since Second World War. But the assistance from the international communities has been criticised on different grounds. Famine relief arrives late, and there are issues of distributing the imported food rather than procuring food from the local markets.[8] Apart from this, the non-emergency food aid is also indirectly related with international political relationship and alliances as well. The recipient country requires to extent her support to the donor country. Hence, a poor international relation may also aggravate the impact of famine rather than curing it.

From the above discussion, it is clear that, asymmetric distribution of income and resources, fluctuations in income, dependence on single source of income, lack of off-farm income or remittances, absence of latent entitlements, disenfranchisement, lack of well-designed famine prevention policies, poor international relations are some of the pertinent reason behind famines. This can be controlled through proper policy. The following section attempts to study the Odisha situation as a case study.

Natural Calamities in Odisha

Literature on the occurrence of natural calamities in Odisha's context is scanty. Most of the available literature prior to 1950s has been narrated as a part of the Odisha's history rather than an exclusive discussion on the natural calamities.[9] However, Bhatta (1997) is one of the rare studies which narrate the 'Natural Calamities in Odisha in the 19th Century'. Citing from earlier studies, Bhatta points out that, there is a high correlation between the frequency and magnitude of natural calamities and economic progress of the state. Frequent natural calamities may lead to a low economic progress and hence widespread hunger and deprivation. There are few historical studies that exist on the different types of natural calamities in Odisha, none of them including Bhatta (1997), follows a well-recognised systematic framework to explain hunger and deprivation. Hence, we attempt a systematic discussion about different natural calamities that have occurred in Odisha.

Drought

Prior to the modern economic policies, agriculture was solely dependent on monsoon, hence, failure in the monsoon led to crop failures, thereby leading to scarcity and famines, which was aggravated by the existing poverty in the state. Hence, it was the failure of the monsoon, or in a larger sense climatic conditions that led to famines. 'In Europe nearly 450 famines occurred over a period of 850 years from 1000 AD to 1850 AD, but after Europe became the leader of commerce and industry it achieved highest material prosperity, famines and scarcity were hardly seen' (Bhatta, 1997). Therefore, Bhatta states that the economically backward and poor are more frequently menaced by famines and the richer nations or richer sections are hardly affected by it.

In Odisha, drought is not a new phenomenon. Drought occurs every year in some region of the state with varying magnitude and proportion. Though the drought of 1866 is one of the severe droughts in Odisha history, the state since then has experienced, as many as 17 times, drought of moderate and severe nature. That implies that, drought of severe and moderate nature occurred on an average in every eight years, duration (Ota, 2001). Ota has points out that drought of severe nature were experienced in 1866, 1919, 1965 and 2000–2001. Bhatta has provided information on famines in Odisha since the Maratha rule, from1751 onwards.

Bhatta states that both the Mughals and Maratha rulers were responsible for weakening the economic conditions of peasants in Odisha, thereby easing the occurrence of famines.[10] During the Maratha rule of 50 years (1752 to 1803), there were five major famines during 1770, 1780 and 1792. Though there are no records available on those famines, it was pointed out that, in India, about one crore people perished out of starvation and different diseases, and one-third of the cultivable land became fallow.

Table 7.2 *Drought affected years in Odisha, 1803 to 2000–2001*

Drought year	*Region Affected*	*Degree of severity*
1803	At least Coastal Odisha	—
1806	At least Coastal Odisha	—
1807	At least Coastal Odisha	—
1809	At least Coastal Odisha	—
1817	At least Coastal Odisha	—
1828	At least Coastal Odisha	—
1834	At least Coastal Odisha	—
1836	At least Coastal Odisha	—
1837	At least Coastal Odisha	—
1839	At least Coastal Odisha	—
1840	At least Coastal Odisha	—
1863	At least Coastal Odisha	—
1866	Most part of Odisha including coastal Odisha	Severe
1868	Western Odisha	Moderate
1884	Western Odisha	Moderate
1897	Western Odisha	Moderate
1900	Western Odisha	Moderate
1920	Some part of coastal and Western Odisha	Severe (people affected by drought coupled with Cholera and Influenza)
1922	Western Odisha	Moderate
1926	Western Odisha	Moderate
1930	Western Odisha	Moderate
1955	Western Odisha	Moderate
1966	Western Odisha	Severe
1971	Western Odisha	Moderate
1975	Western Odisha	Moderate
1986	Western Odisha	Moderate
1997	Western Odisha	Moderate
2000–2001	28 out of 30 districts	Most severe

Source: Bhatta (1997) and Ota (2001).

Note: Prior to 1866 we could get information on only about occurrence of drought in proper Odisha

After, Marathas rule (1751–1803), the British ruled Odisha from 1803 to 1947. In the very first year of British occupation (i.e. 1803), a famine occurred in Odisha (Mahtab, 1960). The frequent occurrence of natural calamities was the most important factor that contributed to the poverty of the agricultural population.

Table 7.2 presents the compiled data on the occurrence of drought in Odisha and its severity from whatever available source. From the scant data available, it can be seen that drought occurred very frequently in the state, though the frequency has come down in the post independent period. Here we discuss in detail about the causes and consequences of two drought years, the 1866 drought and the recent 2000–2001 drought. Both of them were extremely severe in nature though in the former, thousands of people perished.

The Great Famine of 1866

The great famine of 1866 was one of the important events in Odisha's economy and demography. The famine affected an area of 40,240 sq miles, and an estimated population of 1,18,55,543. About 8 per cent of the total population died during the calendar year of 1866 (Mohanty, 1993). Within Odisha, Cuttack, Puri and Balasore were the most affected regions, which constituted an area of 7,649 sq miles and estimated population of around three million. These regions lost at least 27 per cent of their population in this famine (ibid.). Therefore, this famine deserves extensive discussion.

Almost all the causes discussed in the above theoretical framework have been found relevant. Specifically the factors are lack of rainfall, system of land revenue, government policies, market failures, socio-economic, regional and demographic factors, export, poor communication infrastructure and bureaucratic negligence. However, the immediate major causes of the famine were the decline in output, increase in exports, hoarding and the laissez-faire policy of the government, which could not control the price rise and finally the lack of adequate relief also contributed to people's sufferings.

Deficiency of Rainfall and Lack of Irrigation

Deficit of rainfall during 1865 was the main reason for the famine of 1866.

Table 7.3 *Rainfall received by Odisha division in 1865*

Month	*Cuttack*	*Puri*	*Balasore*
January	1.1	—	0.25
February	2.3	—	1.35
March	3.6	—	5.75
April	0.7	—	5.7
May	7.4	13	11.55
June	8.1	5.6	4.25
July	12.2	4.3	5.9
August	7.7	5.6	8.25

Cont'd...

...Cont'd

September	7.5	5.2	9.3
October	0	—	0.3
November	—	—	—
December	0.8	—	0.6
Total	51.4	33.7	53.2

Source: Bhatta, 1997.

As Table 7.3 presents, though there was not much shortfall in the amount of rainfall received, whatever rainfall that had been received came at an unexpected time. Much of the rainfall that year was early and rainfall in the latter period, expected in the end of September and early October failed to arrive. On the other hand, until famine occurred, the river water was not utilised properly for irrigation purpose. In the absence of a better irrigation facility, to get rid of the drought situation there was not much alternative, except the petty tanks and water lodgements on the countryside. Therefore, the success or failure of crops and thereby the lives of millions of people were solely dependent on the timely supply of rainfall.

Defective Trade System and Export of Rice

The laissez-faire policy followed by the government in case of food and lack of any control on the movement of food, allowed the private traders to carry on trade to maximum extent possible, as profit was more important for them. The famine committee of 1866 stated, 'throughout the famine it was symptomatic in character as rather due to scarcity of grain than scarcity of money'. However, they observed that food stock was largely inadequate at the time of failure of crops especially during 1865 October. But the problem was realised only in 1866, when the scarcity of food had reached alarming proportions. Such shortage has been related to two aspects, one there was huge increase in the export of rice and two, people had not been careful in protecting themselves by retaining sufficient stocks in their homes. Table 7.4 presents the amount of rice exported in the preceding years of famine.

Table 7.4 *Amount of rice exported from major divisions of Odisha (1860 to 1865) (in Maunds)*

Years	*Balasore*	*Cuttack*	*Puri*	*Total*	*% Changes*
1860–61	3,50,074	23,049	83,936	4,57,059	—
1861–62	3,27,504	60,880	46,780	4,35,164	–4.7
1862–63	4,07,622	86,696	4,816	4,99,134	14.70
1863–64	5,20,052	29,464	NA	5,49,516	10.09
1864–65	8,06,576	72,128	58,824	9,37,528	70.60

Source: Bhatta, 1997 (As quoted from Famine Committee Report).

Note: 1 Ton equals to around 28 maunds.

The table illustrates that, just before the famine of 1866, there was a huge increase in quantity of rice exported from Odisha division, it increased by 70 per cent. It obviously implies non-availability of rice in the state, conducive for a famine to occur supplemented by the occurrence of drought. Further, low wages and inadequate savings with people (Bhatta, 1997) pushed them into a more severe deprived state.

Problem of Unemployment

Unemployment problem was another aspect which induced the famine of 1866 to happen. Due to the salt monopoly right of the British government, all the workers involved in salt trade had been rendered jobless. Bhatta points out 'at the peak period in salt trade, about 2 lakh people were engaged. The decay of salt trade after the declaration of salt monopoly right brought huge unemployment in the state'.

Lack of Communication

Lack of both hard and soft infrastructure during the time was another cause for spread of famine. There were no proper roads or transport facilities from Kolkata to Odisha to supply food and other items smoothly and quickly. Another major factor was the administrative system of that time which was not aware about the condition of food availability in the state. In fact 'the Commissioner of Odisha made a great blunder when he informed the Government of Bengal and Board of Revenue about plenty of food stock with the Odisha grain dealers without a thorough inquiry' (Bhatta, 1997). Hence, lack of 'communication' was another reason for the outbreak of famine during 1866.

Lack of a Relief to the People

There was no specific agency to inform the government regarding the drought situation. Missionaries played an important role in sending information to England. All the above-mentioned reasons cumulatively led to the outbreak of drought.

Consequences

The famine of 1866, which took away lives of millions of people in Odisha, also brought many other consequences to the economy and society of the state. Mohanty (1993) states some of them–

(i) *Changes in production relation:* Up to 1866, the occupancy rights were not freely transferred, but after the famine, within one year, the sale and purchase of land increased rapidly. About 1,240 deeds of sale of land had been executed in Jajpur subdivision itself. The famine also reduced the bargaining power of the Zamindar's as the supply of labour was reduced due to high mortality and migration of people from Odisha to outside.

(ii) *Increase in Wage for Public Works:* The wages for public works increased because of deficiency in the supply of labour, owing to high mortality of labour in the famine. There was also heavy expenditure in public works of various kinds like, construction of canals, roads, etc. However, the inflow of money to the non-agricultural works could not raise the wage rate of the agricultural labour. As Mohanty (ibid.) pointed out, 'the money wage could not influence the real wage in the agricultural sector, perhaps because it was fixed by custom and the influence of famine was not strong enough to change the rate of real wage'.

(iii) *Agriculture and Irrigation:* Immediately after famine, the area under cultivation diminished due to lack of cultivators and agricultural labourers. Hunter (1877) stated that the area under salt manufacturing, which was converted for rice cultivation, was not cultivated again. In Balasore alone, the area under fallow land increased so much that, it took nearly 10 years for areas under cultivation to get back to the pre-famine period. However, an important change that has been emphasised more after the famine is the introduction of canal irrigation. The plan for irrigation, which was mooted after the heavy floods of 1855, had taken pace and the work had been entrusted to the East India Irrigation Company. Thereafter, irrigation facilities provided protection to crops in Odisha division.

(iv) *Social inequality:* The 1866 famine also gave birth to a new caste in Odisha, the 'chhatra-khia' caste or 'those who ate in the relief kitchen'. Chhatra-khia were treated as untouchables till they purified themselves and got readmitted into their original caste.

Therefore, it can be inferred that the famine of 1866 was due to both natural and man-made causes. While, non-occurrence of timely rainfall was responsible for failure of crop in 1866, the lack of surplus food and non-availability of food to some sections of the people were man-made. The Government could have avoided the famine had there been a conscious effort. It is more than 150 years since the last major drought in Odisha, the planned economy that came into being after India's independence has been able to achieve self-sufficiency in food production. However, there was an outbreak of drought in Odisha during 2000–2001, which is narrated below.

The Severe Drought of 2000–2001

With general improvement in the administrative process and with socio-economic development, there is less possibility of famine in India. But, the occurrences of drought and drought-related hunger and deprivation have not been removed completely. The features of 1866 famines are seen in different forms in some regions of Odisha (especially in Kalahandi) which has lead to an outbreak of drought-like situation in some parts of the state.

There have been instances of severe drought in the post-independence period in Odisha, the drought of 2000–2001 can be described as extremely severe and unprecedented. Abysmally low levels of rainfall along with poor irrigation facilities are the major contributors of drought in the state. Table 7.5 states that, there was on an average, 30 per cent reduction of the rainfall during agricultural season. Further, there was a high variation in rainfall across the districts of Odisha (see Table 7.6).

Table 7.5 *Average rainfall received during June to October 2000*

Month	*Normal rainfall*	*Actual rainfall*	*Per cent deficiency of rainfall from the normal rainfall*
June	219.0	216.4 (98.8%)	–1%
July	372.6	250.9 (67.3%)	–33%
August	357.8	259.6 (73%)	–27%
September	235.1	153.1 (65%)	–35%
October	116.3	34.1 (29%)	–71%
Total	1,300.8	914.1 (70%)	–30%

Notes: The annual average rainfall in Odisha is 1,482 mm; The average rainfall between June to October in Odisha is 1,300.8 mm; In 2000, the State received till October end only 61.68% of the total average rainfall. Source: Ota, 2001.

What can be inferred is that the state is primarily relying on monsoon for the agricultural production. The failure of rainfall during agricultural season leads to a drought situation.

However, in the 21st century, famine has not been perceived as a major calamity (Mishra, 2005). Rather it is a process of economic and biological decline manifested in distress migration, practice of *laag* (credit), expressions of *kalbal* (a state of getting long time torture with acute suffering and agony) and *durbiksha* (the process of decline). Hence, economic-related variables are one way of understanding famine. One of the similarities between the 1866 famine and the drought of 2000–2001 is the lack of irrigation and failure of timely monsoons. Hence, irrigation is necessary to get rid of drought especially in those districts, where there is no alternative livelihood and very less irrigated area.

Table 7.6 *Districtwise actual and normal rainfall status as on 31 October 2000*

Sl. No.	*Name of the district*	*Normal rain fall (in mm) up to end of October*	*Actual rain fall (in mm) up to end of October*	*% of increase or decrease in rainfall compared to normal rainfall up to end of October*
1	Balasore	1,278.7	1,190.9	–07.0%
2	Bhadrak	1,278.7	865.6	–32.0%
3	Bolangir	1,330.1	752.9	–43.0%
4	Sonepur	1,330.1	1,002.5	–25.0%
5	Cuttack	1,266.5	837.6	–34.0%
6	Jagatsinghpur	1,266.5	856.0	–32.0%
7	Jaipur	1,266.5	838.6	–34.0%
8	Kendrapada	1,266.5	712.4	–44.0%
9	Dhenkanal	1,238.1	893.4	–28.0%
10	Angul	1,238.1	817.7	–34.0%
11	Ganjam	1,042.6	882.0	–15.0%
12	Gajapati	1,042.6	832.4	–20.0%
13	Kalahandi	1,259.2	1,143.3	–09.0%

Cont'd...

...Cont'd

14	Nawapada	1,259.2	629.3	–50.0%
15	Keonjhar	1,282.3	993.6	–23.0%
16	Koraput	1,315.8	1,200.9	–09.0%
17	Malkangiri	1,315.8	1,123.7	–15.0%
18	Nawarangpur	1,315.8	1,192.6	–09.0%
19	Rayagada	1,315.8	843.2	–36.0%
20	Mayurbhanj	1,383.0	1,105.5	–20.0%
21	Phulbani	1,391.2	1,064.6	–23.0%
22	Boudh	1,391.2	860.6	38.0%
23	Puri	1,229.8	847.3	–31.0%
24	Khurda	1,229.8	1,061.0	–14.0%
25	Nayagarh	1,229.8	1,096.0	–11.0%
26	Sambalpur	1,405.1	769.4	–45.0%
27	Baragarh	1,405.1	787.5	–44.0%
28	Deogarh	1,405.1	710.6	–49.0%
29	Jharsuguda	1,405.1	587.6	–58.0%
30	Sundargarh	1,489.1	922.5	38.0%
	Total	1,300.8	914.1	–30.0%

Source: Ota, 2001.

Flood

Frequent occurrence of flood is another external source of hunger and deprivation in Odisha. Mahatma Gandhi has rightly said, 'Flood is a permanent disease in Odisha' (Bhatta, 1997). Sarangi and Penthoi, (2005) points out that during 1868 and 1967, there were 262 flood inundations in the state, of which 68 were classified as heavy floods, 77 as medium floods and 117 as low floods. Mahanadi, the longest river in the state experienced highest number of floods (99 times), Brahmani experienced floods about 77 times and Baitarani caused floods 86 times, during this time period. Adding the recent time periods the authors point out that during 1967 and 2003, almost every year flood has occurred in between 1967–1975, 1977, each year between 1980–1982, 1985, 1990, 1992, 1994, 1995, 2001 and 2003. A total of 282 floods occurred in the state between 1886 and 2003.

The topographical reason behind the occurrence of floods is that most of the rivers in the state flow into the Bay of Bengal. But, before they drain into Bay of Bengal, it covers a long distance in the state as some of these rivers also originate beyond the state of Odisha. Besides, the draining system with low channel capacity, low flood slope, sand bank mouths, and the plains area in Cuttack, Puri and Balasore, which are not far away from sea, facilitated flood and flood-led destruction. In other words, when the amount of flow of water in a river or rivulet exceeds the carrying capacity of the channel within its bank, the excess water overflows and causes flood. In southern and western parts of Odisha, floods occur due to low carrying capacity of channels of different rivers. In coastal belt, floods occur because of (1) cyclones accompanied by heavy downpours for prolonged period; (2) tidal waves and hurricane in Bay of Bengal, and Chilika lake bring devastating floods; (3) depression in the Bay of Bengal causes widespread and continuous rainfall; (4) bursting of embankments causes catastrophic floods; (5) deterioration of the shape of river channels; (6) the interconnection of rivers; (7) deforestation in the upper reaches of important rivers and (8) configuration of the catchments basin.

Satapathy (1993) in her paper 'Floods and Flood Control Policies' presented the case study of Mahanadi Delta. She points out that, the extent of flood damage has shown an increasing trend from 1954 to 1982 even though systematic measures for flood control such as (a) multipurpose storage reservoirs, (b) flood-control embankments and (3) channel improvements have been undertaken in India under successive Five-Year Plans. Her study suggested that in spite of the fact that number of factors limiting the effectiveness of Hirakud Dam as a moderator of flood control, the overall frequency as well as average intensity of floods had come down, in the post-dam period. Taking cue from this she suggests the construction of Tikar Pada Dam. Satapathy's work does not pay attention to the displacement dimension.

However, it seems that while relief distribution or providing compensation to farmers in various forms are the short-term measures adopted by the government for the flood-affected people, in the long run there needs to be a permanent solution. In such context, building of multipurpose dam, reservoirs, flood control embankments are necessary, which not only help in controlling the flood damage to the state, but also promotes agricultural development, which is very much essential for the state. In fact, by doing so it may solve the dual problems of drought and flood (however, construction of dams should take care of the problem of displacement).

Cyclone

The term 'cyclone' refers to all tropical storms that begin over the tropical seas. These storms are known as 'hurricanes' in the Atlantic and the Eastern Pacific, and as 'typhoons' in the Western Pacific (Dash, 2002). As Odisha is situated in the eastern coastal belt it is always prone to flood and cyclone, which not only take away the lives of many human beings and animals, it also destroys property and leads to a tremendous loss of material wealth in the economy. One of the main reasons for the occurrence of cyclone in Odisha is the depression arising in the Bay of Bengal, which proceed to the inlands passing over the coast, as the velocity of wind on land is maximum at the coast. Accompanied with sea waves and heavy downpour, it destroys life, property and crops. The intensity of the cyclonic storms depends on the speed of the winds, which sometimes ranges between 90 to 200 km per hour, but a speed of 80 to 90 km per hour can also destroy large number of houses (Bhatta, 1997). It is generally during August to October when the Kharif crops are grown in the field, the cyclone phenomenon arises. In Odisha most cyclone-prone areas are undivided Balasore, Cuttack, Puri and Ganjam. As the frequency of cyclones in Odisha is concerned, during the 19th century, cyclone has occurred as many as 19 times.[11]

During 20th century, the 1999 cyclone was the most disastrous one experienced by Odisha. Some of the precarious consequences of the cyclone have been discussed here. Odisha was first hit by a cyclone on 18 October 1999, which affected the coastal districts of Ganjam, Puri, Jagatsinghpur, Khurda, Gajapati and Balasore. The state was again hit by a Super Cyclone on 29 October 1999 with winds blowing at a speed of more than 250 kms per hour, tidal waves rising 20 feet high and torrential rains. The Super Cyclone caused severe damage to the districts of Jagatsinghpur, Balasore, Cuttack, Puri, Nayagarh, Jajpur, Kendrapada, Bhadrak and Khurda and moderate damage to the districts of Mayurbhanj, Dhenkanal and Keonjhar. Both the cyclones along with tidal waves and floods affected 14 districts, 126 Development Blocks and 46 Urban Local Bodies.

The assessment report on cyclone in Odisha was carried out by the United Nations Resident Coordinator, who is coordinating the UN system response to this disaster through the United Nations Disaster Management Team (UN-DMT),[12] and is in close contact with the State Government as well as with the Central Government to help coordinate assistance from the wider donor community. This report states the damages in three categories; (a) Physical destruction directly caused by the cyclone; (b) Flooding; and (c) Saline inundation. The physical destructions include loss of life and property. Both the cyclones together took away the lives of about 10,000 human beings and about 4,54,000 heads of cattle; destroyed or severely demolished 18,28,532 dwelling units and destroyed over 80% of the standing crops especially the ready-to-be harvested paddy from 18,10,091 hectares of agriculture land. Apart from this individual loss, there has also been extensive loss of physical infrastructure like office-buildings, panchayat 'ghars' or village community halls, schools (about 11,000 local schools have been badly damaged/destroyed), dispensaries, hospitals as well as community wells, tanks, village roads, physical infrastructures like state/national highways, irrigation channels, railway lines and telecommunications, etc. As there was large-scale inundation, there was a widespread contamination of drinking water. Hence, not only drinking water is a serious problem, as there is no electric power, many hand pumps are damaged and numerous wells, tanks, ponds and canals are polluted; this has also led to the outbreaks of diarrhoeal diseases in spite of enormous preventive measures. Though the quantitative assessment of the loss of lives and property is limited, in fact, the cyclone has many indirect impacts and the impacts were very intense among certain groups of people.

Das (2002) narrates the economic effects of the cyclone. According to him, the agricultural sector where majority of population eke out their living has been ruined to such extent that 'it may take more than a decade to achieve the status quo' (ibid.). Fishing industry and shrimp business, which is the third largest revenue earner for the state and important source of livelihood of many is another important sector which has been severely affected by the cyclone. The cyclone affected 300 kms out of 480 kms of the coastal fishing zone and also took away life of more than 7,000 fishermen, their houses, fishing equipment, boats etc. Das (ibid.) provides the magnitude of loss due to the super cyclone of 1999 as revealed by Table 7.7.

Table 7.7 *Impacts of Super Cyclone in Odisha during 29–30 October 1999*

Population affected	1,56,81,072
Villages affected	14,586
Districts affected	12
Human lives lost	9,893
Livestock perished	4,44,531
Houses damaged	1,66,1683
Crop area damaged (ha)	18,43,047

Source: Das, 2002.

Table 7.8 *Value of properties lost and damaged due to national calamities in Odisha since 1971*

Year	*Value of properties (₹ In crores)*	*Average per capita lost and damaged (in ₹)*
1971–73	45.11	6.66
1974–76	34.17	4.81
1977–79	51.39	6.81
1980–82	150.23	18.82
1983–85	302.76	36.04
1986–88	187.28	21.36
1989–91	465.84	49.59
1992–94	2508.55	258.51
1995–97	420.39	41.12
1998–99	478.25	67.21
Total	4643.97	510.93

Source: Sarangi, 2005.

Given the devastating nature and impact of cyclones on the Odisha's economy and society, it is natural to expect the negative effects on the disadvantaged and the poor, including SC and ST communities. These poor people have limited means to protect them from the impact of cyclones. Therefore, the natural calamities like cyclones lead to hunger and deprivation among these disadvantaged social groups. Therefore, preparedness for such external shocks is necessary to get rid of hunger and deprivation in Odisha.

Sarangi (2005) has provided an estimation of the value of property lost or damaged due to the natural calamities (flood, drought and cyclone) during 1971 to 1999, which is presented in Table 7.8. The table shows an increasing trend in the average per capita lost and damaged (i.e. ₹ 6.66 during 1971–73 to ₹ 67.21 during 1998–99). In 1998–99, the total value of the properties lost/damaged was estimated to be ₹ 478 crores. Hence, the state must take the protective and preventive measures to mitigate the harmful impact of natural calamities.

Summary

This chapter attempted to understand the external sources (drought, famine, flood and cyclone) of hunger and deprivation in Odisha. The existing literature suggests that, while the western and southern parts of Odisha are more prone towards famine or drought, the coastal part of Odisha is more prone to flood and cyclone because of the topographical situation of the state. The western and southern part of the state dominate in land resources and hence lack of water creates drought situations, whereas many of the rivers in the state flow from north-western part to the east, into the Bay of Bengal, after covering a long distance within the state. As the coastal area is plain area, whenever there is excess water in the river, it easily gets flooded near river banks. The occurrence of cyclone also is due to the geographical specificity of the state. As one part of the state is surrounded by the Bay of Bengal, the depression in the sea leads to cyclone. Though it is not a regular phenomenon, it brings great harm to the society and economy when cyclone occurs.

The impact of the natural calamities falls disproportionately among the poor and the disadvantaged. Since the poor and the disadvantaged do not have adequate means to protect themselves from the ill-effects of the natural calamities, their well-being is severely compromised. The natural calamities make them more vulnerable and push them into a vicious circle of hunger and deprivation, even if the economy may be growing.

However, though the state faces different kinds of natural calamities, there are very few permanent solutions by either from the state or the central government. Hence, the occurrence of the natural calamities not only intensifies the problem of the existing poor people but also creates additional hunger and deprivation. Therefore, there is a need for the state to protect especially the poor and the disadvantaged.

Natural calamities is one external source of hunger and deprivation; another external source, which produce hunger and deprivation in the state is the displacement in the name of development, which has been dealt with in the next chapter.

Notes

1. However, considering the fact that the world produces surplus food, the same can be supplied to local market but the main problem may be the failure of effective demand.
2. It is understandable that flood and cyclone are different by their nature from famines, but since the aim here is to find out their combined role in producing hunger and deprivations, the same framework can be applied.
3. As Sen (1982: 448) describes Malthus's analysis with the arithmetic and geometric progressions is 'an attempt to get profound insight from elementary mathematics – a tendency not altogether unknown in modern economics as well'. But Malthus was opposed to an earlier viewpoint put forward by the physiocrats that larger the population the larger the aggregate surplus.
4. Against the impact of population on famines several evidences have been given viz., western Europe is several times more densely populated than Africa, and has only fraction of its population working in agriculture, yet it produces enormous agricultural surpluses and is not famine prone (except during war). Similarly, Japan is also not a famine-prone country, despite having to import the bulk of its food (Devereux, 1993).

5. In a discussion of potential famine regions, Borlaug, predicted that 'the critical area for immediate future is South Asia where the greatest population density exists'. But this proved to be factually inaccurate. With hindsight, it now appears that Asia had just seen its last major famine (Bangladesh, 1974). So that Africa rather than South Asia became the main famine arena of the 1980s. While it is true that Bangladesh (at 663 persons/km^2 in 1983) has the highest concentration of people in the world, compared with Ethiopia (33 persons/km^2) and Sudan (just 8 persons/km^2)' (cited Devereux, 1993).
6. 'Dutch children though born into one of the most densely populated nations are less likely to experience famine, than, pastoralists in sparsely populated Mauritania' (Devereux, 1993: 54).
7. Sen's famous example in this regards is barbers in Ethiopia during 1970s famine.
8. Procuring food from the local market and distributing it in the drought-prone areas has another implication that, it may keep the production process by ensuring adequate demand.
9. Some of the studies in this regards are – Stirling, 1825; Hunter, 1872; Toynbee, 1873; Banerjee, 1898; Maddox, 1900; Senapati, 1917; Banerjee, 1939; Mahtab, 1960; Bhatia, 1963; Mukherjee, 1964; Srivastav, 1968; Das, 1970; Samal, 1978; Patnaik, 1980; Das, 1989, Mishra, 1991 (See Bhatta, 1997 and Sen 1981).
10. Bhatta states, 'the Marathas not only contributed to the deterioration of economic condition of the province but also destroyed the basic foundation of the economic structure. People were afraid that if they produced more, the Marathas might come and loot their property. There was no question of accumulation of wealth and that might invite problems. Though, there are many who praise the Maratha for their liberal views, in other parts of the country, their rule in Odisha, as a matter of fact was sad and unfortunate'.
11. Cyclone occurred in the following years – 1823, 1831, 1832, 1833, 1840, 1842, 1848, 1850, 1851, 1864, 1867, 1872, 1874, 1885, 1887, 1888, 1890, 1891 and 1893.
12. There were also three members of the UNDAC team, WHO, UNICEF, WFP and a representative of USAID who have been part of the assessment report.

8

Coping with External Shocks

Displacement in the Name of Development

Land is the only source of livelihood in most of the agrarian economy. Given the structural specificity of Odisha, SCs/STs who mostly live in the forest area depend to a large extent on land and forest for their livelihood. For the poorer section including SCs/STs, land is more than a livelihood as it serves the purpose of collateral, fallback resource, economic and social power to claim citizen's rights in various circumstances and so on. But in a neoliberal strategy, in the name of development, industrialisation and modernisation these lands were taken away by the profit seeker capitalist, and forests were destroyed to extract iron ore for business. In this process the state also works like an agent to collect the land and hands it over to the capitalist. In the process, mainly the SCs/STs, among whom the incidence of hunger and deprivation is already high, suffered the most by losing habitance and source of livelihood, compared to others who do not depend directly on forest for their livelihood.

This is a common phenomenon in many states of India. But in Odisha, since it already faces high incidence of hunger and deprivation, the forced displacement makes the situation worse. The displacement not only affects the current agrarian situation by reducing output and increasing unemployment, it also reduces the capacity of the growth of industry in the state in future as the raw materials, which can help industrialisation is draining away. So far, there are not many instances in Odisha that the already settled mines company might have established manufacturing firm with vertical integration. Rather, the main motive of the company remains at just extraction of iron ore or coal and then drain these to another country to develop the manufacturing sector like tractor, automobiles and so on. In the absence of such vertical integration, the state becomes the victim and the displaced unemployed also move towards lumpinisation.

This chapter examines how the displacement in the name of development is producing hunger and deprivation in Odisha. The benefits of development do not often reach displaced people. In the process of development, the local inhabitants are the victims of irrigation projects, hydro and thermal electricity projects, mineral-based industrial projects, etc. Yet, there is also a long list of projects, which may displace more people if the projects get approval. In the absence of a sustainable rehabilitation and resettlement policy, the displaced people fall into the net of hunger and deprivation. This chapter examines the issue of displacement and its association with hunger and deprivation. The chapter has three sections. Section one presents the macro-picture of the magnitude of development-induced and disaster-induced displacement and their contributions to existing hunger and deprivation in Odisha. Section two explores the magnitude of hunger and deprivation among the displaced households. This section is primarily from the field survey data from Lanjigarh block in Kalahandi district of Odisha. It also presents some observation from the displacement due to POSCO. Section three concludes the chapter.

Development and Disaster-Induced Displacement: A Macro-Picture

Displacement or forced displacement implies the forced dislocation of people away from their home or home region due to their own government order or due to occurrences of different disasters in the region. The forced displacement of population leads not only to break up of living patterns and social continuity, but it also dismantles the existing modes of production, disrupts social networks, causes the impoverishment of many of those uprooted, threatens their cultural identity and increases the risk of epidemics and health problems (Cernea, 1995; Parasuraman, 1999; Chattergee, 2004). Before we elaborate on the consequences of displacement, below we have narrated the distinction between the development- and disaster- induced displacement.

Development-induced Displacement

Relates to the dominant view of development, i.e. the modernisation theory. During 1950s and 1960s, it was pointed out that, the less developed countries are underdeveloped because of their traditional form of development and hence were encouraged to transform from simple traditional form of production process to a modern complex, westernised production process. Many big dams and reservoirs for irrigation, mining industries, power plants, urban infrastructures, transportation (roads, highways, canals), have become inevitable. Such efforts though facilitated development for the whole economy, also brought poverty among those few who forcefully were displaced and left out from the rehabilitation process. Such displacement leads to vulnerability of the people to various dimensions and is known as development-induced displacement. In Odisha, the Hirakud, Balimela, Machhkund, Salandi, Rengali, Kolab and Indravati dams have displaced large numbers of people in the name of development. Similarly, the industries, especially the mines-led industries, such as, Rourkela, Daitari steelplants and NALCO, etc. have led to development-induced displacement.

Disaster-induced Displacement

The United Nations has defined disaster as 'a serious disruption of the functioning of a society, causing widespread human, material, or environmental losses which exceed the ability of the affected society to cope using its own resources.' When such disaster leads to displacement, it is considered as disaster-induced displacement. For example, in the cyclone of 1999 in Odisha, the tragedy of the living was compounded with damages amounting to over ₹ 7,000 crores, 13,50,000 homes were destroyed, 10,000 km of roads were damaged,

30,00,000 people were unemployed, and 24,00,000 hectare of farmland were inundated (Suri, 2000). Such incident is also included in the disaster-induced displacement. However, as per the above definition, fire, earthquake, drought, epidemic, or industrial accidents are called disasters, for which the society needs external support to cope with. The disaster-induced displacement may occur mainly from two grounds, the natural and the man-made. The natural disasters further may be broken down into two sub-categories – sudden impact and slow onset. The human-made disasters include two sub-categories namely industrial/technological disasters and complex emergencies. The sudden impact disasters include floods, earthquakes, tidal waves, tropical storms, volcanic eruptions, and landslides. All these disasters may affect the habitat and livelihood of people and force them to migrate away from their own habitat. The slow-onset disasters include droughts, famine, environmental degradation, deforestation, pest infestation, and desertification (conversion of arable lands to deserts), which are usually the result of adverse weather conditions combined with poor land use. In Odisha, there are a number of cases, where due to drought conditions, people leave their original villages and migrate to wet areas and settle down there. Though these cases are not registered in government records as displaced persons; they form a part of disaster-induced displacement.

The industrial/technological disasters result out of the society's industrial and technological activities lead to pollution, spillage of hazardous materials, explosions and fires. Such disasters may occur due to poor planning and construction facilities or from neglect of safety procedures.

In Odisha the main disaster-induced displacement may include, drought in south-western part of Odisha and flood and cyclone in the coastal part of Odisha. However, in Odisha, the development-induced displacement is better known than the disaster-induced displacement. Given the brief discussion on various kinds of displacement, the following sections examine details on the impact of these displacements in Odisha.

Impacts of Displacement in Odisha

Odisha is endowed with plenty of natural resources like forests, inland water, mineral deposits and raw materials, which are the most essential elements required for pursuing the goal of modern development. It has long coastline combined with potentially viable ports that are suitable for developing Special Economic Zones (SEZ). Odisha is now developing into an industrial hub. Paradoxically, Odisha is generally known as a poor, traditional, backward and non-industrialised state with highest incidence of poverty in India. It is observed that although rapid industrialisation has taken place, the traditional industrial base has been eroded leading to a number of people being displaced and rendered out of work. The history of development and displacement in the state is presented in the following sections.

There are mainly two kinds of development-induced displacement that have occurred in Odisha. The construction of dams for irrigation and power purposes that led to displacement of large number of people[1] and the other is due to industrialisation through mining and land acquisition, which also displaced large numbers of people. Both kinds of displacement are examined here.

River Dam Project and Displacement

Table 8.1 *Major and medium irrigation projects in Odisha till 2004*

Major Irrigation Projects in Odisha			
	Completed projects		*Ongoing Projects*
1	Baitarani System	1	Rengali Irrigation Project
2	Hirakud System	2	Subarnarekha Irrigation Project
3	Mahanadi Delta System	3	Upper Indravati Irrigation Project
4	Rusikulya System	4	Kanupur Irrigation Project
5	Salandi System	5	Upper Kokab Irrigation Project
6	Potteru	6	Lower Indira Project
7	Anandapur Barrage	7	Lower Suktel Project
8	Machhkumd	8	M.C.I.I. Projects
Medium Irrigation Projects in Odisha			
	Completed projects	33	Remal
1	Aunli	34	Saipal
2	Bagh Barrage	35	Salia
3	Baghalati	36	Salki
4	Baghua	37	Sarafgarh
5	Bahuda	38	Satiguda
6	Balidiha	39	Sundar
7	Bankabahal	40	Sunei
8	Bhaskel	41	Talsara
9	Badnalla	42	Titilagarh
10	Budhabudhiani	43	Upper Suktel
11	Dabaraghati	44	Uttei
12	Daha	45	Gobardhanpur Barrage
13	Dahuka	46	Upper Jonk
14	Derjang	47	Harabhangi
15	Dhanei	48	Hariharjore
16	Dumerbaha	49	Birupa Genguti
17	Ghodahada		*Ongoing Projects*
18	Gohira	1	Baghua Dhanei Doab
19	Hiradharbati	2	Chheligarh
20	Jayamangal	3	Deo
21	Jharabandha	4	Manjore
22	Kalo	5	Rajua D/W
23	Kanjhari	6	Ret
24	Kansbahal	7	Rukura
25	Khadkhai	8	Telengir
26	Kuanria	9	Upper Jonk (Kharkhara)
27	Nessa	10	Sapua-Badjore
28	Ong		*Extension Projects*
29	Pillasalki	1	Bagh Barrage
30	Pitamahal	2	Baghlati
31	Ramnadi	3	Titilagarh
32	Ramial		

Source: www.Odishagov.nic.in

The river dam project in Odisha is meant to facilitate irrigation and hydro-electricity. Table 8.1 presents some of the large and medium dams in the state. Table 8.2 presents the capacity of these large dams in power supply and irrigation. Some of the major water-related projects (which is meant for both irrigation and power generation) are Machhkund, Hirakud, Balimela, Rengali, upper Kolab, Indravati, Potteru, and Salandi. The major producers of power are Indravati (600 MW), Balimela (360 MW), Hirakud (I and II) (347.5 MW) and Rengali (250 MW). In terms of providing irrigation, Hirakud dam is one of the major sources. However, these dams, which were producing power for the state and supplying water to many agricultural fields are the major sources of displacement in the state (see Table 8.3).

Table 8.2 *Major projects according to the capacity of power generations*

Station/Dam	*River Name*	*District*	*Year (foundation and inauguration)*	*Capacity*
Machhkund (Odisha shares)	Duduma	Koraput	1946–1955	34.5 MW
Hirakud I (Burla)	Mahanadi	Sambalpur	1948–1957	275.5 MW
Hirakud II (Chiplima)	Mahanadi	Sambalpur	1956–1963	72 MW
Balimela	Sileru	Malkanagiri	1963–1973	360 MW
Rengali	Brahmani	Angul	1976–1986	250 MW
Upper Kolab	Kolab	Koraput	1976–1993	320 MW
Upper Indravati	Indravati and Hati	Bhawanipatana	1979–2001	600 MW
Potteru	Canals	Malkanagiri		6 MW

Source: http://www.ohpcltd.com

Table 8.3 *Number of families displaced and amount of land acquired due to different development projects in Odisha, from 1950 to 1995*

Types of Project	*Number of villages displaces or affected*	*Number of families displaced or affected*	*Total land acquired in hectare*	*% of displaced family*	*% of land acquired*
Mines	79	3,143	2,427.03	3.87	0.39
Industries	113	10,704	21,963	13.19	3.52
Thermal Power	73	2,426	3,155.31	2.99	0.51
Irrigation and hydel power (Dam)	1,181	64,903	5,95,918.6	79.95	95.58
Total	1,446	81,176	6,23,463.9	100	100

Note: Parida (2006) gives another 142 and 4792 families being affected due to wildlife sanctuary and urban slums. So total families displaced increase to 86,110.

Sources: Dalua, 1993, Fernandes (1997), Reddy (1993), Parida, 2006.

Table 8.3 presents the number of families displaced[2] during 1950 to 1995 due to different types of development projects. These data do not cover all kinds of displaced people as development projects in Odisha have started prior to 1950 and continued even after 1995. This implies that the magnitude of displacement in the state must be higher than whatever has been presented in the table. During these 45 years, there are as many as 81,176 families that have been displaced or affected, of which most of them (79.95%) are displaced due to irrigation and hydro-power projects, and due to industrial projects (13.19%). Similar pattern is seen in case of land acquisitions. Most of the total acquired land (95.58%) is due to irrigation and hydel projects.

However, incidence of such displacement is not equal for all social groups. Since these dams are located in the highland or hilly area, where often the ST and SC people live, it is likely that more of this population are displaced. In this regard, Pati (2006) presents a conservative estimate from Fernandes (1994). As Table 8.4 shows the proportion of tribals displaced due to different types of projects is 40.9 per cent, which implies a higher magnitude of displacements among the tribals. Though the ST were materially well off than SC, they might have become resource poor due to displacement. However, if higher percentage of displaced tribals would have been rehabilitated, probably the possibility of poverty among them might have been checked to certain extent. The proportion of displaced persons who were resettled remained almost similar in both cases. This may also imply that more of the benefits of rehabilitation have gone to non-tribal displaced persons. In the Odisha context, distribution of displaced persons, according to their social group is not available for all projects. Table 8.5 presents some of the figures available which depicts the incidence of displacement among different social groups. It differs from project to project, yet the incidence has remained higher among the ST, when compared to their share in total population and their share in displaced families.

Table 8.4A *Conservative estimate of persons and tribals displaced by development projects in India 1951–90 (in lakh)*

Types of Project	*All DPs (Lakh)*	*% of DPs*	*DPs Resettled (Lakhs)*	*% of Resettled DPs*	*Backlog (Lakhs)*	*Backlog (%)*	*Tribals Displaced (Lakhs)*	*% of All DPs*	*Tribals DPs and Resettled (Lakhs)*	*% of Tribal DPs*	*Backlog of Tribal DPs*	*% of Backlog*
Dam	164.0	77.0	41.00	25.0	123.00	75.0	63.21	38.5	15.81	25.00	47.40	75.0
Mines	25.5	12.0	6.30	24.7	19.20	75.3	13.30	52.20	3.30	25.00	10.00	75.0
Industries	12.5	5.9	3.75	30.0	8.75	70.0	3.13	25.0	0.80	25.0	2.33	75.0
Wildlife	6.0	2.8	1.25	20.8	4.75	79.2	4.5	75.0	1.00	22.0	3.50	78.0
Others	5.0	2.3	1.50	30.0	3.50	70.0	1.25	25.0	0.25	20.2	1.00	80.0
Total	213	100	53.80	25.0	159.20	75.0	85.39	40.9	21.16	25.0	64.23	79.0

Note: DP denotes displaced persons.

Source: Fernandes, 1994, pp. 22–32 in Pati, 2000.

Table 8.5 *Conservative estimate of families displaced by development projects in Odisha across social groups 1951–95*

Station/Dam	*Gen*	*SC*	*ST*	*Total family affected*	*Number of family*
Machhakund	38	10	52	100	2,938
Hirakud	NA	NA	NA	—	22,144
Balimela	NA	NA	NA	—	2,000
Salandi	NA	NA	NA	—	569
Rengali	73.55	15.69	10.76	100	10,897
Upper Kolab	39.00	14.18	46.82	100	3,067
Subarna Rekha	30.29	5.14	64.58	100	3,427
Upper Indravati	42.89	15.27	41.85	100	5,568
Medium projects	41.72	11.59	46.69	100	4,892
Total	27.92	7.11	17.95	52.98	52,564
Total	52.70	13.42	33.88	100	27,851*

Note: * implies the figure is without including the displaced families of Hirakud, Balimela and Salandi for which social groupwise displacement is not obtainable.

Source: Record of irrigations department, Government of Odisha, Parida, 2006.

Table 8.6 provides the magnitude of displacement by different dams. The regional dimension of displacement is also revealed through this data.

Machhkund hydro project built on the Duduma River in the Koraput district is the earliest development project of this kind. The project work has started during 1946. The displacement and rehabilitation aspects of the project was not much clear (Dalua, 1993). But as Mahapatra (1989: 96) said, out of the total 2,938 families that were displaced by the project, 1,500 (51%) were ST, 300 (10.21%) were SC and the remaining were from other castes. Mahapatra (1989: 96) pointed out that in the Machhkund project, out of 2,938 displaced families, only 600 families (450 tribal and 150 others) were rehabilitated. Satnely (1996: 1533–1534) pointed out that the displaced people were rehabilitated with 'land for land' provision. Stanely also pointed out that there were two villages, which were not willing to leave their birth place and have given their memorandum to the Chief Minister and the local government officials. But ultimately, the villagers settled for higher compensation and better rehabilitations than originally envisaged. But there is no further information available on the people's response towards the project.

Hirakud dam is the first multipurpose project of independent India. It is situated in the Sambalpur district of Odisha. The scheme for the Dam was conceived by M. Visveswaraya in 1937.[3] Hence, it seems the displacement might have occurred during this period, i.e.1949 to 1956. The second phase of the dam was the Chiplima scheme, which was sanctioned during July 1956 and was inaugurated on 14 September 1963. Therefore, it seems the displacement due to Hirakud I and Hirakud II might have continued from late 1940s to early 1960s. Therefore, if the displaced people might have joined as agricultural labour, then there might be also chances that the percentage of agricultural labour in the state or in the district Sambalpur might have increased during that period. The third phase expansion of the Hirakud Dam had taken place during 1982 to 1990. Therefore, here too the displacement effects may get reflected in the census data. However, as direct information is concerned, Hirakud Dam has displaced 22,144 families from 249 villages in Odisha (Dalua, 1993).

Table 8.6 *Damwise and regionwise displaced persons and submerged areas in some of the notable multipurpose hydro-electricity project in Odisha*

Station/ Dam	*River Name*	*District*	*Time period*	*Submerged*		*Number of family*		*Population affected*
				V	*A*	*D*	*R*	
Machhkund	Duduma	Koraput	1946–1955	225	20,794	2,938	600	14,690
Hirakud	Mahanadi	Sambalpur	1948–1957	285	1,82,593	22,144	1,879	1,60,000
Balimela	Sileru	Malkanagiri	1963–1973	89	48,000	2,000	NA	9,600
Rengali	Brahmani	Angul	1976–1986	265	1,05,905	10,897	6,639	54,235
Upper Kolab	Kolab	Koraput	1976–1993	149	24,794	13,095	525	50,771
Upper Indravati	Indravati and Hati	Bhawanipatana	1979–2001	99	32,530	5,568	3,585	26,505
Total				1,112	4,14,616	56,642	13,228	3,15,801

Percentages

Station/ Dam	*River Name*	*District*	*Time period*	*Submerged*		*Number of family*		*Population affected*
				V	*A*	*D*	*R*	
Machhkund	Duduma	Koraput	1946–1955	20.23	5.02	5.19	4.54	4.65
Hirakud	Mahanadi	Sambalpur	1948–1957	25.63	44.04	39.09	14.20	50.66
Balimela	Sileru	Malkanagiri	1963–1973	8.00	11.58	3.53	NA	3.04
Rengali	Brahmani	Angul	1976–1986	23.83	25.54	19.24	50.19	17.17
Upper Kolab	Kolab	Koraput	1976–1993	13.40	5.98	23.12	3.97	16.08
Upper Indravati	Indravati and Hati	Bhawanipatana	1979–2001	8.90	7.85	9.83	27.10	8.39
Total				100	100	100	100	100

Notes: V – villages, A – Area in acres, D – Displaced, R – Rehabilitated. These table some of the displaced families and not all due to paucity of data.

Source: Compiled from Tripathy, 2003.

In case of the displaced people of the Hirakud Dam as well, there was no specific policy guidelines for rehabilitation. In the absence of that, the affected people were paid compensation for the land property only. It was pointed out that only 11 per cent of the land owners have been settled with some land since the land was located on a hilly slope and forest area was not suitable for cultivation (Tripathy, 2003). The situation was even worse when some of the people have become the victim of multiple displacements due to mines project in the resettled place. For Hirakud Dam also, there was opposition by the local people and leaders, but such protest was not having much impact on the construction of the project.

Balimela project is the second largest hydroelectricity project in Koraput district. The project commenced during 1962–63 and was completed in 1977. The project has displaced 2,000 families (Dalua, 1991). The Resettlement and Rehabilitation (RR) policy followed for the project is similar to that of Hirakud Dam.

Rengali Dam project started around mid-1970s and was completed by mid-1980s. The Rengali Dam displaced as many as 10,847 families. For this also similar RR policy has been followed. The Indravati Hydro project dam started in 1978–79. The four dams across the river Indravati and its tributaries Patogoda, Kapur and Muvan would form a single reservoir submerging 44 villages in Koraput district and 51 villages in Kalahandi district. Mahapatra (ibid.) pointed out that 578 families have been displaced. The Upper Kolab hydroelectric project located in the district of Koraput (Odisha) was initiated in 1976 by the Irrigation and Power Department, Government of Odisha to produce power and supply irrigation. This project utilises the water potential of river 'Kolab' a tributary of river Godavari. It generates 320 MW of power and provides irrigation facilities to 47,985 hectares by lift canal irrigation and supply of drinking water to Damonjodi, Koraput, Sunabeda and Jeypore towns.

Table 8.6 is compiled from different sources, due to lack of proper records. The data does not include all displaced people. However, the table reveals that the share of displacement that occurred due to Hirakud Dam in Sambalpur district was the largest. It has displaced around 39 per cent of the total families displaced. Followed by this, the second highest share of families have been displaced by the Upper Kolab Dam in Koraput district, which displaced 23 per cent of the total displaced families. In Koraput district, the number of families displaced may increase as the 44 villages displaced by Indravati project were also from Koraput district. In terms of population displaced, these two dams hold the major shares. However, the impacts of displacement do not confine only to these numbers.

Many sociologists and anthropologists have also documented various qualitative consequences of the displacement in Odisha. Ota (1996) found that in Rengali irrigation project, the percentage of landless families after relocation has doubled. Reddy (1997) found that in the coal mining displacement around Singrauli, the proportion of landless people skyrocketed from 20 per cent before displacement, to 72 per cent after. Pandey (1998) studied seven projects causing displacement during 1950–94 in Odisha and found the problem of common property resources (CPR) in the post-displacement period.

However, due to paucity of data, to measure the magnitude of hunger and deprivation, a popular method used is to collect evidence from the village survey. These are micro-evidences and have their own limitations. They may not state more on the conditions (especially human rights related, for which there is legal support) of the displaced persons. Therefore, a primary survey has been attempted. Though displacement occurred regularly in undivided Koraput and recently in Kalahandi district, keeping the highest proportion of agricultural labour in Kalahandi, the district was chosen for primary survey. The following section examines the mines-led industrialisation and displacement.

Mines-led Industrialisation and Displacement

Most of the contemporary discussion on the development of Odisha give an impression that the state is really backward in terms of industrialisation, but the actual scenario shows that Odisha has been there in the industrial and mining map of the country ever since the inception of mining and industry in India. The modern-day mining in Odisha can be traced back to 1909, when coal was first excavated in the Rampur area of Ib valley. The TISCO mines for iron ore at Gorumohisani and manganese mines in Goriajhar were started during 1910. Followed by this, during 1914, the dolomite and limestone mining has started at Panposh and Bisra respectively. In the pre-independence period, mining of chromites in Baula area had started during 1942. In the post-independence period, a number of initiatives have been taken to establish public sector industries in Odisha. Industries like Rourkela Steel Plant, Hindustan Aeronautics Ltd (HAL) and National Aluminium Company (NALCO) came up in the state and the mining sector expanded significantly. Bauxite mining was started at Panchpatmali hills in Koraput in 1984. During 1985–2000, while the mineral development in the state was very slow yet the production of bauxite by NALCO in Koraput and coal mining by MCL in Ib valley and Talcher area were notable. Until late 1950s, the mines were small and mostly manual, hence the industrial growth in the state was low, the environmental impact was also low, and the industrial led displacement was low. But post-1984 the mineral production has increased very rapidly and during this period, displacement has increased as well. During 2007–08, about 1785.81 lakhs tonnes of minerals valued ₹ 10, 636.70 crore were produced in the state. In 1991–92, the mineral production was estimated at 3,72,00,000 tonnes, which reached to more than 7,49,00,000 tonnes by 2001–02, (Economic Survey of Odisha, 2008). However, this rapid increase in the mineral production cannot be taken as synonymous with industrialisation as the latter has remained at low level. Moreover, despite the emphasis in the successive industrial policies of Odisha for large industries and mining sector, which has resulted an expansion of industries, yet its contribution in terms of poverty reduction has not been much as shown by the data. Hence, the mines-led development has brought negative impacts on the native population, the local economy and the state in the form of displacement. Table 8.7 presents the reserve of ores and minerals in Odisha.

However, the nature of manufacturing activities in the state has been primarily concentrated on Natural Resource-Based Industries, particularly the mineral based industries. As the Annual Survey of Industries (1997–98) pointed out, industries like electricity, basic metal and alloys, non-metallic minerals and food products accounted for about 75 per cent of the total industrial employment and 80 per cent of the net value added. However, on the other hand, these industries also led to large-scale deforestation. The Economic Survey of Odisha reveals that mining and industrial sectors have accounted for nearly 39 per cent of total conversion of forest area to non-forest area and it led to acquisition of 6,22,463.94 hectares of land (Vasundhara, 2005). Since the majority of large industries and mineral resources are located in the Northern and Southern Tribal dominated regions, the incidence of displacement is more likely to happen among the tribals, who lose their major source of livelihood like forest produce, forests, common property resources and revenue lands, etc. Therefore, though tribal people constitute 22.21 per cent of total population

(Census, 2001), the incidence of poverty among them in southern and northern areas are 85.5 per cent and 79.1 per cent respectively. On the one hand, displacement for mining hardly helped the natives; on the other hand, stagnation in agriculture shrunk their base of employment.

Table 8.7 *Data on reserves of ores and minerals in Odisha (in Million Tonne by 1.1.2000)*

Mineral	*Location/District*	*Resources*
Bauxite	Koraput, Rayagada, Kalahandi and Balangir and Small deposits in Phulbani, Nuapada and Sundargarh	1733 (3075)
Beach Sand	Ganjam and Puri	82* (366)
China Clay	Mayurbhanj and small deposits in Malkanagiri	314 (2353)
Chromite	Keonjhar and Jajpur	183* (186*)
Coal	Sambalpur, Jharsuguda and Angul – Talcher	60,987 (2,45,692)**
Dolomite	Sundargarh and Koraput	882 (7349)
Fireclay	Cuttack, Sambalpur and Jharsuguda	177 (706)
Graphite	Bargarh, Sambalpur, Balangir, Kalahandi, Phulbani and Rayagada	4.6 (16)
Iron Ore	Sundargarh, Keonjhar, Mayurbhanj, Jajpur	4,177 (12,317)
Limestone	Sundargarh, Bargarh, Koraput and Malkanagiri	2,224 (1,69,941)
Manganese	Sundargarh, Keonjhar and Rayagada	116 (406)
Nickel	Jajpur and Keonjhar	174 (188)

Notes and Sources: Figures in brackets are the reserves of India, * as per latest figures of Directorate of Geology Odisha, ** as per latest figures of Ministry of Coal, rest of the data are from Indian Bureau of Mines.

In the post-1990s, Odisha adopted three successive industrial and mining policies (1992, 1996 and 2001). These policies envisage the growth of private investment in the sectors of infrastructure development, mineral-based industries and large industrial projects. What is the relevance of such industrialisation in producing hunger and deprivation? What are the consequences of these displacement that need to address from the grassroots level data to find the alterations in the land acquisition act and rehabilitation and resettlement policies? Table 8.8 presents the number of persons displaced due to a few industry and mining projects in Odisha. The results show that around 44,000 of people have been displaced, due to the above-mentioned projects. Apart from this, there are many other projects, which might have displaced people in various parts of the state for which there are not much information. In the absence of proper rehabilitation and resettlement policy, displacement may push the people into hunger and deprivation. Some of the evidences from the primary survey carried on at various levels are presented below:

The survey carried out by Pandey (1996) among tribal households in five villages at Talcher found an increase in unemployment from 9 per cent to 43.6 per cent, accompanied by a large shift from primary to tertiary occupation, and reported reduction in the level of earning up to 50 to 80 per cent among tribes and scheduled castes. Downing (2002) pointed out in some of the survey villages, the proportion of landlessness became almost twice due to displacement by some of the mining projects. As we would show the evidence below, the impact of displacement is more among the SCs/STs who loose their livelihood.

Table 8.8 *Displacement due to some of the industry and mining projects in Odisha*

Projects	*Year*	*Land Acquired (acres)*	*No. of villages*	*Nr. Families displaced*	*Population Displaced*
Hindustan Aeronautic Ltd. in Koraput	1962	7,200	10	1,200	6,000
Bauxite mining in Koraput	1981	5,936	41	788	3,104
Bauxite mining in Sambalpur	1985	2,379	24	1,396	9,111
Steel plants at Rourkela in Sundargarh	1962	19,557	30	2,467	12,335
Steel plant at Gopalpur in Ganjam	1996	7,598	11	2,912	13,892
Aluminum smelter plant at Angul	1985	3,828	3	0	0
Total		46,498	119	8,763	44,442

Source: Tripathy, 2003.

The Structural Inequality and Displacement in Odisha

The previous chapters have shown that structural inequality is one of the perennial features of Odisha's economy. The SC/ST are the most disadvantaged in access to education, health, better paid job, land or even to food and other capabilities enhancing facilities provided through public provisioning. Such precarious situation affects them directly by displacing them and indirectly by destroying the forest, the source of livelihood of most of the SCs/STs. Incidentally, it is found that the projects were undertaken and people were displaced in those area where SCs/STs concentrations are high. Table 8.9 presents the sectorwise diversion of forest area to non-forest use. Out of the total forest area of 2,26,622 hectares in 2003, about 27,274 hectares were diverted for various development projects particularly after the enactment of Forest Conservation Act, 1980. Of the various development projects for which forest land has been diverted, mining (35%) and irrigation (22%) constitute the most. This has a serious implication for hunger and deprivation, especially among the SCs/STs.

Table 8.9 *Sectorwise diversion of forest land to non-forest use (under Forest Conversion Act, 1980)*

Name of the sector	*Number of projects*	*Forest area diverted as on 31 March 2003*	*Percentage to total area diverted from different project*
Irrigation	59	6,092.4829	22.34
Industry	5	2,406.086	8.82
Mining	77	9,469.4384	34.72
Transmission	44	2,723.4265	9.99
Roads and Bridges	25	197.8122	0.73
Railways	5	1,965.0287	7.20
Others	32	4,420.179	16.21
Total	247	27,274.45	100

Source: Vasundhara, 2005.

Table 8.10 presents a conservative estimate of displaced and partially affect people due to displacement during 1951–55. It shows that the incidence of displacement is the highest among the SCs/STs. Out of the number of displaced people, SCs/STs together account for 56 per cent (43% for SCs and 13% for SCs).

Yet as the report by Vasundhara (2005) reveals, only 25 per cent of the displaced persons have been rehabilitated. The resettlement and rehabilitation policies in Odisha have been discussed below in details with micro-evidences.

Table 8.10 *Percentages distribution of displaced people by their social groups in Odisha 1951–1995*

Project	*DPS*				*PAPs*				*Grand total*
	Tribal	*Dalits*	*Others*	*Total*	*Tribals*	*Dalits*	*Others*	*Total*	
Water	1,53,540	42,335	2,14,125	4,10,000	1,54,830	37,986	1,97,184	3,90,000	8,00,000
Percentage	37.45	10.33	52.22	100	39.7	9.74	50.56	100	54.57
Industries	25,937	11,045	34,812	71,794	14,719	13,286	58,270	86,275	1,58,069
Percentage	36.13	15.38	48.49	100	17.06	15.5	67.54	100	10.78
Mines	50,000	15,000	35,000	1,00,000	1,00,000	30,000	70,000	2,00,000	3,00,000
Percentage	50	15	35	100	50	15	35	100	20.47
Wild life	840	Nil	Nil	840	80,250	16,050	10,700	1,07,000	1,07,840
Percentage	100			100	75	15	10	100	7.36
Others	25,000	5,240	1,97,699	50,000	11,000	7,500	31,500	50,000	1,00,000
Percentage	50	10.46	39.52	100	22	15	63	100	6.82
Total	2,55,317	71,620	3,03,697	6,32,634	3,60,799	1,04,822	3,67,654	8,33,275	14,65,909
Percentage	40.36	11.64	48.01	43.16	43.14	12.53	43.96	56.84	100

Source: Development-induced displacement and rehabilitation in Odisha 1951–1995; database on its extent and nature, by Fernadez and Asif.

Rehabilitation and Resettlement Policy

A Macro Picture

Rehabilitation and Resettlement (RR) is not a recent phenomenon in Odisha. The resettlement and rehabilitation programme can be traced back to the Land Acquisition Act of 1894. However, the extent of benefit from the RR policy is very less. Table 8.11 presents the percentages of displaced people who are resettled in Odisha.

Table 8.11 *Percentage displaced people who are resettled, 1954–1995*

Category	*Displaced*	*Resettled*	*% Resettled*	*Back log*	*% Unsettled*
Water resource	3,25,000	90,000	27.69	2,35,000	72.31
Industries	71,794	27,300	38.03	44,494	61.97
Mines	1,00,000	60,000	60.00	40,000	40.00
Miscellaneous	5,46,794	1,92,840	35.27	3,53,954	64.73
Total	5,46,794	1,92,840	35.27	3,53,954	64.73

Source: Fernadez and Asif, 1997.

The table presents only the conservative method, the actual might be higher than the figure presented here. It shows, during 1954 to 1995 around 65 per cent of the displaced people unsettled. It indicates the necessity of an effective and sustainable RR policy.

It shows that only 65 per cent of the displaced people are resettled during 1954 and 1995. The incidence of resettled people is the lowest (40%) among the displaced people in mines.

It is also explored from the Mining and Industrial Update, Odisha, 2006 that till 20 July 2005, there were already 593 mining projects working in the state of which around 100 are located in Kalahandi, Koraput, Rayagada, Malkanagiri, Nuapada and Kandhamal (the southern region), where there is concentration of tribal people. The same source also depicts that till November 2006, there are additional 78 projects pending for the state approval. Such large scale displacement produces the fear more than that of a fatal diseases.

However, after several public pressures, protests and resistance from communities, the Rehabilitation and Resettlement policies of 1994, 2005 and 2006 have evolved. The main features of the policies are presented below:

Features of Land Acquisition Act (LAA) of 1894

Some of the important features of LAA have been discussed here. The LAA of 1894 authorises the government to acquire land compulsorily for 'public purposes' and for which the individual hardship can never outweigh the question of public demand. However, if the involved people object to such forceful acquisition, the district collector may intervene in it, mainly to decide the compensations award and then a notice would be given to the people concerned. Thereafter, the collector would take possession of the land. This may imply that though there is legal right for land acquisitions or displacement, there was no legal right for rehabilitations and this might be the reason that the earlier land acquisition act might have contributed to hunger and deprivation. However, there was provision of compensation in the LAA of 1894. The Act (sections 23 and 24) suggests for the valuation of the land to be acquired and payment of adequate compensation to the legal owner of the property. The valuation may include (1) the market value of land, (2) damage sustained by the person interested and (3) if any standing crops trees or other type of damage suffered by the individual aggrieved.

Therefore, the Act suggests the right of the state over land is superior to the right of the individual and hence the former can take away the property of any individual at any time, of course for public purposes. However, as the term 'public purpose' has not been defined in the Act, the applications of the whole Act becomes arbitrary (Reddy, 1995). Anything put by the Government in the name of 'public purpose' is eligible for acquisition of land. Further, it is the district Collector who has the absolute authority to determine what public purpose is and what is not, and the affected person cannot even question the decision of the collector on the element of public purpose though he can go to the court questioning the amount of compensation. In brief, it may be said that though there was an Act, there are lots of arbitrariness when such act is implemented in practice.

In 1973, Odisha came up with its first Resettlement and Rehabilitation policy. The first mining sector and the revised resettlement policy were introduced in 1994.

The Resettlement and Rehabilitation (R&R) Policies of 1994

During 1994, the Department of Water Resources of Government of Odisha came out with a most comprehensive policy on resettlement and rehabilitation of the project affected person due to water resource development. As documented (Mahapatra, 1995), the 1994 policy is the outcome of the persistent agitations and political pressure. The new policy emphasised, 'before 1994, the widows, divorcees, orphans, physically and mentally handicapped without independent means of livelihood, in addition to the landless, were recognised as displaced persons in the submergence zone. Cash in lieu of homestead land, house construction assistance, maintenance allowance for one year and rehabilitation assistance (2 acres of irrigated land or 4 acres of unirrigated land) or cash grant in its lieu will now be paid jointly to husband and wife in the family concerned'. The new R&R policy refers to the displaced persons including: (i) the landless labourers, (ii) if some persons loose a proportion of their only holding in the village and thereby compelled to lead an uneconomic living (iii) if in a village more than 75 per cent of the total agricultural land is acquired or it is found on due enquiry that partial acquisition of land property in the submergence zone may lead to a socially and economically unviable living, the entire village may be treated as fully submerged village, even if homestead land is not affected or is partially affected; and all the villagers (including those who do not have landed property) will be treated as equivalent to displaced persons. Another special feature of this policy is the construction of the compensatory afforestation, plantation and similar other uses. The displaced persons include the landless labourers in the submergence zone, but the landless person loosing only his house due to acquisition for the project, but located in the village but not displaced, will be entitled to only an equal area of the house site, subject to the maximum of 0.20 acre of land along with house building assistance as applicable. Similarly 'if landless person loses his house in the submerged zone because of the project, he should be considered in all senses to be displaced person and entitled to all benefits a displaced person is to receive for rehabilitation and resettlement'. The contradiction that has been pointed out is 'for the purpose of displacement, the village should be taken as the unit' so that everyone in village gets the benefits.

Odisha government further improved the rehabilitation and resettlement policy during 2006, which is better than the National Rehabilitation and Resettlement policy in many respects (Jena, 2006). But how effective is the policy in the field need to be examined from specially conducted survey.

The objective of the survey is to examine the impacts of displacement. Does displacement result in hunger and deprivation, and thereby results in violation of human rights? If yes, how does it violate? Further, what are the effects of the recent policies of Odisha on the displaced people? To what extent has the reduction of hunger and deprivation been achieved due to new policy?

Micro Study

Here we have presented two case studies (1) Displacement due to the Vedanta Alumina Limited and the (2) Displacement due to POSCO. From these the primary survey for the first one is conducted by us.

Displacement due to Vedanta Alumina Limited: The case study was carried out in the Lanjigarh block of Kalahandi district, where the Vedanta Alumina Limited, a subsidiary of M/S Sterlite Industries (India) Limited (SIIL), plans to exploit the Niyamgiri bauxite (for commercial purposes) reserve located on top of Niyamgiri Hill. The company is to mine the bauxite deposit from the Hill jointly by the Odisha Mining Corporation Limited (OMC) and VAL (Vedanta Alumina Ltd.) as per the lease agreement signed between OMC and VAL in October 2004. Vedanta required 723.343 ha of land for setting up of an Alumna Complex and another 721.323 ha of land was required for bauxite mining. Most of this land are categorised as Reserved Forest and Revenue land. The forest land was composed of forest and grasslands ecosystems of extremely high value as wildlife habitat, whereas the revenue lands were the areas of the settlement of human being and agricultural field.

Vedanta Aluminium has evacuated three villages. Some of these displaced villagers had to leave the place and there is no account where did they go and some others are living in nearby villages and some others are resettled in the rehabilitated village provided by the Vedanta. We could meet the latter two categories of displaced people and took their personal interview.

Experiences of the Displaced People Living in Nearby Villages

There are four categories of displaced people:

(a) People who lost their land but not home, hence got negligible compensation and are not eligible to get a house in the rehabilitated colony. Hence they have to find a place of living near Lanjigarh (main village), where the real estate price is very high.

(b) People who lost their home and received some compensation but did not like to stay in the rehabilitated colony as it is located around 6 to 7 kilometres away from the main village, where all other facilities (health, education, police and postal etc.) are available.

(c) People who lost both land and home and received partial compensation and did not get a house in the rehabilitated colony.

(d) People residing were adjacent to the Alumina factory. They had to leave the place because of the negative externality of the factory and were not paid anything. This group also include the landless labourers, who were completely depending on the villagers as agricultural or non-agricultural labourers.

The loss of food security and livelihood, especially of the marginal farmers and landless agricultural labourers, is the key problem. The farmers have taken generations together to convert the bush and forest cover land into agricultural land but they had to leave their land (rice bowl) and became landless. With the compensation money they received, they could not buy the same quality and amount of land they possessed earlier as demand for land increased. The case of landless labourers are even worse, who were working in the agricultural field are now jobless. As a result, some are migrating to distant places like Rajkot, Kerala to earn their livelihood. In the absence of regular livelihood, the common property resources would have supplemented to an extent but the common property resources are also depleted by the company's work. As people have been displaced and there is no proper account of all the habitants, it has affected the already established community services and social integration.

Experiences of the Displaced People Living in the Rehabilitated Colony

For the company purposes, almost four villages were displaced (Barbhatta and Kinare villages were displaced completely and Sindhi Bahal and Kotdwar were displaced partially). There is also speculation by the local people that another five villages (Khajur Padar, Tangankana, Kasi gadi, Goiputa and Bandguda) are going to be displaced soon as the company discover more mines storage in that region. The company started the survey during 2002 and the displacement began during 2003. Then the people were hardly aware about the aim and objective of the company. However, they were given assurance by the Company that the displaced people will get home for home, land for land and job. However, these promises by the Company were partially met.

Our enquiry found that, around 300 households have been displaced so far, out of which around 150 households have been resettled. The resettlement has started during 2006. Table 8.12 provides some information about the resettled households in the rehabilitated colony. The results show that the cost of access to almost all the facilities including, health, education and other civil rights have increased after displacement. However, unlike the earlier category of case study, a positive thing is that almost all the rehabilitated households have received some amount of compensation. However, the beneficiaries themselves do not know how to use the money in a sustainable manner. Some of these beneficiaries have already spent their money and every now and then are agitating before the company for a permanent job. However, the company is unable to accommodate all the displaced people in jobs. Moreover, the company also choose its workforce in such a manner that it would help the interest of the company rather than be a threat. Perhaps, therefore, even the daily workers were being brought from outside the state.

Another thing that came out clearly in our investigation is that almost all the displaced persons (including those who are living in the rehabilitated colony) are unhappy and agitating against the company. Our investigation reveals that, the company received full support from the local politicians and police authority. So it is not merely the interest of the company but the interest of the local politician, bureaucrats and police authority together went against the displaced persons and, therefore, the company could sustain and function. Therefore, despite existence of RR policy, displacement is contributing to existing hunger and deprivation in Odisha.

Table 8.12 *Nearest education and health institutes*

Education	*Old village*	*New village*	*Health*	*Old village*	*New village*
LP School	1 KM	1 KM	PHC	2 KM	7 KM
UP School	2 KM	7 KM	Medical	2 KM	7 KM
High School	2 KM	7 KM	ICDS centre	1 KM	1 KM

Civil rights and others

Nearest	*Old village*	*New village*		*Old village*	*New village*
Police station	3 KM	7 KM	Electricity	No	Yes
Post office	3 KM	7 KM	Main road	2.5 KM	3 KM
Participate in voting	3 KM	1 KM	Drinking water	1.5 KM (From River Banshi Dhara)	From tube well but polluted water
PDS distribution point	3 KM	7 KM	Bathing water	Niyam giri Stream	—
Bust stop	3 KM	7 KM	Pond	2 KM	0
Forest	1 KM	2 KM	River	Banshi Dhara	Getting polluted
Daily Market	2 KM	7–8 KM	Cultural programme	2 KM	7–8 KM
Bank Services	2 KM	7–8 KM			

Source: Field survey.

Displacement Due to POSCO

On 22 June 2005, the state government of Odisha signed a memorandum of understanding (MoU) with the Korean steel giant, the Pohang Steel Company (POSCO). This is India's largest Foreign Direct Investment (FDI), which also involves a 12-million tonne integrated steel plant, and over 4,500 acres of land for the construction of a new port. The steel plant is expected to affect seven villages in three gram panchayats, namely Dhinkia, Nuagaon and Gadakujang, and will allow POSCO to extract 600 million tonnes of iron ore over the next 30 years. As per the MoU, the state government has agreed to provide almost all facilities required by the company including free encumbrances to POSCO, mining and prospecting licenses, clearances related to matters of forest and environment, and permits for drawing water from the Mahanadi river, defending in various recommendations made in favour of POSCO in the eventuality of litigation in the appropriate judicial, quasi judicial areas and extended similar facilitation of clearances by the state government for all aspects of the project – the steel plant, mines, roads, railways and port. Finally, in 2006, the project was also granted 'Special Economic Zone' (SEZ) status, and the Prime Minister issued a statement saying that land acquisition by POSCO must be expedited. However, the project has great impacts on the people of affected areas.

The Lives and Livelihood Issue

The report of the government and POSCO stated that about 400 families will be displaced by the project, which will be around 2,000 population with the assumption of 5 members per household. But according to the 2001 census, the three panchayats have 3,350 households which add up to 22,000 people who will be affected and displaced. The difference may be due to anomalies in the Government records, which recognise only 438 acres out of 4,000 acres as being privately owned. Other lands are though under betel, cashew and other cultivation by the STs communities for several generations may not be having legal entitlement for the land. This land is not only their living place but also the source of livelihood as an average family involved in cashew farming. About 50 per cent of the families are also involved in pisciculture (mostly prawns). There are many landless families who depend on ancillary employment like making baskets for packaging Paan leaves grown in the area and people living adjacent to the mines (Banspal block) may also suffer in curtailing their source of livelihood and may also face various health related problems.

Negative Externality to Agricultural Growth

It is reported that the project will require 286 million litres of water per day, which will be supplied from nearest Jobra and Naraj barrages of river Mahanadi, as a result of which the farmers who were using the irrigated water from the canals of Taldanda, Machhagaon, Birupa of Cuttack, Jagatsinghpur, and Kendrapada may suffer a great loss and agricultural productivity may come down.

Impact on Environment

Odisha has some of the largest mass-nesting areas in the world for the Olive Ridley Turtle, an endangered species. The proposed POSCO port site lies between the mass nesting beaches of Gahirmata and Devi River mouth where the offshore waters have a high density of turtles from November to April. In 2007, 130,000 of these animals nested at the beaches along the Gahirmata Marine Sanctuary. The port itself will lie just inland from beaches on which turtle nesting occurs, and less than 30 km from the mass nesting beaches of Devi. Lighting, marine pollution, ship traffic and ancillary development of what is currently largely untouched coastline will pose a long-term threat to the species. Similarly, the proposed port will endanger several species of fish in the Jatadhari estuary, additionally impacting the livelihoods of fishing communities.

The project involves the felling of over 2,80,000 trees which will directly affect the dense forest covers in the Gandhamardhan and Malangatoli areas. Mining will also affect the Kandadhar waterfall, a famed tourist destination. The new port might entail erosion , thereby threatening the existing government port in Paradeep. POSCO has applied for environmental clearance for different parts of the project separately rather than as a whole in the hope of expediting clearance.

Implications of SEZ Status

POSCO gets a 10-year tax break as a result of its being granted SEZ (Special Economic Zone) status by the Central Government. Land will also be sold to POSCO at a lower than market price. These impose significant costs to the exchequer. The SEZ status of this project also removes it from the purview of the local panchayat governments, thus further decreasing the control of the villagers over their local environment. Further, the SEZ status also grants immunity to POSCO from adherence to hard-won labour and environmental laws designed to protect employees.

The number of jobs created due to the steel plant is supposed to number 13,000 and it is claimed that another 48,000 indirect jobs are likely to be created. These numbers do not justify the subsidies granted to the project, including the discounted price of ₹ 2,000 ($50) per tonne of iron ore.

Relief and Rehabilitation

As the government record recognises only a small fraction of the total number of affected people are compensated and most of the villagers did not receive the compensation. Instead, the state sponsored violence (through 15 platoons of armed paramilitary forces) against the villagers who opposed the POSCO steel plant. However, the tussle between the local inhabitants and the state police is on. It is unfortunate but not hard to imagine that the POSCO struggle may cost lives of poor and the disadvantaged people in the future as the evidence of many deaths of tribal people in police firing in Kashipur and Kalinganagar show.

Summary

The chapter attempted to enquire how displacement (especially the development-induced displacement) has contributed to the existing hunger and deprivation of Odisha. Development-induced displacement in Odisha had started at least from the later part of the 1940s and is continuing till date. In fact, in almost all decades there have been displacements in the state due to one or the other developmental project. The Export Promotion Zone (EPZ) or the Special Economic Zone (SEZ) has primarily two objectives – the economic growth and the employment expansion. In case of Odisha, there might be some contribution of the EPZ and SEZ to state economic growth and employment but it has brought a huge social cost by directly displacing people and by destroying the common property resources on which people were directly relying for their livelihood. There was no proper rehabilitation and resettlement policy for the displaced people of the state until a revised R&R policy was introduced during 1994. The new R&R policy has some effects in ensuring some amount of compensation for the displaced people but the policy has failed to provide a sustainable compensation to the displaced people and hence the displacement has added to the existing hunger and deprivation in the state, which is also depicted by the micro-level case study.

The political economy aspects of displacement have shown that the SCs/STs are the most affected victims of displacement as most of the projects undertaken are found largely on SC/ST populated areas. Further, as the development project destroyed forests (the major and perhaps the only source of livelihood for the poor and tribal), the ultimate burden of displacement threatened the food security and might be pushing these disadvantaged groups into hunger and deprivation.

The state aligning with 'capitalist' would like to extract the material resources from Odisha, which hits the poor most. The increasing number of developmental projects not only created a fear in the minds of the people, rather in the absence of a sustainable R&R policy, the displaced and unemployment people also become refugee and urban proletariats. These may give birth to lumpinisation in the state. In fact, whatever little was done to help the poor through public provisioning is snatching away by displacement.

So far the present study has attempted to understand the problem of hunger and deprivation in interdisciplinary political economy approach. It examined the problem not only through the dominant political economy approach of distribution of forces of production and relations of production and economic growth; it also incorporates Sen's political economy approach of entitlement failure and capability failure. Further, we have brought the social history of the people of the state to explain the problem. The contributions of social structure and inappropriate state policies to the existing situation of hunger is very high, which may be true in any other Indian states. But in Odisha, it is the social structure that is the fundamental cause of hunger than any other factors. The next chapter conclude the study.

Notes

1. In Odisha, since 1901, there have been as many as 149 medium and large dams, out of which 18 are under construction, ranging in height from 10 to 71 metres and length from 45 to 6245.85 metres (CWC, 1990: 87–90).
2. The area displaced or the number of households or people displaced may slightly differ from study to study. However, such difference would not affect to elicit the larger picture of displacement.
3. The official processing was taken till 1946 when the foundation laying ceremony of the Dam was done but the foundation was laid by Sri Jawaharlal Nehru on 13 April 1948 and the construction of the project commenced since February 1949. The first phase of the Dam was completed during August 1956 and the project was inaugurated by Sri Jawaharlal Nehru on 13 January 1957.

9

Summary and Conclusion

The prime objective of this study is to understand the problem of hunger and deprivation (nature, extent and causes) in the state of Odisha and finding out a suitable policy for reducing the impact of the same in the state. As it is perceived from the experiences, observations and literature, in the contemporary society (21st century) studying hunger independently (which has multiple dimensions) is a difficult task, because hunger is not confined only to food deprivation but contagion with many other non-food deprivations. Hence, to study hunger, it requires studying some other types of basic non-food deprivations as well.

Given that conceptual clarification, the problem of hunger has been examined by analysing the trend and pattern of some outcome indicators, such as calories, protein and fat intakes, income and perception-based poverty, and anthropometrics-based poverty, such as stunting, wasting, underweight and Body Mass Index etc. The problem of basic non-food deprivation has been examined by analysing the outcome indicators pertaining to housing, health and education. To study the causal factors of hunger and deprivation, the study adopted an interdisciplinary political economy approach, as hunger and deprivation are related with many disciplines, such as economics, sociology, political science and so on, but within that the two major causes of hunger and deprivation are (a) inadequate amount of growth, which may be a necessary but not a sufficient condition to reduce hunger and deprivation; and (b) unequal distribution of wealth, which are the product of political economy (the institutional set-up and the process of structural changes in the society). Therefore, the dominant approach adopted in this study is the political economy approach, which relates the existing hunger and deprivation of the state with the process of her structural and institutional changes. The study is focused on Odisha – a state with the highest incidence of below poverty line population among all Indian states. Further the incidence of poverty has not reduced since 1993–94 (as per the estimation of Planning Commission of India).

Apart from conceptual clarification, Chapter 1 examined the trend and pattern of hunger and deprivation in India and Odisha. Higher adversity of hunger and deprivation has been observed in Odisha compared to any other Indian states. The results show that the position of Odisha in terms of hunger and deprivation is worse than almost all other Indian states. At the end, the chapter presented the objectives, hypotheses, theoretical framework, methodology, and data sources for the study. It also set the theoretical framework to study the problem of hunger and deprivation in the context of Odisha, where the conventional political economy of class relations or the role of state in public provisioning, as well as the social hierarchy aspects of hunger and deprivation have been brought in.

First Objective: To Examine the Mass (Inadequate Growth) and Class (Unequal Distributions) Character of Hunger and Deprivation in Odisha

Chapters 2, 3 and 4 dealt with this objective.

Chapter 2 presents the structure of hunger and deprivation within the state of Odisha. In Odisha, the political economy aspects of the problem has been seen through (i) the history of the accumulating and de-accumulating groups; (ii) the higher and lower strata of the society and their access to wealth, health, education and housing; and (iii) the gender and regional dimensions.

This chapter inferred that the levels of both hunger and deprivation are not only high, they have also been the regular features of Odisha economy and society. The reasons are (a) failure of agriculture, (b) the institutional and structural rigidities, and (c) the colonial rule. However, a single most important variable that can explain the higher incidence of hunger and deprivation in Odisha to a greater extent, is the social structure and social hierarchy prevailing in the state, where the SCs/STs women and children faced most disadvantages. The results also show a higher variation of hunger and deprivation across social groups, geographical regions, occupations, educational status, landholding, and dependency ratio. In other words, hunger and deprivation are higher for those who reside in rural areas, who have low human capabilities, who belong to low social groups, who possess less assets (specifically land) and who have less earning members who contribute to employment and livelihoods.

Agriculture has been the major source of income and employment in the state, where 80 per cent of the manpower was engaged but the sector largely depends on uncertain monsoon. The agricultural sector is characterised by disguised unemployment, low labour productivity, and low wages. Moreover, the topology and geography of the state are such that they accentuated the mass hunger and deprivation in Odisha through various natural calamities, such as drought in southern-western part of Odisha and flood and cyclone in the coastal part of Odisha. Agriculture being the main source of livelihood, uncertainty in it leads to a mass hunger and deprivation in the state.

The high levels of hunger and deprivation become a multiple burden on certain sections and communities of the Odisha economy and society due to the existence and functioning of certain institutions. The caste-based institution is one of the dominant mechanisms, the operation of which brings unequal share to the agents of the factors of production and the least share to the labourers. Caste as a social institution works to such an extent that irrespective of the place of living, the SCs/STs are more prone to hunger and deprivation. In that social system, there was also no place for SCs/STs in the traditional power structure. Thirty-eight per cent of the people of the state belong to SCs/STs, who were traditionally excluded from the state affairs and continued to face various exclusions. Hence, they got marginalised over a period of time and became victims of hunger and deprivation. Further, in the caste-based institution, the division of work was such that the least remunerated 'occupation' has been assigned to lower strata of the society (broadly the SCs and STs). The findings here suggest that irrespective of place of living, incidence of hunger and deprivation among SCs/STs are higher. Historically, these communities did not possess adequate wealth (notably land), neither they had access to any valuable socio-cultural institutions like school, temple and other public places, which might have been the reasons for higher levels of hunger and deprivation among them. Therefore, these communities are in need of special attention from a developmental state but as it will be seen in later chapter that the problem of social exclusion is pervasive in the state. So is the case of women in the Odisha society, who have less access to education, land and other public spheres. Therefore, within the mass problem of hunger and deprivation in the state, these sections/communities had to face multiple incidences of hunger and deprivation.

Geographically too, these communities concentrate in specific regions (southern and part of northern Odisha), far away from the centre of decision-making of the state. Lack of awareness among the people of this region on one hand, and lack of interest of the state to this region on the other, together failed to bring development in a sustainable manner to reduce hunger and deprivation. Instead, certain investments were made (mostly in mines and coal), which are highly extractive and unsustainable in nature. That added to the hunger and deprivation through displacement and destroying the main and sometimes the only source of livelihoods of these people.

Apart from the state's systemic failure, which was responsible for the creation of hunger and deprivation, the British land revenue system was an additional breeder of hunger and deprivation through their various exploitative land revenue policies historically.

In the post-independence era the problem could have been ameliorated to a great extent but so far the social elites are the political elites and are the principal agents of policy matters. They had little political will or desire to bring the disadvantaged sections to the mainstream. This systematic negligence might be one of the reasons that the incidence of hunger and deprivation continued to remain high in Odisha and in fact, remained relatively the highest in the country.

In brief, it seems the strong social hierarchy, which existed in the socio-cultural history of Odisha is still persisting in different forms even now. This remains an important barrier to reduce hunger and deprivation in the state.

Chapter 3 analysed the growth and structural transformation aspects of hunger and deprivation in Odisha. The analysis shows a high degree of product transformation from primary to service sector, but the secondary sector remained passive in this process. Further, the employment transformation did not follow the output transformation. The growth rate of output since 1970s made the structural transformation (in output) to happen, however, that growth has failed to trickle down and reduce hunger and deprivation. In other words, the rate of change in output has not much effect on the rate of change in poverty, as there is less employment growth in the primary sector.

The rate of growth in output was higher than the rate of growth in employment during the previous 27 years in Odisha. Further, the rates of annual compound growth rates of both output and employment were found to be higher in service sector than in primary sector since the 1990s. Since most of the workers are dependent on primary sector (63%) and the rate of growth is lower in the primary sector, poverty situation has worsened. Further, the study of the employment transformation reveals that there is negligible shift of workforce from primary to secondary to service sectors. The relatively more dependency on primary sector might be due to the less impressive growth and employment in the secondary sector.

The low and negative employment elasticity exhibits the lack of absorptive capacity or displacement of employees from the particular sector. Hence, the rate of change in employment did not bring about reductions in poverty. However, the growth in employment during the late 1970s to early 1990s perhaps accounts for some reduction in poverty in the state. But the momentum in growth in employment in subsequent period did not keep pace. However, the analysis of output and employment shows that the poverty reduction is more sensitive to the primary sector growth.

In brief, the analysis of the output shows that the primary sector has a scope to grow and the analysis of employment shows that the poverty reduction is more sensitive to the primary sector growth. Hence, the primary sector development ought to be the primary focus of the policy makers *in Odisha* to tackle the worsening situation of poverty and hunger *in Odisha*.

While the empirical exercise has been seen in terms of dual model of the development economies, it shows that there is almost no transfer of labour from rural to urban sector. There is transfer of some labour force from agriculture to non-agriculture, yet the percentage of workers in the agricultural sector still remains high. In other words, contrary to what one would expect from theory, the labour productivity in the agricultural sector or in the primary sector has not increased in Odisha. Therefore, it can be suggested that, the Odisha's economy is still waiting for a big push, especially in the secondary sector for overall development of the state and for poverty reduction. Chapter 3 responded to the first objective and part of the first hypothesis that hunger and deprivation are due to the problem of economic growth. The analysis suggests that hunger and deprivation in Odisha can be related more to the growth (both employment and output) of primary sector rather than overall growth rate.

The regional analysis within Odisha reveals interesting results. For most of the indicators analysed, Southern region of Odisha remained at a disadvantaged position and Coastal region remained the best performer. The Northern region remained in the middle of the other two regions in terms of overall performance. In Southern Odisha, poverty level is the highest, casual labour is much higher, share of STs population is much higher and literacy level is much lower. Due to these interrelated factors, hunger and deprivation are much higher in Southern region of Odisha.

Among all the major factors, social group is the most disadvantaged. The STs population has the highest level of poverty regardless of regions. Of course, poverty level among STs population in the Southern region is much higher as compared to their counterparts in the other two regions. Next in order is the SCs population who are also disadvantaged, especially in Southern region followed by Northern region. On the whole, Southern region of Odisha and STs and SCs communities play a dominant role in the overall deteriorating situation of hunger and deprivation in Odisha.

Given this background and growth exercise, Chapter 4 examines to what extent the land distributions contributed to the existing hunger and deprivation. No clear relationship has been found between the operational holdings and income poverty, which might be due to lack of overall agrarian reforms. It was inferred that, land reforms or landholdings are necessary but not sufficient for poverty reduction. While one would disagree with the concentration of land, the point derived from the case studies of Odisha, Kerala and West Bengal is that ensuring equitable landholding alone is not a panacea for poverty reduction; rather poverty reduction through land reforms is conditional on complementary economic and social policies of the state.

Key findings from this chapter are presented below:

(1) First, the significant growth in agricultural output in Kerala, Odisha, West Bengal, Punjab, Haryana and Uttar Pradesh has not contributed to poverty reduction during the 1960s. Chapter 3 also shows that during the 1990s, Odisha had healthy growth but poverty has not reduced much.

(2) Second, the social justice approach to poverty reduction has been implemented in all the three states of Kerala, West Bengal and Odisha. In Odisha, though the share allocated to the social sector is comparable with that of Kerala and better than that of West Bengal, the per capita assistance in Odisha remained very low. And the large dependence on public provisioning, together with insufficient growth is a threat to the sustainability of such schemes.

(3) Third, in Odisha, it becomes extremely difficult for the households with high incidence of poverty to even access public support. Therefore, adopting the alternative approach to social justice, by providing land entitlements to the landless through land reforms is important from a poverty reduction point of view. In Kerala and West Bengal, however, many other conditional factors played an important role in facilitating land reforms which led to poverty reduction.

In Kerala, the overall rural transformation (particularly social and economic transformation) was made possible not by land reforms alone but rather by combining land reforms with supportive measures in labour and credit markets including the agricultural investment (both private and public). In West Bengal also, it was not land reforms which directly reduced poverty, but rather, as pointed out by the Economic Review of West Bengal (cited in Government of West Bengal, 2004) and Roy and Pal (2002), agricultural investment (public investment for minor irrigation and private investment for bore wells), non-land inputs such as credit, seed, fertiliser, and irrigation facilities which played a significant role in bringing about agrarian change and reducing rural poverty in West Bengal (see Lieten, 1996: 121). Although Odisha had an advantage over Kerala and West Bengal in terms of both occupational distribution and land distribution at the initial point, this has not only failed to reduce poverty but the pattern of land distribution has gradually worsened. Such conditions can be attributed to deficiencies in rural credit systems (Sarap, 1991), agricultural investment and irrigated areas. These may have contributed to keeping the agricultural wages at a low level in Odisha, unlike in Kerala and West Bengal. In other words, comprehensive agrarian reforms may have been able to save the smallholder from becoming landless. Further, the failure of the institutional system to retain the land with the smallholder leads one to infer that the cause of poverty in Odisha lies not only with weak policy but also in its institutional structure.

From these lessons, we can make the following concluding policy points:

(i) As Odisha's experience of land possession/landlessness has shown, landholding alone does not ensure poverty reduction. Rather, irrigated and productive landholding is important as shown by West Bengal and Kerala. Therefore, efforts should be made by both private and public agents to transform the land into productive assets.

(ii) In a developing economy concentration of land is not desirable, despite the fact that it may facilitate economies of scale and increase growth of agricultural output with the help of technology. This is because the introduction of technology can cause unemployment, and therefore, in the absence of strong and adequate redistributive measures, worker households may be driven to a state of poverty. An examination of the NSSO data also shows that concentration of land is occurring in all the three states but it is most visible in Odisha. Therefore, a minimum amount of landholding should be considered as a basic need for every agricultural household in an economy. This may justify institutions stipulating that, for the welfare of the people, if a household (or person) possesses only the specified minimum amount of land, then legally, neither he/she be allowed to sell nor any other household (or person) be allowed to buy the piece of land. This study argues that the scenario 3 type of agrarian structure is better, at least from a poverty reduction perspective.

Raj (1974: 11) stated 'Even a toe-hold on land means a great deal to those who are seriously handicapped because they have none', but the attempt has been a failure in case of Odisha due to lack of other supportive measures. Hence, Raj's statement should be qualified by saying that other factors – at least a minimum amount of irrigation (or an adequate amount of water) is necessary to make even a toehold land an effective instrument for poverty reduction.

Second Objective: To Examine Social Exclusion Dimensions of Hunger and Deprivation in Odisha

The conventional approach to political economy mostly deals with the distribution of forces of production and relations of production, which apparently study the history of class conflicts between the bourgeoisies and proletariats. It is the persistent class struggle that determines the political and economic power of a person and subsequently state whether a person will face hunger and deprivation. But in India, the socio-cultural history is such that caste as an institution divided the society into different groups in a social hierarchical order and that order determines the political economy power of a person. Apparently, the higher is the order in a social hierarchy, the higher is the wealth, education, occupation, income and well-being, and hence lower levels of hunger and deprivation and vice versa. Therefore, hunger and deprivation among the socially lower order bear a historical reason. In the social history of Odisha some groups have been excluded from various mainstream activities and this social exclusion reinforces the hunger and deprivation among them.

Chapter 5 examined how the caste-based social exclusion contributes to the existing hunger and deprivation in Odisha. From the review of literature and empirical data analysis, we can infer that social exclusion (caste based) does prevail in the state to a great extent and that might be the reason why the incidence of hunger and deprivation among the SCs and STs are higher. As shown below, the macro-level analysis reveals the precarious socio-economic and demographic conditions of SCs and STs households in Odisha:

1. As regards caste composition of the population in Odisha, there are 22 per cent of STs population, 17 per cent of SCs population and 61 per cent of 'other' social group. The occupational structure shows that the share of agricultural labourer among the STs is 47 per cent as compared to 46 per cent among SCs and 26 per cent among 'other' caste group.
2. During 2004–05, the incidence of poverty is the highest among the STs households (75%) followed by SCs households (53%) and 'other' category of households (25%).
3. In education too, the STs and SCs remained backward with 37 per cent and 56 per cent of literacy rate respectively compared to 74 per cent among 'other' category.
4. In terms of housing too, most of the dilapidated houses are found among the SCs with 16 per cent followed by other castes (10%) and STs (8%).
5. The infant mortality rate is also the highest among the SCs/STs group as compared to 'other' castes in Odisha.

The social exclusion is so deep that even after 60 years of constitutional efforts a segment of people remained backward in terms of almost all basic necessities. It is also inferred from the micro-level case studies that there is prevalence of exclusion (in explicit or implicit form) of Dalits and SCs/STs from various socio-economic spheres, such as exclusion in the socio-cultural institutions like educational institutions, religious institutions like temple and economic institutions like labour market, etc. And that might be the reason why Dalits and SCs/STs are more prone to hunger and deprivation compared to any other community of other castes.

So far the study has inferred that neither the growth (output and employment) nor the distributive justice is enough to reduce mass hunger and deprivation in the state. Further, both types of employment and possession of land have shown a pattern where the SCs/STs are more vulnerable to hunger and deprivation. And this chapter further confirms that within SCs/STs, there are socially excluded groups such as Ganda, Ghansi, Chamar, Hadi, etc. whose access to capability enhancing facilities are restricted by the structural factors of caste system. While the SCs/STs remain at lower level both in economic power and social ladder, the women mass from these communities are further pushed down by the cultural norms, where there are restricted access of women to education, health and other capability enhancing programs. Therefore, the role of public provisioning is crucial here to reduce hunger and deprivation among such disadvantaged groups.

Third Objective: To Examine the Role of State and Institution (Public Action) to Reduce Hunger and Deprivation and Formulate Policy for the Same for the State

Chapter 6 aims to understand 'to what extent functioning of the various public provisioning schemes are helpful in reducing the intensity of hunger and deprivation in Odisha'. A large number of schemes have been introduced in the state – most of them are central schemes, some are state assisted schemes and recently some schemes are also introduced by some voluntary and non-governmental organisations. These schemes have been broadly classified into five groups, such as nutritional support, educational support, health support, housing support, support for the old person and women. Despite the functioning of these schemes, the incidence of hunger and deprivation has remained at a high level in Odisha in general, and higher among SCs/STs of the state in particular. At the macro-level, some of the factors responsible for the ineffectiveness of these schemes are the inadequate entitlements and inadequate coverage of these public provisioning leading to a slow pace of ameliorating hunger and deprivation.

A micro-survey was also conducted to understand the details of the functioning of these schemes. The analysis is based on two levels – the institutional and the household. At the institutional level, it found that various institutions do exist, though less in numbers but their functioning to achieve the desired goals are very poor. Such poor performances are due to several factors – inadequacy of the institution (such as inadequate infrastructure and manpower) to manage the task, influence of the few groups for the functioning of the institution and lack of regular government attention, etc. From the household perspective, some of the reasons for the ineffective functioning of the schemes are inadequate coverage, irregular and insufficient entitlement, wrong targeting of household, wrong targeting of commodities and meagre assistance (subsidy). Therefore, when the support-led development has been considered an alternative to achieve development quicker, there must be adequate coverage with adequate amount.

The study concludes that, at the initial period itself, the level of hunger and deprivation remained higher in Odisha than any other Indian states. In the absence of effective policies, the problem is getting accentuated. The incidence became more visible in certain regions, such as among some social groups, occupation groups and among the uneducated. In certain cases, all these variables (backward region, lower social groups, agricultural labourer and uneducated household) are found to be existing together and hence they bear the multiple incidence of hunger and deprivation. Their counterparts (the developed region, higher social groups, capitalist and higher educated household) get rid of hunger and deprivation in multiple ways. The investigation also suggests that the society and economy are ruled mostly by higher castes, which hold the political, economic and social power and make decisions favouring the non-poor. Hence, the state interest gets undermined when their self interest gets hampered and therefore, the state might have shown a tardy progress in reducing hunger and deprivation. However, some improvements have been achieved due to the contribution of democracy, which also needs to be well developed in every sphere for the rapid amelioration of structural and institutional changes in the state.

Fourth Objective To: Examine the External Sources of Hunger and Deprivation in Odisha

As mentioned in the previous sections, hunger and deprivation in Odisha are largely due to absence of distributional justice, which makes the socially excluded groups more vulnerable. In addition, the socially excluded groups do not benefit from the public provisioning. This further gets intensified with the external shocks like natural calamities and displacement. Therefore, a greater incidence of hunger and deprivation is found among them. This then leads to a vicious circle for the vulnerable populations even if the economy may be growing.

Chapter 7 examined the drought (famine), flood and cyclone and their impacts. The descriptive analysis found that, while the western and southern part of Odisha are more prone to famine or drought; the coastal Odisha is more prone to flood and cyclone. While drought and flood are the regular phenomenon, cyclone, though not a regular phenomenon, can do much harm to the society and economy. In fact, the reviews of some studies and empirical data reveal an increasing trend in loss of per capita value (in rupees) due to natural calamities. In spite of that, there are no preventive or permanent solutions to the problem. Similarly, though it has been observed that the coastal Odisha is regularly hit by flood, the efforts are too less either for preparedness or for rehabilitation measures. In recent years, the occurrences of famines have been avoided due to general improvements in the administrative system and timely flow of information that reaches people faster. However, the bureaucratic delays led to forced migration and even to starvation deaths in the drought affected areas in the western and southern parts of Odisha. However, such problem can be considered as class problem – the economically backward people are more affected by draught than the economically better-off people. Moreover, the coping mechanism and survival strategy for the draught-affected people become more difficult in the absence of assets. In Odisha, it was also observed that lack of adequate physical infrastructure, and educational backwardness still determine the incidence of hunger and deprivation in the drought-affected region, which the Government alone can ameliorate. The chapter infers that external shock is responsible not only for the creation of hunger and deprivation in the state, it also remains a hindrance in state's economic progress. The natural calamities disproportionately affect the already vulnerable, and push them into vicious circle of hunger and deprivation even if the economy may be growing. The chapter suggests that the constructions of small and medium dams/reservoirs may partly solve the dual problems of drought and flood.

Fifth Objective To: Examine the Displacement Induced Hunger and Deprivation

Chapter 8 examines the issues related to development-induced displacement in Odisha. The analysis suggests that in Odisha there have been huge social costs of major development projects. The impact is felt disproportionately among the SCs/STs, Dalits or untouchables. The question of land and displacement directly affect these communities due to greater demand for land as well as mineral and forest resources located in regions inhabited by these population groups. Due to development-induced displacement, these communities loose natural resource-based livelihoods and land. Under the new policies of economic liberalisation, there has been rapid increase in land alienation by the state for private industries and mining companies, including the promotion of SEZs. This has further aggravated the situation leading to heightened levels of hunger and deprivation among the poor, mostly the socially excluded groups.

The displacement not only affects the current agrarian situation by reducing output and increasing unemployment, it also reduces the capacity of the growth of industry in the state in future as the raw materials needed for industrialisation is draining away. So far, there are not many instances in Odisha that the already settled mines companies might have established manufacturing firms with vertical integration. Rather, the main motives of the company is just extraction of iron ore or coal and then drain these to other countries to develop the manufacturing sector like tractor, automobiles and so on. In the absence of such vertical integration, the state becomes the victim and the displaced unemployed also move towards lumpenisation.

This chapter examines how the displacement in the name of development is producing hunger and deprivation in Odisha. The benefits of development do not often reach displaced people. In the process of development, the local inhabitants are the victims of irrigation projects, hydro- and thermal- electricity projects, mineral-based industrial project, etc. Yet there is also a long list of projects, which may displace more people if the projects get approval. In the absence of a sustainable rehabilitation and resettlement policy, the displaced people may fall into the net of hunger and deprivation.

The results suggest that development-induced displacement in Odisha had started at least from the later part of 1940s and is continuing till date (in almost all decades there have been displacement in the state due to one or the other developmental project). In the last decade, the promotion of SEZs led to huge social costs by directly displacing people and by destroying the common property resources on which people were directly relying for their livelihoods. There was no proper rehabilitation and resettlement (R&R) policy for the displaced people of the state until a revised R&R policy was introduced during 1994. The new R&R policy has some effects in ensuring some amount of compensation for the displaced people but the policy has failed to provide a sustainable compensation to the displaced people and hence the displacement has added to the existing hunger and deprivation in the state, which is also depicted by the micro-level case study.

The political economy aspects of displacement shows that the SCs/STs are most affected victims of displacement as most of the projects undertaken are found largely on SCs/STs populated area. Further, as the development project destroyed forests (the major and perhaps the only source of livelihoods for the poor and the tribal), the ultimate burden of displacement threatened the food security and might be pushing these disadvantaged groups into hunger and deprivation.

So far, the study has attempted to understand the problem of hunger and deprivation in interdisciplinary political economy approach. It examined the problem not only through the dominant political economy approach of distribution of forces of production and relations of production; it also incorporated Sen's political economy approach of entitlement failure and capability failure. Further, we have discussed the social history of the people of the state to explain the problem. The contribution of social structure to the existing hunger is very high, which may be true in any other Indian states. But in Odisha, the social structure is the fundamental cause of hunger.

Since there are regional and social disparities, it is required to reduce the disparities to reduce the higher incidence of hunger and deprivation in the state. The disparities can be reduced through land redistribution followed by other agrarian reforms (as landlessness is increasing and as there is low productivity and large dependents in agriculture), through effective implementation of public provisioning and through positive discriminations.

Appendix Tables

Appendix A1.1 *Calorie intakes by the people of Odisha and her position compared to major Indian states in rural and urban sector, 1973–74 to 2004–05*

	Rural						*Urban*					
State	*1972–73*	*1977–78*	*1983*	*1993–94*	*1999–00*	*2004–05*	*1972–73*	*1977–78*	*1983*	*1993–94*	*1999–00*	*2004–05*
ORI	2,494	2,674	2,624	2,740	2,635	2,512	2,845	3,033	2,735	2,754	2,802	2,596
AP	94.8	89.8	94.8	107.1	105.3	101.49	106.2	98.0	109.8	112.2	111.7	106.00
ASM	96.2	96.7	104.0	113.9	112.5	99.88	106.6	100.1	111.4	108.3	106.5	100.12
BIH	89.7	91.4	94.9	103.9	99.9	98.13	105.0	101.4	104.5	103.3	105.9	96.76
GUJ	93.2	96.6	100.2	110.9	107.1	105.55	104.8	111.5	111.3	110.6	111.3	106.57
HAR	62.0	76.7	82.1	88.1	87.0	91.75	94.7	61.4	98.7	105.3	105.1	104.38
H.P	67.5	81.4	78.8	94.0	86.2	NA	76.9	104.8	93.4	94.5	87.1	NA
J. K	63.3	77.6	82.1	86.9	80.7	NA	92.3	93.5	100.8	93.4	47.1	NA
KAR	90.6	88.5	92.7	106.4	105.1	110.37	118.2	108.9	104.2	110.8	112.3	108.85
KER	127.9	101.5	111.3	111.8	105.9	98.55	132.1	118.4	107.0	112.6	112.2	102.45
MP	82.3	92.1	89.8	101.6	102.9	105.28	102.1	104.3	103.9	107.7	107.6	108.30
MAH	105.2	97.8	97.8	112.9	105.4	104.45	115.5	116.3	109.9	113.2	112.8	114.82
PUN	57.1	74.2	78.7	91.1	89.2	90.92	81.8	106.6	105.3	107.2	105.1	99.31
RAJ	73.1	73.0	86.0	88.7	87.0	92.56	96.6	107.8	98.2	101.8	97.7	100.39
TN	102.0	103.0	112.5	116.7	116.3	109.50	123.6	110.5	102.9	116.4	111.7	108.44
UP	77.5	81.9	87.0	94.5	90.3	91.58	105.3	111.1	108.2	105.3	107.4	99.92
WB	103.9	94.4	103.7	100.3	102.4	98.70	109.4	83.4	109.8	106.5	107.9	105.23
IND	87.3	89.8	94.4	102.1	98.8	98.90	108.0	103.7	106.3	108.3	106.3	104.89

Notes: Odisha's figured have been expressed in K.Calories. Rest of the figures are Odisha's consumption as percentages of other state's consumption.

Source: Various NSSO reports.

Appendix A1.2 *Protein intake among all the Indian states in rural and urban sector, 1972–73 to 2004–2005*

	Rural						*Urban*					
State	*1972–73*	*1977–78*	*1983*	*1993–94*	*1999–00*	*2004–05*	*1972–73*	*1977–78*	*1983*	*1993–94*	*1999–00*	*2004–05*
ORI	62.38	68.2	65	65.6	62.1	59.9	69.87	89.6	68	69.7	70.5	67
AP	93.3	79.5	94.2	103.5	101.5	96.93	107.7	92.1	109.7	113.9	113.7	107.37
ASM	93.3	82.0	103.2	109.2	106.5	93.45	99.6	93.9	111.5	107.9	103.2	99.11
BIH	76.8	77.6	80.2	87.4	85.1	82.85	90.5	93.4	90.7	93.1	94.8	88.04
GUJ	86.0	87.2	89.0	95.2	92.5	90.90	97.4	116.1	101.5	103.3	105.4	95.58
HAR	55.0	66.0	66.3	67.0	66.8	69.98	83.3	41.1	80.0	89.7	91.9	90.54
H.P	57.4	69.1	64.4	74.1	68.2	NA	65.2	107.7	79.1	82.6	77.6	NA
J. K	62.4	63.4	71.4	69.1	66.6	NA	89.2	90.9	90.7	81.8	58.8	NA
KAR	87.1	86.0	91.5	95.9	92.5	99.50	120.5	123.1	100.0	107.1	108.1	104.69
KER	129.3	107.4	112.1	103.5	94.4	85.45	126.8	160.3	107.9	106.7	102.0	93.06
MP	73.1	79.4	75.6	83.6	85.9	82.39	91.5	238.3	86.1	94.8	95.1	93.84
MAH	91.2	83.2	86.7	95.6	88.5	86.44	101.1	124.1	95.8	102.7	103.5	105.02
PUN	58.6	63.6	65.7	70.7	69.9	72.78	79.7	109.9	88.3	91.7	89.6	87.01
RAJ	59.0	58.7	69.1	66.0	64.6	69.09	79.1	106.3	76.4	84.7	81.5	85.68
TN	101.4	94.3	112.1	112.5	111.5	107.16	125.3	103.0	121.4	116.2	110.3	110.20
UP	65.6	68.8	71.4	74.1	71.0	72.96	89.0	109.3	87.2	89.1	92.8	84.07
WB	98.8	78.8	100.0	96.9	97.9	93.59	95.9	74.7	103.0	101.5	104.3	99.11
IND	79.6	78.1	84.4	87.5	84.6	84.60	98.6	100.8	97.1	99.3	98.6	95.85

Notes: Odisha's figures are expressed in grams. Rest of the figures are Odisha's consumption as percentages of other state's consumption. *1972–73 estimated by using conversion factor of (per capita = .8 of PCU).*

Source: Various NSSO reports.

Appendix A1.3 *Fat intakes among all the Indian states in rural and urban sector, 1972–73 to 2004–2005*

	Rural						*Urban*					
States	*27th*	*32nd*	*38th*	*50th*	*55th*	*61st*	*27th*	*32nd*	*38th*	*50th*	*55th*	*61st*
ODISHA	11.13	16	18	18.5	20.2	22.1	29.625	49.2	28	34.2	33.4	34.4
AP	42.2	37.0	60.0	54.4	55.3	53.13	76.0	73.9	73.7	79.5	65.88	65.03
ASSAM	57.4	43.4	85.7	72.5	74.3	68.21	93.3	88.8	96.6	92.2	71.37	77.30
BIHAR	49.7	66.7	72.0	64.5	61.2	62.25	91.9	109.1	90.3	85.9	80.1	69.49
GUJARAT	21.8	29.2	32.1	31.5	30.3	35.08	40.3	69.2	45.2	48.0	40.73	44.27
HARYANA	18.9	29.6	30.0	27.7	27.7	32.45	56.2	83.1	46.7	56.6	48.34	51.65
HP	19.5	34.6	32.1	33.1	29.7	NA	34.3	88.0	41.8	50.4	38.52	NA
J&K	25.6	29.9	39.1	32.0	28.3	NA	71.6	74.5	56.0	47.3	40.53	NA
KARNATAKA	37.9	35.6	56.3	52.1	44.6	52.87	72.7	103.6	65.1	74.0	60.84	64.78
KERALA	46.1	27.3	46.2	45.3	41.5	42.83	87.5	100.4	63.6	74.3	62.2	60.25
MP	41.4	53.7	54.5	52.4	52.1	50.92	69.1	93.2	66.7	69.1	62.78	64.66
MAH	36.5	40.0	50.0	44.2	41.0	42.83	56.6	89.1	93.3	58.4	52.11	56.12
PUNJAB	17.6	25.7	28.1	24.9	27.7	30.48	45.0	83.2	46.7	51.7	47.51	46.36
RAJ	19.2	25.9	34.0	28.0	30.2	34.91	49.9	93.4	48.3	53.5	44.24	49.86
TN	48.4	47.8	69.2	22.9	55.2	59.89	100.4	80.0	82.4	81.8	62.55	67.72

Cont'd...

...Cont'd

UP	30.9	48.0	48.6	41.4	42.9	47.22	66.8	128.8	68.3	67.1	59.86	60.99
WB	68.5	43.2	81.8	69.8	68.0	67.79	75.2	59.7	80.0	82.2	68.3	71.67
ALL	36.3	41.9	52.9	47.3	45.0	50.23	64.6	84.1	65.1	66.3	55.02	59.11

Notes: Odisha's figured have been expressed in grams. Rest of the figures are Odisha's consumption as percentages of other state's consumption. 1972–73 estimated by using conversion factor of (per capita = .8 of PCU). 1972–73 estimated by using conversion factor of (per capita = .8 of PCU).

Source: Various NSSO reports.

Appendix A1.4 *Percentage Population in Income Poverty by States 1973–74 to 2004–05*

State		*1973–74*	*1977–78*	*1983*	*1987–88*	*1993–94*	*1999–00*	*2004–05*
ORI	R	67.3	72.4	67.5	57.6	49.7	48.01	46.8
	U	56.3	53.6	50.6	44.1	41.6	42.83	44.3
	C	66.18	70.07	65.29	55.58	48.6	47.2	46.4
AP	R	48.1	38.11	26.53	20.92	15.92	11.05	11.2
	U	50.61	43.55	36.3	40.11	38.33	26.63	28.0
	C	48.86	39.31	28.91	25.86	22.19	15.77	15.8
ASM	R	52.67	59.82	42.6	39.35	45.01	40.04	22.3
	U	36.92	32.71	21.73	9.94	7.73	7.47	3.3
	C	51.21	57.15	40.47	36.21	40.86	36.09	19.7
BIH	R	63	63.3	64.4	52.6	58.2	44.3	42.1
	U	51.8	48.76	47.33	57.7	34.5	32.9	34.6
	C	61.91	61.55	62.22	52.13	54.6	42.6	41.4
GUJ	R	46.35	41.76	29.8	28.67	22.18	13.17	19.1
	U	52.57	40.02	39.14	37.26	27.89	15.59	13.0
	C	48.15	41.23	32.79	31.54	24.21	14.07	16.8
HAR	R	34.23	27.73	20.56	16.22	28.02	8.27	13.6
	U	40.18	36.57	24.15	17.99	16.38	9.99	15.1
	C	35.36	29.55	21.37	16.64	25.05	8.74	14.0
HP	R	27.42	33.49	17	16.28	30.34	7.94	10.7
	U	13.17	19.44	9.43	6.29	9.18	4.63	3.4
	C	26.39	32.45	16.4	15.45	28.44	7.63	10.0
J & K	R	45.51	42.86	26.04	25.7	30.34	3.97	4.6
	U	21.32	23.71	17.76	17.47	9.18	1.98	7.9
	C	40.83	38.97	24.24	23.82	25.17	3.48	5.4
KAR	R	55.14	48.18	36.33	32.82	29.88	17.38	20.8
	U	52.53	50.36	42.82	48.42	40.14	25.25	32.6
	C	54.47	48.78	38.24	37.53	33.16	20.04	25.0
KER	R	59.2	51.5	39	29.1	25.8	9.4	13.2
	U	62.2	59.5	48.7	43.4	24.6	20.3	20.2
	C	59.79	52.22	40.42	31.79	25.4	12.7	15.0
MP	R	62.7	62.5	48.9	41.9	40.6	37.1	36.9
	U	58.3	62.1	54.6	48.2	48.3	38.4	42.1
	C	61.78	61.78	49.78	43.07	42.5	37.43	38.3
MAH	R	57.71	63.97	45.23	40.78	37.93	23.72	29.6
	U	43.87	40.09	40.26	39.78	35.15	26.81	32.2
	C	53.24	55.88	43.44	40.41	36.86	25.02	30.7
PUN	R	28.2	16.4	13.2	12.6	11.9	6.4	9.1
	U	27.7	27.6	23.9	12.9	11.4	5.8	7.1
	C	28.15	19.27	16.18	13.2	11.8	6.2	8.4
RAJ	R	44.76	35.89	33.5	33.21	26.46	13.74	18.7
	U	52.13	43.53	37.94	41.92	30.49	19.85	32.9
	C	46.14	37.42	34.46	35.15	27.41	15.28	22.1

Cont'd...

...Cont'd

TN	R	57.43	57.68	53.99	45.8	32.48	20.55	22.8
	U	49.4	48.69	46.96	38.64	39.77	22.11	22.2
	C	54.94	54.79	51.66	43.39	35.03	21.12	22.5
UP	R	56.4	47.6	46.5	41.1	42.3	31.2	33.4
	U	59.5	57.1	50.3	45.2	35.4	30.9	30.6
	C	57.07	49.05	47.07	41.45	40.4	31.2	32.8
WB	R	73.2	68.3	63.1	48.3	40.8	31.9	28.6
	U	34.5	38.7	32.2	32.8	22.4	14.9	14.8
	C	63.43	60.52	54.85	44.72	35.6	27	24.7
IND	R	56.4	53.1	45.6	39.1	37.3	27.1	28.3
	U	49.2	47.4	42.2	40.1	32.4	23.6	25.7
	C	54.88	51.32	44.48	38.36	35.9	26.1	27.5

Source: Complied from Planning Commission of India and www.indiastat.com.

Appendix A1.5 *Percentage Distribution of Households by the type of response on the question whether all members of the household get two square meals a day by states1983 to 2004–05*

	Rural								*Urban*							
	Some months				*No month*				*Some months*				*No month*			
States	*1983*	*1993–94*	*1999–00*	*2004–05*	*1983*	*1993–94*	*1999–00*	*2004–05*	*1983*	*1993–94*	*1999–00*	*2004–05*	*1983*	*1993–94*	*1999–00*	*2004–05*
ORI	31.8	14.9	6.3	4.8	5.02	0.5	1.5	1.3	11.33	3.3	0.9	0.4	0.84	0.7	1.6	0.2
AP	14.7	1.7	1.7	0.6	0.8	1.2	0.2	0.0	6.9	0.9	0.6	0.1	0.04	0.7	0.1	0.0
ASM	12.5	6.1	3.0	1.7	3.6	3.0	4.1	3.6	7.2	1.2	0.6	0.3	1.79	0.6	0.9	2.1
BIH	31.8	5.1	4.8	2.0	5.4	1.5	1.3	0.8	9.1	1.1	2.0	0.8	2.89	0.9	1.1	1.1
GUJ	2.8	0.9	0.4	0.3	0.1	0.4	0.1	0.0	1.1	0.1	0.1	0.0	0.09	0.7	0.7	0.0
HAR	0.7	0.8	1.3	0.0	0.1	0.0	0.2	0.0	0.5	0.6	0.7	0.0	0	0.0	0.1	0.2
KAR	17.8	2.7	0.9	0.3	0.9	0.8	0.0	0.0	10.4	1.0	0.1	0.1	0.58	0.3	0.1	0.0
KER	15.3	7.4	2.3	2.2	3.7	0.4	0.6	0.3	11.7	4.8	1.5	1.7	1.85	1.1	0.2	1.0
MP	13.4	2.5	2.9	1.5	1.7	0.3	0.2	0.5	4.9	0.9	1.1	0.4	0.37	0.3	0.1	0.0
MAH	13.4	4.1	1.7	1.0	0.7	0.4	0.2	0.0	5.0	1.1	0.3	0.5	0.42	0.4	0.2	0.0
PUN	1.3	0.1	0.2	0.8	0.2	0.0	0.0	0.1	1.8	0.1	0.1	0.2	0.82	0.1	0.0	0.0
RAJ	3.1	0.6	0.2	0.0	0.7	0.0	0.1	0.0	1.2	0.2	0.2	0.0	0.27	0.0	0.0	0.0
TN	16.1	1.5	0.4	0.1	1.4	0.9	0.3	0.1	7.2	1.0	0.2	0.4	1.05	0.6	0.1	0.0
UP	10.4	2.9	1.4	1.4	0.6	0.5	0.7	0.3	3.9	0.8	0.7	0.7	0.71	0.2	0.3	0.1
WB	31.0	11.1	8.2	10.6	8.6	3.0	2.4	1.3	6.1	2.0	1.1	0.7	1.55	1.5	0.9	0.6
IND	16.2	4.2	2.6	2.0	2.4	0.9	0.7	0.4	5.6	1.1	0.6	0.4	0.77	0.5	0.4	0.1

Sources: NSSO, various calorie adequacy reports.

Appendix A1.6(a) *Body mass index of women across the Indian states, 1998–99 and 2005–06*

Name of the states	*Mean BMI and percentages of less than 18.5 of women*			
	1998–99		*2005–06*	
	Mean BMI	*< 18.5*	*Mean BMI*	*< 18.5*
AP	20.3	37.4	20.9	33.5
ARU	21	10.7	21.1	16.4
ASM	20.1	27.1	20.0	36.5
BHI	19.4	39.3	19.4	45.1

Cont'd...

...Cont'd

Goa	21.6	27.1	21.5	27.9
GUJ	20.7	37	20.8	36.3
HAR	21.3	25.9	21.1	31.3
HP	20.8	29.7	20.8	29.9
J & K	21	26.7	21.4	24.6
KAR	20.4	38.8	20.7	35.5
KER	22	18.7	22.6	18.0
MP	19.8	38.2	19.7	41.7
MAH	20.2	39.7	20.6	36.2
MNP	21.1	18.8	21.5	14.8
MGH	20.3	25.8	21.0	14.6
MZR	20.4	22.6	21.2	14.4
NAG	20.9	18.4	20.8	17.4
ORI	19.2	48	19.7	41.4
PUB	23	16.9	22.9	18.9
RAJ	19.9	36.1	20.1	36.7
TN	21	29	21.6	28.4
UP	20	35.8	20.1	36.0
WB	19.7	43.7	20.2	39.1
DEL	23.7	12	22.9	14.8
IND	20.3	35.8	20.5	35.6

Notes: The Body Mass Index is the ratio of the weight in kilograms to the square of the height in metres (kg/m²).

Source: NFHS, 1998–99, 2005–06.

Appendix A 1.6(b) *Nutritional Status of children across Indian states, 1992–93 to 2005–06 (All are in percentages)*

Name of the states	*Weight for age*						*Height for age*						*Weight for height*					
	% <3SD			*% <2SD*			*% <3SD*			*% <2SD*			*% <3SD*			*% <2SD*		
	1	*2*	*3*	*1*	*2*	*3*	*1*	*2*	*3*	*1*	*2*	*3*	*1*	*2*	*3*	*1*	*2*	*3*
AP	15.6	10.3	9.9	49.1	37.7	32.5	U	14.2	18.7	U	38.6	42.7	U	1.6	3.5	U	9.1	32.5
ARU	14.5	7.8	11.1	39.7	24.3	32.5	27.9	11.9	21.7	53.9	26.5	43.3	3.6	2	6.1	11.2	7.9	32.5
ASM	18.7	13.3	11.4	50.4	36	36.4	26.3	33.7	20.9	52.2	50.2	46.5	1.7	3.3	4.0	10.8	13	36.4
BHI	31.1	25.5	24.1	62.6	54.4	55.9	39.5	33.6	29.1	60.9	53.7	55.6	4.1	5.5	8.3	21.8	21	27.1
Goa	8.9	4.7	6.7	35	28.6	25.0	11	4.8	10.2	32.5	18.1	25.6	2.4	0.7	5.6	15.3	13	14.1
GUJ	17.6	16.2	16.3	50.1	45.1	44.6	25.3	23.3	25.5	48.2	43.6	51.7	3.7	2.4	5.8	18.9	16	18.7
HAR	9	10.1	14.2	37.9	34.6	39.6	19.3	24.3	19.4	46.7	50	45.7	0.6	0.8	5.0	5.9	5.3	19.1
HP	12.9	12.1	11.4	47	43.6	36.5	U	18.1	16	U	41.3	38.6	U	3.3	5.5	U	17	19.3
J&K	13.8	8.3	8.2	44.5	34.5	25.6	18.6	17.3	14.9	40.8	38.8	35	3.4	1.2	4.4	14.8	12	14.8
KAR	19.4	16.5	12.8	54.3	43.9	37.6	22.7	15.9	20.5	47.6	36.6	43.7	2.6	3.9	5.9	17.4	20	17.6
KER	6.1	4.7	4.7	28.5	26.9	22.9	9	7.3	6.5	27.4	21.9	24.5	1.3	0.7	4.1	11.6	11	15.9
MP	22.3	24.3	27.3	57.4	55.1	60	U	28.3	26.3	U	51	50.0	U	4.3	12.6	U	20	35.0
MAH	21.3	17.6	11.9	54.2	49.6	37.0	23.5	14.1	19.1	48.5	39.9	46.3	4.2	2.5	5.2	20.2	21	16.5
MNP	7.2	5.3	4.7	30.1	27.5	22.1	16	11.2	13.1	33.6	31.3	35.6	1.2	1.8	2.1	8.8	8.2	9.0
MGH	17.2	11.3	27.7	45.5	37.9	48.8	38.4	24.5	29.8	50.8	44.9	55.1	4.8	1	19.9	18.9	13	30.7
MZR	5.3	5	5.4	28.1	27.7	19.9	16	13.9	17.7	41.3	34.6	39.8	0.6	2.8	3.5	2.2	10	9.0
NAG	7.6	7.4	7.1	28.7	24.1	25.2	13.2	11.7	19.3	32.4	33	38.8	2.3	2.4	5.2	12.7	10	13.3
ORI	22.7	20.7	13.4	53.3	54.4	40.7	25.2	17.6	19.6	48.2	44	45.0	3.6	3.9	5.2	21.3	24	19.5
PUB	14.2	8.8	8.0	45.9	28.7	24.9	15.7	17.2	17.3	40	39.2	36.7	2.8	0.8	2.1	19.9	7.1	9.2
RAJ	19.2	20.8	15.3	41.6	50.6	39.9	26.6	29	22.7	43.1	52	43.7	5.2	1.9	7.3	19.5	12	20.4
TN	13.3	10.6	6.4	48.2	36.7	29.8	U	12	10.9	U	29.4	30.9	U	3.8	8.9	U	20	22.2

Cont'd...

...Cont'd

UP	24.6	21.9	16.4	59	51.7	42.4	35.6	31	32.4	59.2	55.5	56.8	2.7	2.1	5.1	16.1	11	14.8
WB	18.4	16.3	11.1	56.8	48.7	38.7	U	19.2	17.8	U	41.5	44.6	U	1.6	4.5	U	14	16.9
DEL	12	10.1	8.7	41.6	34.7	26.1	19.3	18	20.4	43.2	36.8	42.2	2.7	4.1	7.0	11.9	13	15.4
IND	20.6	18	15.8	53.4	47	42.5	28.9	23	23.7	52	45.5	48	3.2	2.8	6.4	17.5	16	19.8

Notes: Each index of the children nutritional status is expressed in standard deviation units (SD) from the median of the international references population.

Source: NFHS, 1992–93, 1998–99, 2005–06.

Appendix A1.7 *Housing conditions across the Indian states*

State	*Region*	*1973–74*		*1988*			*1993*			*2002*	
		Pucca	*Katcha*	*Pucca*	*Katcha*	*% of KHHB*	*Pucca*	*Katcha*	*% of KHHB*	*Pucca*	*Katcha*
ORI	R	5.97	78.78	9.36	71.43	51.76	12.06	59.59	43.07	18	53
	U	49.53	34.33	53.15	29.46	63.92	55.42	16.78	58.26	68	11
AP	R	16.61	57.5	29.52	45.91	39.12	37.47	39.2	36.67	46	22
	U	51.77	31.56	61.71	21.86	51.55	68.06	16.61	41.53	80	5
ASM	R	7.71	84.86	8.13	75.39	48.86	7.5	70.88	35.6	15	55
	U	45.06	39.9	38.68	33.93	58.33	48.67	20.48	45.63	65	10
BIH	R	13.91	48.23	18.35	53.17	41.45	20.76	42.04	39.64	28	38
	U	62.97	15.06	66.5	15.79	49.29	65.98	8.8	45.57	70	11
GUJ	R	29.64	32.21	34.14	17.97	40.34	39.94	21.69	25.6	35	8
	U	76.16	8.91	81.77	3.12	63.86	81.22	6.64	71.25	91	1
HAR	R	54.9	32.61	76.14	13.82	34.02	79.42	14.34	51.32	83	4
	U	82.5	8.24	93.47	2.38	74.18	86.97	5.61	55.38	90	0
KAR	R	15.15	39.01	24.32	24.45	49.62	26.31	17.9	55.43	21	11
	U	51.08	14.26	62.66	10.64	56.65	63.55	13.28	69.98	64	3
KER	R	31.28	43.8	50.07	26.87	64.44	55.3	18.96	57.27	36	8
	U	57.15	23.21	68.66	19.63	70.11	73	10.72	66.57	53	3
MP	R	13.02	17.68	12.07	18.56	44.55	13.35	8.35	34.57	19	8
	U	54.54	13.39	55.06	9.73	48.53	58.65	4.65	57.93	66	2
MAH	R	20.64	29.27	32.58	18.17	51.93	42.54	9.53	38.82	37	6
	U	67.05	8.19	74.69	5.64	74.68	81.8	2.89	64.17	83	1
PUN	R	36.12	46.81	67.08	18.1	51.76	79.99	8.73	49.42	71	5
	U	80.63	7.47	88.97	3.25	50	91.96	3.34	65.17	85	1
RAJ	R	26.71	49.17	38.39	36.27	33	46.08	30.37	18.41	54	23
	U	65.89	14.97	87.11	5.95	37.05	87.45	4.25	37.5	87	4
TN	R	22.32	60.89	31.15	46.12	42.53	35.16	41.33	29.94	26	27
	U	59.84	24.91	64.06	18.56	53.74	69.25	13.53	43.2	63	8
UP	R	18.11	52.42	28.16	44.46	40.5	37.19	31.61	33.35	51	23
	U	71.62	10.44	74.93	9.95	54.26	77.16	7.17	55.55	89	2
WB	R	10.5	65.87	13.38	53.86	37.52	16.59	45.97	29.09	21	26
	U	68.42	10.11	73.27	6.76	53.33	70.93	6.54	58.87	65	2
IND	R	18.42	49.17	27.15	39.35	41.88	32.28	31.72	34.91	36	21
	U	64.52	15.84	71.08	10.92	55.71	73.84	8.27	52.63	77	3

Notes: KHHB – Katcha households in bad conditions, the semi-Pucca households is the almost equal with the 100 minus Pucca and Katcha households.

Source: NSSO various rounds.

Appendix A1.8 *Percentages distribution of the conditions of houses across the major Indian states, 2001*

States		Good	Liveable	Dilapidated	Total	Good	Liveable	Dilapidated	Total
ORI	R	14,45,152	43,79,769	6,68,275	64,93,196	22.26	67.45	10.29	100
	U	4,92,622	4,87,756	83,344	10,63,722	46.31	45.85	7.84	100
	C	19,37,774	48,67,525	7,51,619	75,56,918	25.64	64.41	9.95	100
AP	R	67,74,327	50,23,028	5,50,779	1,23,48,134	54.86	40.68	4.46	100
	U	29,30,718	10,53,348	89,973	40,74,039	71.94	25.86	2.21	100
	C	97,05,045	60,76,376	6,40,752	1,64,22,173	59.10	37.00	3.90	100
ASM	R	10,17,579	26,87,782	4,51,249	41,56,610	24.48	64.66	10.86	100
	U	3,65,308	2,91,716	45,837	7,02,861	51.97	41.50	6.52	100
	C	13,82,887	29,79,498	4,97,086	48,59,471	28.46	61.31	10.23	100
BIH	R	39,90,293	67,33,105	13,47,303	1,20,70,701	33.06	55.78	11.16	100
	U	6,97,601	4,82,335	1,02,501	12,82,437	54.40	37.61	7.99	100
	C	46,87,894	72,15,440	14,49,804	1,33,53,138	35.11	54.04	10.86	100
GUJ	R	23,98,093	27,90,981	1,63,379	53,52,453	44.80	52.14	3.05	100
	U	19,78,755	9,80,956	65,338	30,25,049	65.41	32.43	2.16	100
	C	43,76,848	37,71,937	2,28,717	83,77,502	52.25	45.02	2.73	100
HAR	R	10,23,979	12,93,824	1,16,652	24,34,455	42.06	53.15	4.79	100
	U	6,24,186	3,98,462	41,240	10,63,888	58.67	37.45	3.88	100
	C	16,48,165	16,92,286	1,57,892	34,98,343	47.11	48.37	4.51	100
KAR	R	23,74,555	38,11,564	3,82,201	65,68,320	36.15	58.03	5.82	100
	U	20,42,033	13,35,753	11,6871	34,94,657	58.43	38.22	3.34	100
	C	44,16,588	51,47,317	4,99,072	1,00,62,977	43.89	51.15	4.96	100
KER	R	25,89,112	18,70,667	4,40,271	49,00,050	52.84	38.18	8.99	100
	U	10,59,344	4,79,821	92,806	16,31,971	64.91	29.40	5.69	100
	C	36,48,456	23,50,488	5,33,077	65,32,021	55.85	35.98	8.16	100
MP	R	39,61,811	37,64,788	3,04,062	80,30,661	49.33	46.88	3.79	100
	U	17,49,501	9,19,925	86,805	27,56,231	63.47	33.38	3.15	100
	C	57,11,312	46,84,713	3,90,867	1,07,86,892	52.95	43.43	3.62	100
MAH	R	49,07,931	52,46,567	6,89,279	1,08,43,777	45.26	48.38	6.36	100
	U	49,87,367	27,34,716	2,42,438	79,64,521	62.62	34.34	3.04	100
	C	98,95,298	79,81,283	9,31,717	1,88,08,298	52.61	42.43	4.95	100
PUN	R	14,32,259	11,71,867	1,25,806	27,29,932	52.47	42.93	4.61	100
	U	9,31,417	4,68,445	55,599	14,55,461	63.99	32.19	3.82	100
	C	23,63,676	16,40,312	1,81,405	41,85,393	56.47	39.19	4.33	100
RAJ	R	30,28,068	37,90,637	2,43,191	70,61,896	42.88	53.68	3.44	100
	U	13,40,113	7,58,453	49,056	21,47,622	62.40	35.32	2.28	100
	C	43,68,181	45,49,090	2,92,247	92,09,518	47.43	49.40	3.17	100
TN	R	71,08,341	9,98,603	74,834	81,81,778	86.88	12.21	0.91	100
	U	52,00,932	5,37,684	42,215	57,80,831	89.97	9.30	0.73	100
	C	1,23,09,273	15,36,287	1,17,049	1,39,62,609	88.16	11.00	0.84	100
UP	R	86,00,883	1,03,68,609	12,86,127	2,02,55,619	42.46	51.19	6.35	100
	U	27,84,666	20,49,766	2,35,554	50,69,986	54.92	40.43	4.65	100
	C	1,13,85,549	1,24,18,375	15,21,681	2,53,25,605	44.96	49.03	6.01	100
WB	R	41,69,520	58,06,165	10,52,713	1,10,28,398	37.81	52.65	9.55	100
	U	24,18,715	17,88,524	2,91,327	44,98,566	53.77	39.76	6.48	100
	C	65,88,235	75,94,689	13,44,040	1,55,26,964	42.43	48.91	8.66	100
IND	R	6,06,25,573	66,01,1484	84,14,211	13,50,51,268	44.89	48.88	6.23	100
	U	3,33,03,924	1,68,18,398	18,90,143	5,20,12,465	64.03	32.34	3.63	100
	C	9,39,29,497	8,28,29,882	1,03,04,354	18,70,63,733	50.21	44.28	5.51	100

Source: Census of India, 2001.

Appendix A1.9 *Expectation of life at birth among the major Indian states, 1970–75 to 1997–2001*

		1970–75	*1976–80*	*1981–85*	*1986–90*	*1991–95*	*1996–2000*	*1997–2001*	*1999–2003*
ORI	R	45.1	48.5	52.4	53.8	55.7	57.3	57.5	58
	U	55.4	58.2	59.6	61.7	64.4	65.3	65.4	65.8
	C	45.7	49.4	53	54.4	56.5	57.9	58.1	58.7
AP	R	47.3	51.7	57.1	58.3	60.7	62.1	62.2	62.5
	U	57.2	60.6	63.8	62.1	65.1	66.7	66.9	67.2
	C	48.8	53.1	58.4	59.1	61.8	63.3	63.4	63.7
ASM	R	44.9	50.4	51.2	53	55.1	56.8	56.9	57.2
	U	55.7	60.8	60.2	60.9	64.1	66.3	66.4	66.7
	C	45.5	51.1	51.9	53.6	55.7	57.5	57.7	58
BIH	R	NA	NA	52.1	54.3	58.5	59.5	59.7	60.1
	U	NA	NA	61	61.2	65.8	66.4	66.6	66.9
	C	NA	NA	52.9	54.9	59.3	60.4	60.6	61
GUJ	R	47.4	51	56.2	56.4	60.1	61.8	62	62.3
	U	55	57.1	60.7	61	63.2	65	65.3	65.8
	C	48.8	52.4	57.6	57.7	61	63.1	63.3	63.5
HAR	R	62.4	62.4	66	65.7	67.4	64	64.2	64.6
	U	56.4	53.6	58.9	60.9	62.6	68.2	68.4	68.7
	C	57.5	54.8	60.3	62.2	63.4	64.8	65	65.4
KAR	R	52.6	53.9	58.7	60.3	60.9	62.4	62.6	62.9
	U	64.2	64	67.1	64.9	66.6	68.6	68.7	69
	C	55.2	56.3	60.7	61.1	62.5	64.2	64.3	64.6
KER	R	61.7	65.5	68.5	69.6	73	73.4	73.4	73.5
	U	63.5	65.7	67.6	68.6	73.6	74	74.1	74.2
	C	62	65.5	68.4	69.5	72.9	73.5	73.6	73.6
MP	R	45.7	47.6	50	51.5	53.2	55.2	55.3	55.9
	U	56.6	58.3	60.3	61.1	62.8	64	64.1	64.4
	C	47.2	49	51.6	53	54.7	56.5	56.6	57.1
MAH	R	51.9	54	59	60.7	62.5	64	64.2	64.5
	U	58.8	62.2	64	66.6	69.1	70	70.1	70.5
	C	53.8	56.3	60.7	62.6	64.8	66	66.1	66.4
PUN	R	57.1	59.8	61.7	64.3	66.5	67.4	67.5	67.9
	U	61.1	63	68.1	67.4	70.1	70.9	70.9	70.6
	C	57.9	60.5	63.1	65.2	67.2	68.2	68.3	68.6
RAJ	R	46.6	50.3	52	54	57	59.1	59.4	59.9
	U	59.6	60.3	61.1	61.9	64.2	65.7	65.8	66.3
	C	48.4	51.9	53.5	55.2	59.1	60.7	60.9	61.3
TN	R	46.2	51	54.6	58.7	61.7	63.5	63.7	63.9
	U	59.7	59.7	62.5	64.7	67.1	68.5	68.7	69.1
	C	49.6	53.4	56.9	60.5	63.3	64.8	65	65.4
UP	R	42	45.2	48.7	52.4	56	57.2	57.9	58.4
	U	51.5	54.9	57.8	58.6	61.2	62.7	62.9	63.3
	C	43	46.2	50	53.4	56.8	58.6	58.8	59.3
WB	R	NA	NA	55.1	59.2	60.6	62	62.2	62.8
	U	NA	NA	64.9	65.7	67.6	68.8	68.9	69.2
	C	NA	NA	57.4	60.8	62.1	63.6	63.7	64.1
IND	R	48	50.6	53.7	56.1	58.9	60.7	61	61.4
	U	58.9	60.1	62.8	63.4	65.9	67.3	67.6	68.1
	C	49.7	52.3	55.5	57.7	60.3	61.9	62.2	62.7

Note: R – Rural, U – Urban, C – Combined.

Source: Registrar General of India, (1999 and 2006).

Appendix A1.10 *Infant Mortality Rate for selected years among the major Indian states from 1973 to 2005*

		1971	*1976*	*1981*	*1986*	*1991*	*1996*	*2001*	*2005*
ORI	R	131	130	140	127	129	99	90	78
	U	84	81	68	75	71	65	56	55
	C	127	127	135	123	124	96	87	75
AP	R	115	127	93	87	77	73	71	63
	U	65	97	52	59	56	38	35	39
	C	106	122	86	82	73	65	62	57
ASM	R	144	126	107	111	83	79	73	71
	U	73	100	76	69	42	37	38	39
	C	139	124	106	109	81	74	70	68
BIH	R			124	104	71	73	62	62
	U			60	68	46	54	50	47
	C			118	101	69	71	61	61
GUJ	R	155	159	123	124	73	68	68	63
	U	110	100	89	66	57	46	37	37
	C	144	146	116	107	69	61	60	54
HAR	R	74	122	108	91	73	70	64	64
	U	58	54	52	58	49	60	51	45
	C	72	112	101	85	68	68	62	60
KAR	R	105	99	77	82	87	63	65	54
	U	54	60	45	47	47	25	25	39
	C	95	89	69	73	77	53	55	50
KER	R	60	58	40	28	17	13	11	15
	U	48	47	24	20	16	16	8	12
	C	58	56	37	27	16	14	10	14
MP	R	144	145	152	124	125	102	89	80
	U	79	88	80	82	74	61	56	54
	C	135	138	142	118	117	97	85	76
MAH	R	111	91	90	73	69	58	52	41
	U	88	61	49	44	38	31	34	27
	C	105	83	79	63	60	48	45	36
PUN	R	109	115	88	72	58	54	55	49
	U	76	75	51	55	40	40	35	37
	C	102	108	81	68	53	51	51	44
RAJ	R		152	118	113	84	90	81	75
	U		73	53	71	50	60	55	43
	C		142	108	107	79	85	78	68
TN	R	127	121	104	93	65	60	50	39
	U	77	81	55	54	42	39	32	34
	C	113	110	91	80	57	53	44	37
UP	R	173	184	157	140	102	88	83	77
	U	119	121	97	88	74	67	58	54
	C	167	178	150	132	97	85	80	73
WB	R			98	75	76	58	52	40
	U			44	55	47	44	36	31
	C			91	71	71	55	49	38
IND	R	138	139	119	105	87	77	69	64
	U	82	80	62	62	53	46	40	40
	C	129	129	110	96	80	72	63	58

Note: R – Rural, U – Urban, C – Combined.

Source: Registrar General of India, (1999 and 2006).

Appendix A2.1 *Nutritional Status of children (based on Stunting (H/A)) by region, across the type of locations, castes, standard of living, education and occupations of the households and respondent*

	Coastal		Northern		Southern		Total	
	–2 SD	*–3 SD*	*–2 SD*	*–3 SD*	*–2 SD*	*–3 SD*	*–2 SD*	*–3 SD*
				Based on place of location				
Rural	45.0	20.4	42.7	15.3	47.5	17.0	37.0	18.0
Urban	—	—	—	—	—	—	44.8	14.0
Total	43.3	19.43	42.7	15.3	47.2	17.1	44.0	17.5
				Based on caste				
SC	47.59	27.78	55.13	20.78	52.24	14.71	50.69	22.84
ST	60.00	26.67	44.44	15.28	53.33	23.81	49.46	19.71
OBC	47.31	22.16	40.25	13.84	45.45	18.18	44.28	18.29
Other	34.60	10.38	30.65	12.90	19.35	0.00	32.24	9.84
Total	43.36	19.44	42.89	15.38	47.21	17.41	44.0	17.61
				Based on economic class				
Low	50	24.8	79.3	15.1	55.6	21.0	50.6	20.4
Medium	41.7	16.9	38.7	18.7	24.5	7.5	38.7	16.4
High	14.5	1.9	30.0	7.5	22.2	5.6	21.2	4.5
Total	43.0	19.2	42.7	15.6	47.2	17.3	43.8	17.6
				Based on occupation of the child of mother				
NW	41.6	18.1	40.5	14.8	46.6	16.6	42.1	16.8
Agl	38.9	23.5	38.1	14.3	37.9	3.4	38.2	12.5
Non Agl.	59.3	30.5	52.8	18.0	51.9	23.4	54.2	23.1
Total	43.4	19.6	42.8	15.4	47.2	17.1	44.0	17.6
				Based on education of the mother				
0	51.0	28.9	49.4	18.3	50.0	19.8	50.1	22.5
1	45.4	14.2	35.6	11.9	43.2	8.3	42.1	12.8
2	27.0	8.0	34.8	13.0	42.9	7.7	31.1	9.9
3	12.0	4.0	20.0	0.0	0.0	0.0	12.8	2.1
Total	43.3	19.4	43.0	15.4	47.4	17.2	44.0	17.6
				Based on occupation of the child's father				
NW	35.7	23.1	45.5	27.3	0.0	0.0	35.7	22.2
Agl.	38.4	14.5	39.1	11.8	39.2	11.3	38.8	13.0
Non Agl.	47.7	23.8	45.0	17.3	52.7	21.3	47.8	20.9
Total	43.3	19.6	42.9	15.6	47.2	17.5	44.0	17.8
				Based on education of the father				
0	54.7	29.7	49.7	18.7	51.1	20.6	51.8	23.0
1	48.1	21.2	41.8	14.8	49.3	17.8	46.3	18.5
2	36.1	13.5	42.2	13.9	40.4	9.8	38.9	13.1
3	18.0	3.3	22.7	9.1	28.6	7.1	21.0	5.9
Total	43.4	19.4	42.8	15.4	47.4	17.1	44.0	17.5
				Number of observations				
Total	572	571	443	442	269	270	1284	1283

Notes: For education, 0 – illiterate, 1 – literate and up to primary, 2 – higher than primary and up to high school, 3 – secondary and more.

Source: NFHS, unit level data.

Appendix A2.2 *Nutritional status of children (based on wasting (W/A)) by region, across the type of locations, castes, standard of living, education and occupations of the respondent*

	Coastal		*Northern*		*Southern*		*Total*	
	–2 SD	*–3 SD*	*–2 SD*	*–3 SD*	*–2 SD*	*–3 SD*	*–2 SD*	*–3 SD*
				Based on place of location				
Rural	50.0	16.7	61.7	23.5	57.3	27.2	55.5	21.2
Urban	—	—	—	—	—	—	45.3	16.6
Total	49	16.6	60.2	23	57	26.3	54.4	20.7
				Based on Caste				
SC	57.93	20.83	62.82	23.38	58.21	32.84	59.31	24.31
ST	50.00	20.00	61.81	26.90	58.10	27.62	59.14	26.43
OBC	49.73	18.38	62.26	22.01	62.12	24.24	56.59	20.73
Other	42.18	11.32	48.39	14.75	37.50	12.90	42.95	12.17
Total	49.04	16.46	60.27	22.85	56.67	26.39	54.52	20.75
				Based on economic class				
Low	57.41	19.63	66.14	28.29	63.08	31.79	62.01	25.98
Medium	47.33	16.05	56.00	14.67	35.85	15.09	48.88	15.47
High	12.96	3.70	38.46	20.51	50.00	5.56	27.93	9.91
Total	48.85	16.58	60.23	22.95	56.77	26.69	54.44	20.90
				Based on Occupation of the child of mother				
NW	48.48	17.58	57.88	20.26	58.28	25.77	53.15	19.81
Agl	44.44	16.67	46.51	16.28	50	20	47.25	17.58
Non Agl.	55.17	6.78	75.28	35.23	55.26	30.26	63.23	26.01
Total	49.04	16.43	60.27	22.85	56.51	26.39	54.48	20.73
				Based on education of the mother				
0	55.5	20.5	66.9	28.8	60.4	29.2	60.9	26.0
1	50.3	14.8	53.9	13.9	43.2	21.6	50.6	15.3
2	38.6	11.9	53.6	15.9	50.0	7.7	45.1	13.1
3	12.0	8.0	20.0	13.3	28.6	0.0	17.0	8.5
Total	49.0	16.6	60.3	22.9	56.7	26.4	54.5	20.8
				Based on Occupation of the child of father				
NW	28.6	0.0	36.4	27.3	0.0	0.0	28.6	10.7
Agl	44.7	13.7	52.5	15.6	50.0	19.8	48.1	15.4
Non Agl.	53.6	19.5	66.1	26.9	61.5	30.2	60.0	24.7
Total	49.0	16.4	60.4	22.9	56.7	26.1	54.6	20.7
				Based on education of the father				
0	62.2	21.6	67.7	29.9	61.8	29.0	64.1	26.8
1	50.5	17.8	60.7	21.6	54.8	27.4	54.6	20.7
2	44.5	14.2	57.8	18.3	51.0	19.2	50.2	16.5
3	24.6	6.6	38.6	13.6	38.5	21.4	31.4	10.9
Total	49.1	16.6	60.3	22.9	56.7	26.3	54.6	20.8
				Number of Observations				
Total	572	571	443	442	269	270	1,284	1,283

Notes and Source: Same as 2.2.

Appendix A2.3 *Nutritional status of children (based on under weight (W/H)) by region, across the type of locations, castes, standard of living, education and occupations of the respondent*

	Coastal		*Northern*		*Southern*		*Total*	
	–2 SD	*–3 SD*	*–2 SD*	*–3 SD*	*–2 SD*	*–3 SD*	*–2 SD*	*–3 SD*
				Based on Place of location				
Rural	17.1	—	32.2	—	27.2	—	24.4	4
Urban	—	—	—	—	—	—	23.9	3.6
Total	17.65	—	31	—	27.8	—	24.3	4.0
				Based on caste				
SC	17.24	2.07	26.92	5.19	31.34	2.99	23.10	3.11
ST	20.00	3.33	36.81	8.28	24.76	2.86	30.47	5.71
OBC	16.67	3.23	32.08	6.29	28.79	4.62	24.57	4.63
Other	18.48	1.42	19.67	4.84	29.03	6.25	19.80	2.61
Total	17.66	2.27	31.00	6.55	27.88	3.72	24.40	4.05
				Based on economic class				
Low	20	2.59	36.25	7.57	31.12	5.13	28.73	5.03
Medium	16.9	2.47	24.00	4.00	20.75	0.00	19.73	2.69
High	11.1	0.00	23.08	7.50	16.67	0.00	16.22	2.68
Total	17.8	2.29	30.91	6.35	28.09	3.76	24.49	4.00
				Based on occupation of the child of mother				
NW	19.0	2.4	27.6	7.4	27.0	4.3	23.1	4.3
Agl	16.7	0.0	21.4	0.0	13.8	0.0	18.0	0.0
Non Agl.	5.2	1.7	47.2	6.7	35.1	3.9	32.1	4.4
Total	17.5	2.3	30.9	6.5	27.9	3.7	24.3	4.0
				Based on education of the mother				
0	18.3	1.9	35.0	8.2	26.9	4.2	26.6	4.8
1	15.3	2.2	26.5	5.0	35.1	2.8	21.1	3.1
2	20.0	3.0	26.1	1.4	30.8	0.0	23.1	2.2
3	16.0	0.0	13.3	6.7	14.3	0.0	14.9	2.1
Total	17.5	2.1	30.9	6.3	27.9	3.7	24.3	3.9
				Based on occupation of the child of father				
NW	0	0	0	0	33.3	0.0	3.6	0
Agl	15.2	2.7	23.1	3.7	23.7	1.0	19.3	2.7
Non Agl.	20.5	2.0	36.9	8.5	30.2	5.3	28.7	5.1
Total	17.6	2.3	31.0	6.5	27.9	3.7	24.4	4.0
				Based on education of the father				
0	21.6	4.1	39.4	6.5	29.8	4.5	30.4	5.1
1	15.0	1.4	29.1	8.1	21.9	2.8	20.8	3.9
2	18.7	1.3	25.7	4.6	31.4	3.9	23.2	2.9
3	13.1	3.2	20.5	4.5	28.6	0.0	17.6	3.3
Total	17.5	2.3	31.0	6.3	27.9	3.7	24.3	4.0
				Number of observations				
Total	572	571	443	442	269	270	1,284	1,283

Notes and Source: Same as Table 2.2.

Appendix A2.4 *Percentages distribution of the conditions of houses across the districts of Odisha, 2001*

	Total houses	*Good*	*Livable*	*Dilapidated*
Odisha	75,56,918	25.64	64.41	9.95
Bargarh	2,71,498	21.59	70.28	8.13
Jharsuguda	1,02,635	28.97	66.02	5.01
Sambalpur	1,94,560	29.37	63.86	6.77
Debagarh	55,650	15.04	75.51	9.45
Sundargarh	3,81,865	29.99	64.75	5.26
Kendujhar	3,23,511	23.12	68.05	8.84
Mayurbhanj	4,62,527	20.33	70.35	9.32
Baleshwar	3,76,989	23.47	61.40	15.13
Bhadrak	2,27,321	20.80	65.93	13.27
Kendrapara	2,78,766	21.61	63.70	14.69
Jagatsinghapur	2,22,394	22.45	55.58	21.98
Cuttack	4,48,133	31.78	51.88	16.34
Jajapur	3,31,149	23.42	58.25	18.33
Dhenkanal	2,08,888	24.12	63.65	12.23
Anugul	2,19,065	29.82	60.32	9.86
Nayagarh	1,66,212	22.01	67.89	10.10
Khordha	3,28,811	38.17	51.54	10.29
Puri	2,74,520	24.13	59.78	16.09
Ganjam	6,48,741	33.74	57.45	8.81
Gajapati	1,11,710	24.10	71.28	4.62
Kandhamal	1,45,185	25.70	70.17	4.13
Baudh	83,460	20.04	74.71	5.25
Sonapur	1,15,965	19.61	74.62	5.76
Balangir	3,30,591	23.12	70.91	5.98
Nuapada	1,27,293	20.72	75.36	3.92
Kalahandi	3,14,763	21.95	71.58	6.47
Rayagada	1,87,892	24.31	72.73	2.96
Nabarangapur	2,25,045	20.88	73.66	5.46
Koraput	2,83,421	28.39	67.42	4.19
Malkangiri	1,08,358	21.28	72.23	6.50

Source: Census of India, 2001.

Appendix A2.5 *Level of literacy across the districts of Odisha, 2001*

Districts	*Total*			*Rural*			*Urban*		
	P	*M*	*F*	*M*	*F*	*T*	*P*	*M*	*F*
Anugul *	68.8	81.4	55.4	66.1	79.7	52.0	85.1	91.4	77.5
Balangir	55.7	71.7	39.5	52.7	69.5	35.8	78.0	87.1	68.2
Baleshwar	70.6	81.7	58.9	69.5	81.2	57.4	78.8	85.8	71.2
Bargarh *	64.0	77.4	50.3	62.7	76.5	48.6	79.2	87.7	70.1
Baudh *	57.7	76.2	39.0	56.4	75.3	37.4	82.0	92.9	70.5
Bhadrak *	73.9	84.7	62.8	74.3	85.4	63.0	70.4	78.9	61.4
Cuttack	76.7	85.8	66.9	73.6	84.3	62.5	84.7	89.7	78.9
Debagarh *	60.4	73.3	47.2	59.0	72.4	45.6	76.3	84.1	67.7
Dhenkanal	69.4	80.6	57.9	68.0	79.5	56.1	84.1	90.8	76.7
Gajapati *	41.3	54.7	28.4	37.6	51.4	24.5	71.1	81.2	61.1
Ganjam	60.8	75.2	46.4	56.5	72.2	41.3	79.7	88.2	70.6
Jagatsinghapur *	79.1	88.6	69.3	78.7	88.6	68.8	82.2	88.1	74.5

Cont'd...

...Cont'd

Jajapur *	71.4	81.9	60.8	71.0	81.6	60.1	81.7	88.3	74.3
Jharsuguda *	70.7	82.2	58.5	66.7	79.5	53.6	77.5	86.6	67.4
Kalahandi	45.9	62.7	29.3	43.5	60.7	26.5	74.4	84.6	63.4
Kandhamal	52.7	69.8	35.9	50.1	67.8	32.8	85.4	93.6	76.6
Kendrapara *	76.8	87.1	66.8	76.5	87.0	66.3	82.6	89.6	75.3
Kendujhar	59.2	72.0	46.2	56.9	70.2	43.6	73.7	82.8	63.6
Khordha *	79.6	87.9	70.4	74.1	85.0	63.0	86.7	91.4	80.9
Koraput	35.7	47.2	24.3	27.3	39.2	15.6	74.9	83.4	65.9
Malkangiri *	30.5	40.1	20.9	27.9	37.4	18.4	65.3	74.9	54.9
Mayurbhanj	51.9	65.8	37.8	49.5	63.8	35.0	82.6	89.1	75.2
Nabarangapur *	33.9	47.0	20.7	31.3	44.6	18.0	73.7	83.1	63.5
Nayagarh *	70.5	82.7	57.6	69.8	82.2	56.7	85.6	92.4	78.2
Nuapada *	42.0	58.5	25.8	40.3	57.0	23.8	70.4	81.6	59.0
Puri	78.0	88.1	67.6	77.3	88.2	66.3	81.9	87.5	75.9
Rayagada *	36.1	48.2	24.6	29.9	42.1	18.3	72.2	81.8	62.4
Sambalpur	67.3	79.0	55.2	62.8	75.9	49.5	79.0	86.9	70.6
Sonapur *	62.8	78.9	46.2	61.7	78.1	44.7	77.3	88.6	65.0
Sundargarh	64.9	75.3	53.9	55.3	67.4	43.1	82.5	89.0	75.0

Source: Census of India 2001.

Appendix A2.6 *Land utilisation pattern across the districts of Odisha, 1955–56*

Districts	*Total Area**	*Under Forest*	*NAC*	*CWOFL*	*Fellow land*	*NSA*	*Sown more than once*	*Total cropped area*
Balasore	100	6.48	11.18	9.56	8.43	64.51	1.30	65.80
Bolangir	100	15.84	8.03	33.60	3.51	38.91	2.49	41.52
Cuttack	100	8.17	9.36	10.37	15.23	57.06	16.24	73.21
Dhenkanal	100	47.40	7.58	6.39	2.19	36.44	2.10	38.54
Ganjam	100	17.08	16.18	29.98	9.07	27.61	0.33	27.94
Kalahandi	100	12.28	24.08	16.44	8.70	38.39	7.35	45.74
Keonjhar	100	31.80	13.22	19.33	3.49	32.29	0.50	32.79
Koraput	100	12.31	29.63	12.94	7.35	37.72	0.04	37.76
Mayurbhanj	100	22.17	20.25	5.47	21.40	32.92	0.48	33.30
Phulbani	100	44.36	8.84	15.78	1.53	29.49	0.81	30.21
Puri	100	29.32	11.65	15.19	2.87	40.97	9.65	50.72
Sambalpur	100	25.80	6.56	28.60	7.76	29.22	1.14	30.37
Sundargarh	100	36.60	16.92	10.09	6.32	29.97	—	29.97

Source: Statistical Abstract of Odisha, 1955–56.

Appendix A2.7 *Annual rainfall (in inches) across the districts of Odisha (1931–1950)*

	Districts												
	Cuttack	*Balasore*	*Puri*	*Sambalpur*	*Ganjam*	*Koraput*	*Dhenkanal*	*Keonjhar*	*Phulbani*	*Bolangir*	*Sundargarh*	*Kalahandi*	*Mayurbhanj*
1931	59.01	67.09	54.23	60.61	49.79	65.87	58.45	66.07	51.21	50.57	61.08	38.34	64.62
1932	56.35	58.55	58.78	54.29	50.99	53.56	58.14	63.21	61.96	66.72	51.77	59.05	60.92
1933	75.89	83.2	65.57	74.64	72e	66.3	73.75	63.54	91.98	80.92	63.99	67.92	78.63
1934	57.8	58.36	52.58	61.84	44.7	64.81	54.18	59.05	65.53	57.29	74.2	59.39	51.9
1935	48.86	47.47	48.04	51.48	35.53e	48.92	44.49	49.82	53.64	39.34	56.46	42.77	56.92

Cont'd...

...Cont'd

1936	81.14	75.55	82.04	74.48	61.65	73.2	70.09	66.06	88.68	74.18	82.39	71.35	65.93
1937	60.53	63.69	59.62	71.01	47.99	57.48	56.65	71.51	69.31	64.16	67.5	63.76	70.45
1938	51.38	56.54	60.62	56.8	50.32	65.7	47.65	58.4	53.89	45.57	57.04	57.69	52.18
1939	47.81	51.29	48.51	75.51	46.32	58.97	65.16	62.93	80.46	69.68	72.26	58.41	74.58
1940	65.01	73.36	67.43	62.23	62.59	60.46	60.66	81.46e	69.9	65.08	60.8	58.7	83.97
1941	60.39	70.32	49.23	64.81	44.12	51.79	54.66	72.91	59.93	46.37	70.37	38.49	74.75
1942	53.92	54.33	49.5	57.25	41.74	47.57	49.07	58.33	56.13	58.92	63.34	49.4	70.33
1943	57.21	60.28	68.41	76.88	—	57.51	65.65	63.97	77.5	61.61	83.59	70.51	83.22
1944	67.97	59.62	70.69	67.71	58.46	63.59	68.51	62.17	74.75	60.34	59.89	68.82	69.22
1945	51.13	63.35	57.04	60.86	47.91	39.07	51.09	61.9	64.57	52.77	62.01	59.11	64.62
1946	66.39	77.28	67.31	67.79	54.32	72.17	64.93	77.18	57.68	45.4	74.91	48.93	81.4
1947	64.73	51.03	56.42	52.61	51.53	63.58	47.96	57.03	63.66	48.18	49.38	58.83	56.62
1948	58.64	56.66	57.63	48.91	45.45	55.15	50.34	57.51	41.53	52.07	63.97	—	69.1
1949	46.21	61.91	52.04	62.88	50.62	101.2e	—	—	59.72	52.92	—	—	52.83
1950	66.62	59.58	59.88	57.24	45.26	50.34	45.96	59.75	64.42	51.85	—	—	—
ANR 1950	58.51	59.41	57.38	60.56	51.93	59.58	55.91	62.13	63.65	58.1	65.65	54.76	63.46

Notes: e-outliers (excluded), ANR – Annual Normal Rainfall of 1931 to 1950.

Source: Census of India, 1951, Volume XI, Odisha, Part-I Report, e-excluded (outlier).

Appendix A2.8 *Level and variability of annual rainfall (in inches) across the districts of Odisha (1931–1950)*

Districts	*Coefficient of variation*	*Annual normal rainfall 1950*	*Period average*	*N*
Phulbani	18.78	63.65	65.32	20
Bolangir	18.2	58.16	57.20	20
Kalahandi	17.67	54.76	57.15	17
Dhenkanal	15.04	55.91	57.23	19
Mayurbhanj	14.99	63.46	67.48	19
Cuttack	14.83	58.51	59.85	20
Balasore	14.73	59.41	62.47	20
Koraput	14.62	59.58	58.74*	19
Puri	14.52	57.38	59.28	20
Sundargarh	14.19	65.65	65.28	18
Sambalpur	13.09	60.56	62.99	20
Ganjam	11.66	51.93	50.22**	17
Keonjhar	9.9	62.13	62.85*	18

Note: N implies number of years included. This differs due to dropping of outliers or non-availability of data. * Number of star implies numbers of outlier year. Mean is the simple average after dropping outliers.

Source: Census of India, 1951, Volume XI, Odisha, Part-I Report.

Appendix A2.9 *Percentages distributions of households across their landholding in Odisha, 1951*

State and districts	*Odisha*	*Cuttack*	*Balasore*	*Puri*	*Sambalpur*	*Ganjam*	*Koraput*	*Dhenkanal*	*Keonjhar*	*Phulbani*	*Bolangir*	*Sundargarh*	*Kalahandi*	*Mayurbhanj*
PHHLC	4.20	2.39	3.73	3.33	4.31	1.91	5.95	4.55	5.35	6.12	6.91	6.57	4.82	6.46
0 Acres	36.46	44.53	27.46	38.51	37.75	48.18	31.43	34.27	13.48	34.22	35.13	26.01	41.39	25.36
1Acres	13.99	18.78	18.69	19.16	9.95	13.76	11.43	10.29	17.95	13.56	3.70	7.73	10.30	14.38
2Acres	10.26	10.97	12.47	10.48	8.40	11.41	10.73	11.77	13.50	10.31	6.57	9.67	5.12	10.20

Cont'd...

...Cont'd

3Acres	8.49	7.33	10.96	7.97	8.03	7.21	9.70	10.29	13.83	8.13	6.38	11.27	3.43	11.04
4Acres	6.44	4.56	8.16	5.47	6.55	4.95	7.71	7.38	10.83	7.41	6.15	6.32	3.66	10.54
5Acres	5.29	2.89	6.51	5.38	5.60	3.73	7.84	8.08	9.69	4.48	5.18	6.71	4.02	4.69
6–10 Acres	8.89	3.63	8.93	6.19	10.59	5.33	8.26	11.31	11.50	13.73	14.70	14.79	11.09	15.45
11–32 acre	5.76	1.80	4.78	2.79	9.31	2.80	4.32	5.51	7.32	5.01	15.46	12.48	11.44	6.54
33 acres	0.31	0.21	0.17	0.11	0.38	0.57	0.38	0.04	0.40	0.82	0.14	0.20	0.35	0.47
34–79 Acres	0.96	0.44	0.86	0.54	1.74	0.39	0.66	0.56	0.42	1.04	2.57	1.93	2.56	0.69
80 and over	0.26	0.40	0.21	0.07	0.42	0.08	0.23	0.09	0.12	0.32	0.60	0.39	0.43	0.10
Acres not stated	2.89	4.46	0.78	3.32	1.28	1.58	7.31	0.42	1.06	0.98	3.40	2.50	6.19	0.55
Total	100	100	100	100	100	100	100	100	100	100	100	100	100	100

Source: Census, 1951.

Appendix A2.10 *Livelihood patterns across the districts of Odisha, 1951*

	Agricutural Classes	*Nonagricultural Classes*
Keonjhar	89.9	10.1
Mayurbhanj	87.2	12.8
Sundargarh	86.8	13.2
Balasore	85.2	14.8
Kalahandi	84.4	15.6
Dhenkanal	83.7	16.3
Bolangir	82.7	17.3
Sambalpur	79.2	20.8
Phulbani	79	21
Puri	78.7	21.3
Koraput	77.3	22.7
Cuttack	74.9	25.1
Ganjam	66.1	33.9

Source: Census of India, 1951.

Appendix A2.11 *Livelihood pattern of general population per 10,000*

	I	*II*	*III*	*IV*	*V*	*VI*	*VII*	*VIII*
Odisha	59.53	5.94	12.31	1.5	6.33	2.91	0.53	10.95
Inland O	64.67	4.12	13.17	0.75	5.94	1.92	0.31	9.12
Coastal O	53.39	8.11	11.3	2.4	6.79	4.09	0.79	13.13
Cuttack	54.42	6.63	10.63	3.19	6.48	3.54	0.76	14.35
Balasore	67.11	8.05	8.43	1.65	3.65	2	0.59	8.52
Puri	52.6	10.94	13.42	1.7	7.26	3.94	0.93	9.21
Sambalpur	59.86	4.32	14.26	0.77	9.11	1.94	0.55	9.19
Ganjam	44.94	7.25	11.61	2.3	8.52	6.72	0.76	17.9
Koraput	59.42	6.85	10.38	0.66	4.39	2.44	0.24	15.62
Dhenkanal	67.62	2.3	11.65	2.1	7.5	1.7	0.32	6.81
Keonjhar	76.75	2.33	10	0.81	3.93	0.88	0.09	5.21
Phulbani	63.67	3.3	11.66	0.33	7.38	3.76	0.12	9.78
Bolangir	61.3	3.26	17.86	0.28	6.81	1.71	0.56	8.22
Sundargarh	75.19	2.73	8.65	0.25	4.58	1.41	0.42	6.77
Kalahandi	59.21	5.59	19.25	0.37	4.51	1.81	0.12	9.14
Mayurbhanj	69.29	3.8	13.43	0.71	4.78	1.07	0.23	6.69
Coastal	54.8	8.2	11.0	2.2	6.5	4.1	0.8	12.5
Northern	68.3	3.1	12.6	0.8	6.1	1.5	0.4	7.1
Southern	60.8	5.2	13.8	0.5	5.4	2.7	0.2	11.5

Source: Census of India, 1951.

Appendix A2.12 *Females per 1000 males in agricultural non-agricultural classes, 1951*

	Agriculture				*Non-Agriculture*			
	Total	*SSP*	*NED*	*ED*	*Total*	*SSP*	*NED*	*ED*
Odisha	1,021	182	1,936	1,081	1,025	290	1,868	1,135
Inland O	1,010	187	1,831	1,476	993	250	1,800	1,473
Coastal O	1,036	175	2,071	518	1,053	325	1,922	759
Cuttack	1,022	142	2,157	528	1,032	338	1,845	1,003
Balasore	981	166	1,882	540	1,034	345	1,861	947
Puri	1,038	140	2,056	675	999	229	1,975	1,074
Sambalpur	1,018	158	1,965	2,680	977	286	1,788	3,084
Ganjam	1,109	289	2,143	302	1,121	373	1,994	370
Koraput	985	118	1,825	858	1,006	125	1,964	298
Dhenkanal	1,027	203	1,927	992	971	186	1,943	1,373
Keonjhar	1,004	173	1,995	161	923	228	1,900	186
Phulbani	1,014	166	2,316	218	1,019	235	2,196	197
Bolangir	1,004	157	1,674	1,832	1,042	266	1,609	3,084
Sundargarh	1,023	183	1,739	1,753	948	224	1,693	2,447
Kalahandi	1,012	246	1,634	1,627	1,060	305	1,633	2,116
Mayurbhanj	1,014	298	1,640	1,633	919	385	1,409	1,661

SP – self supporting person, NED – Non-earning dependent, ED – Earning dependent

Source: Census of India.

Appendix A2.13 *Distribution of backward classes population across the districts of Odisha during 1951*

	Categories of Castes					
	NBC	*TBC*	*SC*	*ST*	*SC/ST*	*OBC*
Odisha	33.3	66.7	20.2	18	38.2	28.5
Inland Division	18.3	81.7	33.4	17.5	50.9	30.8
Coastal Division	51.1	48.9	4.6	18.6	23.2	25.7
Kalahandi	9.5	90.5	28.4	20.6	49	41.5
Bolangir	12.6	87.4	20.3	19.9	40.2	47.2
Balasore	46.7	53.3	5.7	19.6	25.3	28.6
Cuttack	52.7	47.3	3.2	19.5	22.7	24.6
Phulbani	13.4	86.6	42.7	18.9	61.6	25
Ganjam	54.1	45.9	6.9	18.6	25.5	20.4
Dhenkal	41.2	58.8	12.4	17.3	29.7	29.1
Sambalpur	17.8	82.2	27	17.1	44.1	38.1
Mayurbhanj	16	84	49.1	16.9	66	18
Koraput	18.1	81.9	41.2	16.5	57.7	24.2
Puri	48.8	51.2	4	16.2	20.2	31
Keonjhar	16.9	83.1	38.8	15.3	54.1	29
Sundargarh	9.4	90.6	56.6	13.2	69.8	20.8

Notes: Odisha Inland division comprises 10 districts (Sambalpur, Koraput, Dhenkanal, Keonjhar, Phulbani, Bolangir, Sundargarh, Kalahandi, Mayurbhanj and Ganjam Agency), Odisha Coastal Division comprises 4 districts (Cuttack, Balasore, Puri, Ganjam Plains). NBC – Non-backward classes, TBC – Total backward classes, SC-Scheduled Tribes, ST – Scheduled Castes, OBC – Other Backward Classes.

Source: Census of India, 1951, Volume XI, Odisha, Part-I Report, pp. 7.

Appendix A2.14 *Percentages distributions of landless households by their social groups across districts in Odisha, 1951*

State and districts	*Odisha*	*Cuttack*	*Balasore*	*Puri*	*Sambalpur*	*Ganjam*	*Koraput*	*Dhenkanal*	*Keonjhar*	*Phulbani*	*Bolangir*	*Sundargarh*	*Kalahandi*	*Mayurbhanj*
Total Avg	36.46	44.53	27.46	38.51	37.75	48.18	31.43	34.27	13.48	34.22	35.13	26.01	41.39	25.36
Ag.classes	25.43	34.96	20.50	29.76	26.05	32.05	19.40	23.49	7.67	25.44	24.87	18.48	30.11	14.22
Backward	25.96	40.55	23.65	35.64	23.87	37.03	23.35	32.86	6.89	27.18	25.05	18.83	24.67	11.03
Nbackward	24.42	29.57	17.29	25.75	34.32	26.81	5.75	11.93	11.20	13.56	24.27	11.47	48.18	31.95
Difference	1.54	10.98	6.36	9.89	–10.45	10.22	17.60	20.93	–4.31	13.62	0.78	7.36	–23.51	–20.92
Non-ag	65.14	59.71	37.59	65.82	72.80	78.27	84.31	72.36	63.43	76.16	72.97	68.41	76.26	54.92
Backward	67.31	63.59	34.18	70.01	70.11	81.59	86.66	74.82	66.63	76.79	70.32	69.03	76.61	58.09
Nbackward	61.83	55.08	41.33	60.66	77.73	7.45	79.38	65.91	54.96	73.29	85.97	66.24	73.25	48.58
Difference	5.47	8.51	–7.15	9.35	–7.62	74.13	7.28	8.92	11.67	3.50	–15.65	2.79	3.36	9.52

Source: Census, 1951.

Appendix A2.15 *Level of literacy across the districts of Odisha during 1951*

	Percentages			
	Total	*Males*	*Females*	
Odisha	15.8	13.5	2.3	38.2
Inland O	10.8	9.6	1.2	50.9
Coastal O	21.7	18.2	3.5	23.2
Balasore	23.7	20.3	3.4	25.3
Cuttack	23.3	19.3	4	22.7
Puri	21.8	18.3	3.5	20.2
Dhenkanal	18.7	16.6	2.1	29.7
Sambalpur	16.3	14.1	2.2	44.1
Ganjam	16.2	13.7	2.5	25.5
Keonjhar	12.1	10.7	1.4	54.1
Sundargarh	10.3	8.8	1.5	69.8
Bolangir	10.1	9.2	0.9	40.2
Phulbani	10	9.3	0.7	61.6
Mayurbhanj	9.6	8.6	1	66
Kalahandi	6.3	5.8	0.5	49
Koraput	5.3	4.6	0.7	57.7

Source: Census of India, 1951.

The earliest estimate ever attempted was in 1822 by Stirling. He calculated the inhabitants of the Mogulbandi (number was 12,96,369). A second estimate was just before and after the famines of 1866. The population estimated was 30,15,826 before and 20,86,288 after the famine. But these figures are reliable only to limited extents. The first official counting of population of Odisha was carried out in 1872, when the Imperial Census Operations in India begun. But at that time since Odisha was fragmented into different territories, the populations counted were also expressed in the name of different territories. And since the number of population based on the territorial adjustment was not available till 1901, here, the analysis is confined from 1901. Table 2.7 presents the rate of growth of population and density of population in Odisha and across the districts of Odisha from 1901 to 1951.

For Odisha, from 1901 to 1911, rate of growth of population was 10.4 per cent. During this decade, although there was a series of floods that resulted in crop failures; outbreaks of cholera and fever, still the population had increases in all the districts of Odisha except Balasore, where the population diminished by 15,820 persons. The contributing factor of this population growth was also credited to development of railways, which facilitated migration.

Appendix A2.16 *Distribution of the rate of growth of population and density of population across the districts of Odisha*

Districts		*Years*					
		1901–11	*1911–21*	*1921–31*	*1931–41*	*1941–51*	*1901–1951*
India	Rate						51.5
Orissa	Rate	10.4	–1.9	11.9	10.2	6.4	42.2
	Density*	189	186	208	229	244	
Inland Orissa	Rate	19.55	1.7	15.86	12.94	8.29	
	Density		122	142	160	173	
Coastal Orissa	Rate	3.01	–5.37	7.96	7.26	4.17	
	Density		393	424	455	474	
Cuttack	Rate	2.42	–2.75	6.42	4.68	3.29	14.6
	Density		522	556	582	601	
Balasore	Rate	–1.38	–6.98	1.29	4.68	–0.25	–2.9
	Density		417	422	442	441	
Pure	Rate	1.95	–10.13	10.87	8.12	7.99	18.6
	Density		300	337	360	389	
Sambalpur	Rate	15.79	4.53	11.75	12.63	8.25	64.9
	Density		141	158	178	192	
Ganjam	Rate	11.18	–3.24	13.24	12.19	4.11	42.3
	Density		260	294	330	344	
Kavaput	Rate	20.21	–3.32	17.88	18.77	12.56	83.1
	Density		82	96	114	129	
Dhenkanal	Rate	3.28	–12.81	23.38	16.72	13.12	46.7
	Density		122	150	176	199	
Phulbani	Rate	27.62	4.05	21.37	15.02	11.07	105.9
	Density		118	143	165	183	
Bolangir	Rate	13.16	–0.44	9.38	9.33	1.52	36.8
	Density		88	96	105	107	
Sundargarh	Rate	39.49	15.5	11.60	8.32	5.28	105.1
	Density		211	236	256	269	
Kalahandi	Rate	30.65	4.22	15.74	12.33	12.53	99.2
	Density		100	115	130	146	
Mayur bhanj	Rate	20.34	1.82	19.05	13.75	8.10	79.4
	Density		116	138	157	170	
	Rate	19.46	3.44	17.94	10.69	4.48	68.6
	Density		187	221	245	256	

Note: Population density represent for the later years, i.e. if population density under 1901–11, then it represent the density for the latter period that is 1911 and so on.

Source: Census Report 1951.

Decade 1911–1921 was a period of disasters – epidemics; wars; scarcity of food and high prices; and high mortality and low birth rate. Although first four years of the decade were healthy, the unhealthy year started from 1915 when the death rate began rising with continuous outbreaks of cholera, dysentery and smallpox. In 1917, there was flood; in 1918, there was draught. There was also four years of World Wars, which altogether increased the price of rice. The cumulative effects of all these unexpected and unhappy circumstance decreased the population of Odisha by 1.9 per cent over the population of 1911. As it is evident from the A2.16, half of the districts were having the negative population growth.

Decade 1921–1931 was the era of prosperity for Odisha, although the blessings were by no means unmixed. Many districts of Odisha such as Sambalpur and Balangir in particular, had bumper crops for the entire period. For the last three years of the decade, all the coastal districts also received good harvests, which brought a large number of rice mills into existence, particularly in Balasore. There was also effective exploitation of forest resources such as timber, bamboo and lac etc. There was advancement in transport as well. Two railways lines were opened, one joining the Talcher coalfields with the main railway line connecting Calcutta (now Kolkata) and Madras (now Chennai) and the other Raipur-Vizianagaram railway line, which passes through western Odisha (completed in 1929). This decade also marks as the beginning of industrialisation in Odisha. Two new iron mines were opened in the Mayurbhanj district and the manganese and limestone quarries were developed in Sundargarh districts. Another feature of this decade was the increase in emigration for Assam Tea Labour. Above all, this decade was considered as the period of prosperity. Population had increased by 11.9 per cent. However, this decade was also seeking the provisions for public health (especially northern and southern Odisha) to escape from malaria, cholera and dysentery.

Decade 1931–41 is memorable for Odisha, as it was constituted into a separate province in 1936. First half of this decade was the time of peace and prosperity but the later part witnessed unhappy circumstances. During this decade, the effects of World War I had completely gone; the state received good crop in most of years except in 1933, 1937 and 1938 due to floods. In 1940 while there was flood in coastal Odisha, there was lack of rain in Sambalpur area. However, the employment of agricultural labour was normal throughout the period. But migration was as usual. People generally used to move towards Calcutta, Jamshedpur and Rangoon for livelihood, especially in the off-season of agriculture. As the health of population was concerned, it was pointed out that, the general level of public health in Odisha was comparatively healthy than previous decades. But at the later period of the decade it was not satisfactory. There were epidemics of cholera, smallpox, malaria, typhoid, diarrhoea and dysentery etc. It was most virulent in the years 1932, 1936 and 1939. In 1936, severe epidemic of malaria spread over 417 villages in Gajam. In 1938, the death rates from this disease was rising to the high figures of 30.0 per mile in Khondmals, 22.1 in Balasore and 22.0 in Angul. Despite this, there was a satisfactory growth of population 10.2 per cent.

Decade 1941–51, although brought the memorable historical events, such as the political freedom of India from British rule and merging of feudatory states with Odisha, it was less favourable than previous two decades. The weather, natural conditions, vital occurrences and the material condition of people, war, political unrest, altogether deteriorated the conditions which resulted in decline in the rate of growth of population to 6.4 per cent from 10.2 per cent in 1931–41 and 11.9 per cent in 1921–31. In the opening years there was flood in Balasore, poor harvest in Kalahandi, Sambalpur and Sundargarh due to insufficient rain. In the following years, the situation deteriorated in all directions. The economic repercussion of the War resulted a sharp rise in price of commodities. The enemy had also thrown two bombs in the coastal districts of Odisha. Again in the same year (October), a severe cyclone visited parts of Bengal; and the northern parts of Balasore districts. In November again, cyclone visited Puri and Ganjam. In 1943, there was Bengal Famine, whose repercussion effects spread to Odisha in the form of emigration. In 1944, there was a high flood. In 1945, rainfall was inadequate at the time of sowing but there was flood when the crops had fairly grown up. In 1946, again there was an abnormal rainfall and high flood. Last three years of the decade also had moderate floods. Overall, this was a decade in which almost all the years there was flood. The public health also deteriorated due to the outbreak of cholera and smallpox. There was small surplus of birth over death in 1943 but in 1944, 1945 and 1947 death exceeded births. The Famine Enquiry Commission of Bengal famine expresses, 'the total recorded number of deaths in Odisha in 1943 was 2,33,584, an increase of 17.9 per cent over the quinquennial average of 1,98,150. In the district of Balasore, the increase in mortality was 40.7 per cent. In this district 1,105 deaths from starvation were recorded; but many of the victims were destitute from Bengal'. on the whole, this decade did not give much happiness.

As the population density was concerned, there was no population pressure in Odisha as a whole (244) in 1951. However, the same has remained higher for coastal Odisha (474) compared to inland Odisha (173). Across the districts, it remained highest for Cuttack (601) and remained lowest for Phulbani (107).

To summarise, the last 50 years of population growth (1901 to 1951) shows population pressure as such has remained lower in Odisha compared to all India. Hence, population pressure as reasons of hunger may not be accepted. But within Odisha, while the population growth for Keonjhar, Balangir and Sudargarh remained high, for Balasore the growth rate was negative and for Cuttack and Puri also, it remained at low level. Yet the density of population remained low among the former group of district. Hence, here too, mere population does not show as a reason of hunger and deprivation, rather working population may be more important. During 1951 Census, these figures can be traced from the livelihood tables, which we have dealt with in livelihood section.

Appendix A3.1 *Percentages distribution of NSDP of Odisha across different sectors, 1950–51 to 2005–06 (1993–94 prices)*

	Sectors	*50s*	*60s*	*70s*	*80s*	*90s*	*2000–01 to 2005–06*
1	Agriculture and Animal Husbandry	54.74	49.24	48.47	47.07	34.29	24.89
2	Forestry and Logging	8.45	11.02	9.34	7.34	3.83	3.18
3	Fishing	1.20	0.42	0.79	1.18	2.28	2.23
1 to 3	Agriculture, Animal Husbandry, Forestry and Fishing	64.39	60.68	58.61	55.59	40.41	30.30
4	Mining and Quarrying	1.00	1.43	1.70	1.96	5.02	8.88
1 to 4	Sub-Total Primary	65.39	62.11	60.31	57.55	45.43	39.18
5	Mnf Regd.	1.31	3.46	4.81	4.63	5.83	5.76
6	Mnf Un Regd.	4.20	4.11	3.50	2.61	2.24	1.28
5 and 6	Manufacturing	5.52	7.58	8.31	7.25	8.06	7.03
7	Electricity, Gas and Water-Supply	0.38	0.55	1.30	1.55	1.89	1.97
8	Construction	4.49	7.25	6.16	5.40	6.68	4.86
5 to 8	Sub-Total – Secondary	10.39	15.38	15.76	14.19	16.64	13.87
9	Trade, Hotel and Restaurants	5.77	7.28	7.88	8.25	10.36	11.40
10	Railways	0.58	0.73	0.84	1.21	2.42	3.01
11	Transport by Other means	0.80	0.97	1.44	1.90	2.73	4.36
12	Storage	0.01	0.01	0.02	0.04	0.07	0.09
13	Communication	0.19	0.22	0.28	0.39	0.67	2.93
10 to 13	Transport, Storage and Communications	1.58	1.93	2.59	3.54	5.90	10.40

Cont'd...

...Cont'd

9 to 13	Sub-Total Tertiary	7.35	9.22	10.47	11.79	16.26	21.80	
14	Banking and Insurance	0.18	0.38	0.70	1.76	3.52	5.76	
15	Real Estate, Ownership of Dwellings and Business Services	9.03	5.61	5.48	5.92	5.84	5.22	
14 and 15	Financing, Insurance, Real Estate and Business Services	9.21	5.99	6.18	7.68	9.35	10.98	
16	Public Administration	1.56	2.41	2.91	4.16	5.24	5.53	
17	Other Services	6.10	4.90	4.37	4.62	7.08	8.64	
16 and 17	Community, Social and Personal Services	7.66	7.31	7.27	8.78	12.32	14.17	
14 to 17	Sub Total – Finance and Services	16.88	13.30	13.46	16.46	21.68	25.15	
9 to 17	Service	24.22	22.51	23.93	28.26	37.94	46.95	
1 to 17	Total NSDP	100	100	100	100	100	100	
	Per capita NSDP of Odisha (99–00 prices, in ₹)	6,270 (50–51)	6,872 (60–61)	7,858 70–71	7,995 80–81	8,303 90–91	10,211 2000–01	16,195 2007–08
	Per capita NNP (99–00 prices, in ₹)	5,708 (50–51)	7,121 (60–61)	8,091 70–71	8,594 80–81	11,535 90–91	16,172 2000–01	22,553 2006–07

Notes: 50s – 1950–51 to 1959–60, 60s – 1960–61 to 1969–70, 70s – 1970–71 to 1979–80, 80s – 1980–81 to 1989–90, 90s – 1990–91 to 1999–2000. Output of all variables has been converted to constant prices of 1993–94 by using the splicing method.

Sources: Compiled from the estimates of State Domestic Product, 1950–51 to 1993–94; Directorate of Economics and Statistics and Central Statistical Organization.

Appendix A3.2 *Percentages change in each sector of NSDP of Odisha over different decades, 1950–51 to 2005–06*

	Sectors	*50s to 60s*	*60s to 70s*	*70s to 80s*	*80s to 90s*	*2000–01 to 2005–06*	*1950–51 to 2005–06*
1	Agriculture and Animal Husbandry	–10.05	–1.56	–2.89	–27.14	–27.43	–54.53
2	Forestry and Logging	30.40	–15.22	–21.43	–47.78	–17.11	–62.41
3	Fishing	–64.80	87.32	49.32	92.84	–2.20	85.69
1 to 3	Agriculture, Animal Husbandry, Forestry and Fishing	–5.76	–3.42	–5.14	–27.31	–25.03	–52.95
4	Mining and Quarrying	43.19	19.53	14.97	156.02	76.97	791.53
1 to 4	Sub-Total Primary	–5.02	–2.90	–4.57	–21.06	–13.76	–40.08
5	Mnf Regd.	164.05	38.83	–3.61	25.71	–1.14	339.11
6	Mnf Un Regd.	–2.18	–14.82	–25.45	–14.35	–42.98	–69.66
5 and 6	Manufacturing	37.34	9.71	–12.81	11.27	–12.75	27.52
7	Electricity, Gas and Water-Supply	44.15	134.17	19.60	22.14	4.14	413.51
8	Construction	61.47	–15.05	–12.37	23.80	–27.19	8.34
5 to 8	Sub-Total – Secondary	48.02	2.52	–9.97	17.22	–16.63	33.51
9	Trade, Hotel and Restaurants	26.27	8.26	4.70	25.54	10.04	97.70
10	Railways	25.29	14.81	45.21	99.29	24.55	418.45
11	Transport by Other means	21.04	48.80	31.18	44.26	59.67	444.22
12	Storage	20.65	48.68	86.01	90.89	20.76	669.22
13	Communication	18.90	28.39	38.16	70.99	337.09	1476.39
10 to 13	Transport, Storage and Communications	22.35	33.67	36.92	66.61	76.31	557.77

Cont'd...

...Cont'd

9 to 13	Sub-Total Tertiary	25.43	13.59	12.66	37.86	34.08	196.68
14	Banking and Insurance	106.66	83.71	152.31	99.39	63.65	3025.56
15	Real Estate, Ownership of Dwellings and Business Services	–37.83	–2.31	7.94	–1.41	–10.50	–42.15
14 and 15	Financing, Insurance, Real Estate and Business Services	–34.94	3.16	24.28	21.75	17.40	19.21
16	Public Administration	54.20	20.76	43.20	25.97	5.47	254.29
17	Other Services	–19.72	–10.83	5.68	53.33	22.08	41.60
16 and 17	Community, Social and Personal Services	–4.67	–0.42	20.67	40.36	15.01	84.91
14 to 17	Sub Total – Finance and Services	–21.19	1.19	22.33	31.68	16.04	49.05
9 to 17	Service	–7.05	6.27	18.10	34.26	23.77	93.83

Source and Notes: Same as Appendix A3.1.

Appendix A3.1 and A3.2 Sub-sectoral Composition of NSDP

As the sub-sectoral share to NSDP is concerned, within primary sector, the major contributors are (1) 'agriculture and animal husbandry' (2) 'forestry and logging' and (3) 'mining and quarrying'. During 1950s, the share of agriculture and animal husbandry was 54.74 per cent of NSDP, which constituted 84 per cent of primary sector. By 1990s, the share of agriculture and animal husbandry has declined to 34.29 per cent of NSDP and 75 per cent of primary sector. By 2000–01 to 2005–06, the share of 'agriculture and animal husbandry' has declined further to 24.89 per cent of NSDP, which was 63 per cent of the primary sector. The share of 'forestry and logging', to NSDP was 8.45 per cent during 1950s, which constituted 13 per cent of primary sector; it has come down to 3.83 per cent of NSDP, which constitutes 8 per cent of primary sector during 1990s and remained at around that level for the subsequent period (both as a share of NSDP and share to primary sector). Mining and quarrying constitutes negligible share to NSDP, but it is the only subsector of primary sector that has shown persistent improvements both in terms of its share to NSDP (from 1% during 1990s to 8.88% during 2000–2005) and in terms of it share to primary sector (from 1.5% during 1950s to 22.6% during 2000–2005).

In the secondary sector, though the share of secondary sector as a whole has not shown much improvement, there is a change in its compositions within the sector – overall, manufacturing sector has shown an improvement (contributing 5.52% to NSDP during 1950s and 8% during 1990s). But within manufacturing, the unregistered manufacturing, which was the major contributors has shown a sharp decline (4.20% of NSDP in 1950s to 1.28% during 2000–01 to 2005–06). The share of registered manufacturing has shown a sharp increase (from 1.31% of NSDP during 1950s to 5.76% during 2000–01 to 2005–06). The other better looking subsector of secondary sector in terms of its share is construction, which has also not shown much changes in its share to NSDP.

Service sector has three broad subsectors (a) tertiary sector, which is composed of (i) 'trade, hotel and restaurants' (ii) 'railways, transport by other means, storage and communication'; (b) finance, which is composed of (i) 'banking and insurance' (ii) 'real estate, ownership of dwelling and business services' and (c) community, personal and social service, which is composed of (i) 'public administrations' and (ii) 'other services'.

In the tertiary sector, the major contributors to NSDP is the 'trade, hotel and restaurant', which altogether were contributing 5.77 per cent to NSDP during 1950s, and has persistently increased to 10.36 per cent during 1990s and 11.40 per cent during 2000–01 to 2005–06. However, the share of this subsector to tertiary sector has remained at around 24 per cent, though there is a little variation. This implies that other subsectors of the service sector have increased either at par with or higher than 'trade, hotel and restaurant'. 'Communication' sector as a whole that was contributing 1.58 per cent to NSDP during 1950s has gone up to 10.40 per cent during 2000–01 to 2005–06 and its share with in the tertiary sector during the same time period has gone up from 21 per cent to 47 per cent. As both 'trade, hotel and restaurant' and 'communication' are the components' of tertiary sector and have shown improvements, the contributions of tertiary sector to NSDP also reflected this – tertiary sector as a whole that was contributing 7.35 per cent to the NSDP during 1950s has gone up to 21.80 per cent of NSDP during 2000–01 to 2005–06. The share of tertiary sector in the service sector also has shown an improvement from 30 per cent during 1950s to around 46 per cent during 2000–01 to 2005–06.

The share of subsector 'finance' shows, while the 'banking and insurance' component has shown a steady increase from 0.18 per cent of NSDP during 1950s to 5.76 per cent during 2000–01 to 2005–06, the 'real estate, ownership of dwelling and business services' has shown a decline from 9.03 per cent of NSDP during 1950s to 5.22 per cent during 2000–01 to 2005–06. Overall, the finance sector showed a marginal increase.

'Community, social and personal service' subsector shows that both of its subsectors have shown improvements in their respective shares to NSDP. Share of public administration to NSDP has increased from 1.56 per cent during 1950s to 5.53 per cent during 2000–01 to 2005–06. Similarly, share of 'other services' to NSDP has increased from 6.10 per cent during 1950s to 8.64 per cent during 2000–01 to 2005–06. Overall the 'community, social and personal services' has increased from 7.66 per cent to NSDP during 1950s to 14.17 per cent during 2000–01 to 2005–06.

Appendix A3.4 *Percentages distribution of NDP of India across different sectors, 1950–51 to 2005–06*

	Years	*50s*	*60s*	*70s*	*80s*	*90s*	*2000–01 to 2005–06*
1	Agriculture and Animal Husbandry	51.13	43.76	39.58	34.94	28.45	20.56
2	Forestry and Logging	6.88	8.40	9.98	8.54	8.67	11.08
3	Fishing	0.97	1.01	1.07	0.99	1.03	0.91
1 to 3	Agriculture, Animal Husbandry, Forestry and Fishing	58.04	49.83	45.05	38.48	30.99	22.54
4	Mining and Quarrying	1.61	1.99	1.88	2.11	2.17	2.40
1 to 4	Sub-Total Primary	59.65	51.82	46.92	40.59	33.15	24.94
5	Mnf Regd.	4.39	6.00	6.89	8.44	9.70	9.77
6	Mnf Unregd.	4.97	5.52	5.64	5.77	5.68	5.40
5 and 6	Manufacturing	9.36	11.51	12.53	14.21	15.39	15.17
7	Electricity, Gas and Water-Supply	0.33	0.57	0.81	1.03	1.37	1.18
8	Construction	4.49	6.16	6.19	5.80	5.54	5.95
5 to 8	Sub-Total – Secondary	14.17	18.24	19.53	21.03	22.29	22.29
9	Trade	8.90	10.71	11.72	12.86	13.90	16.71
10	Hotel and Restaurants	0.52	0.62	0.63	0.68	0.82	1.11
9 and 10	Trade, Hotel and Restaurants	9.41	11.33	12.35	13.53	14.73	17.82
11	Railways	0.90	1.03	1.10	1.10	0.98	0.87
12	Transport by Other means	1.59	1.94	2.50	3.10	3.37	4.10
13	Storage	0.08	0.07	0.08	0.11	0.08	0.06
14	Communication	0.43	0.59	0.76	0.93	1.31	2.28
11 to 14	Transport, Storage and Communications	2.99	3.63	4.44	5.24	5.73	7.31

Cont'd...

...Cont'd

9 to 14	Tertiary	12.41	14.96	16.79	18.77	20.46	25.13
15	Banking and Insurance	1.17	1.54	2.17	3.31	5.96	7.58
16	Real Estate, Ownership of Dwellings and Business Services	3.20	3.10	3.10	3.98	5.43	6.21
15 and 16	Financing, Insurance, Real Estate and Business Services	4.38	4.64	5.27	7.29	11.38	13.79
17	Public Administration	2.54	3.54	4.66	5.55	5.60	5.91
18	Other Services	6.85	6.80	6.83	6.76	7.11	7.94
17 and 18	Community, Social and Personal Services	9.39	10.33	11.48	12.31	12.71	13.85
9 to 18	Sub Total – Finance and Services	26.17	29.94	33.55	38.38	44.55	52.76
1 to 18	NDP	100	100	100	100	100	100

Source and Notes: Same as Appendix A3.1.

Appendix A3.5 *Percentages change in each sector of NDP of India over different decades, 1950–51 to 2005–06*

	Years	*50s to 60s*	*60s to 70s*	*70s to 80s*	*80s to 90s*	*2000–01 to 2005–06*	*1950–51 to 2005–06*
1	Agriculture and Animal Husbandry	–14.42	–9.56	–11.71	–18.57	–27.75	–59.80
2	Forestry and Logging	22.08	18.84	–14.41	1.47	27.85	61.07
3	Fishing	3.90	5.92	–7.14	3.72	–11.80	–6.52
1 to 3	Agriculture, Animal Husbandry, Forestry and Fishing	–14.14	–9.60	–14.57	–19.48	–27.26	–61.17
4	Mining and Quarrying	23.40	–5.60	12.51	2.72	10.89	49.28
1 to 4	Sub-Total Primary	–13.13	–9.45	–13.49	–18.33	–24.76	–58.19
5	Mnf Regd.	36.59	14.86	22.49	14.98	0.64	122.37
6	Mnf Unregd.	11.05	2.20	2.31	–1.43	–4.97	8.76
5 and 6	Manufacturing	23.03	8.80	13.41	8.32	–1.43	62.08
7	Electricity, Gas and Water-Supply	73.00	42.32	25.97	33.54	–14.13	255.63
8	Construction	37.23	0.55	–6.34	–4.52	7.45	32.61
5 to 8	Sub-Total – Secondary	28.69	7.07	7.68	6.01	–0.01	57.27
9	Trade	20.40	9.44	9.66	8.13	20.22	87.82
10	Hotel and Restaurants	19.73	1.66	8.06	21.70	34.53	115.33
9 and 10	Trade, Hotel and Restaurants	20.36	9.02	9.58	8.80	21.02	89.33
11	Railways	15.06	6.57	–0.09	–10.75	–11.67	–3.42
12	Transport by Other means	22.09	28.71	24.19	8.61	21.71	157.96
13	Storage	–9.64	23.08	26.68	–28.30	–22.34	–21.55
14	Communication	36.95	29.22	22.08	40.74	74.73	431.25
11 to 14	Transport, Storage and Communications	21.31	22.39	17.86	9.49	27.50	144.29
9 to 14	Tertiary	20.59	12.26	11.77	9.00	22.84	102.59
15	Banking and Insurance	31.83	40.36	52.99	79.81	27.17	547.33
16	Real Estate, Ownership of Dwellings and Business Services	–3.23	–0.02	28.32	36.37	14.40	93.69
15 and 16	Financing, Insurance, Real Estate and Business Services	6.15	13.40	38.47	56.11	21.08	215.06
17	Public Administration	39.20	31.67	19.14	0.94	5.45	132.42
18	Other Services	–0.78	0.46	–0.92	5.06	11.73	15.93
17 and 18	Community, Social and Personal Services	10.03	11.14	7.21	3.21	8.96	47.45
9 to 18	Sub Total – Finance and Services	14.39	12.05	14.40	16.09	18.43	101.61

Source and Notes: Same as appendix A3.1.

Appendix A3.6 *Annual Compound Growth rates of NSDP of different sectors and of per capita income of Odisha, 1950–51 to 2005–06 (%)*

	Sectors	*50s*	*60s*	*70s*	*80s*	*90s*	*2000–01 to 2005–06*	*1950–51 to 2005–06*
1	Agriculture and Animal Husbandry	–0.98	3.49	1.07	3.08	1.27	4.77	1.43
2	Forestry and Logging	–2.77	6.39	–0.05	–2.72	–0.28	2.00	0.49
3	Fishing	–13.08	15.39	6.91	9.08	8.35	3.90	5.53
1 to 3	Agriculture, Animal Husbandry, Forestry and Fishing	–1.38	4.26	0.99	2.42	1.49	4.40	1.42
4	Mining and Quarrying	8.38	4.74	3.74	8.72	13.63	14.78	7.15
1 to 4	Sub-Total Primary	–1.25	4.27 *	1.07	2.64 *	2.75 *	6.54 *	1.81 $
5	Mnf Regd.	6.13	10.26	9.09	15.16	2.03	22.23	5.77
6	Mnf Unregd.	4.54	1.70	3.12	2.56	1.85	3.85	0.64
5 and 6	Manufacturing	4.90	5.12	6.47	10.35	2.28	18.56	3.33
7	Electricity, Gas and Water-Supply	1.70	8.59	7.88	5.17	–1.26	15.22	6.66
8	Construction	19.90	–5.60	6.68	3.55	1.41	–1.09	3.20
5 to 8	Sub-Total – Secondary	10.30 $	0.40	6.58 $	7.14 $	1.55	10.63 *	3.49 $
9	Trade, Hotel and Restaurants	6.64	5.50	1.56	5.66	6.12	7.81	4.21
10	Railways	4.93	4.96	4.38	12.81	2.88	8.65	6.45
11	Transport by Other means	4.34	7.55	4.65	11.90	6.74	13.79	6.35
12	Storage	4.33	7.39	4.60	9.90	13.02	5.03	7.65
13	Communication	7.09	3.63	5.36	7.89	12.91	21.60	7.57
10 to 13	Transport, Storage and Communications	4.89	6.14	4.63	11.75	6.00	14.30	6.64
9 to 13	Sub-Total Tertiary	6.26	5.64	2.33	7.45	6.07	10.86	4.96
14	Banking and Insurance	13.69	11.70	9.78	14.54	8.75	13.46	10.74
15	Real Estate, Ownership of Dwellings and Business Services	2.29	0.28	3.67	2.80	2.76	3.20	2.04
14 and 15	Financing, Insurance, Real Estate and Business Services	2.51	0.93	4.36	5.42	5.01	8.40	3.44
16	Public Administration	6.20	5.29	4.53	7.55	6.03	2.39	5.67
17	Other Services	5.00	0.45	1.43	7.49	8.61	5.01	3.53
16 and 17	Community, Social and Personal Services	5.25	1.94	2.69	7.51	7.50	3.98	4.21
14 to 17	Sub Total – Finance and Services	3.76	1.48	3.44	6.53	6.45	5.86	3.83
9 to 17	Service	4.54 *	3.16 $	2.96 $	6.92 $	6.29 $	8.15 $	4.25 $
1 to 17	Total NSDP	1.20 *	3.46 *	2.39 *	4.48 $	3.91 $	7.84 $	2.81 $
	Per Capita Income	–0.32	1.16	0.47	2.92 *	2.44 $	6.64 $	0.99 $

Notes: 50s – 1950–51 to 1959–60, 60s – 1960–61 to 1969–70, 70s – 1970–71 to 1979–80, 80s – 1980–81 to 1989–90, 90s – 1990–91 to 1999–2000. All sectors' output and per capita income have been converted to constant prices of 1993–94 by using splicing method. Statistical significance has been shown for major sector only. * and $ implies 't' statistics significant at 5 and 10 per cent level of significance respectively.

Source: Same as Table 3.1.

Appendix A3.7 *Annual Compound Growth rates of NDP of different sectors and per capita NDP of all India, 1950–51 to 2005–06 (%)*

	Years	*50s*	*60s*	*70s*	*80s*	*90s*	*2000–01 to 2005–06*	*1950–51 to 2005–06*
1	Agriculture and Animal Husbandry	2.84	1.12	1.79	3.15	3.33	4.86	2.67
2	Forestry and Logging	0.25	3.27	–0.61	–0.84	0.68	4.05	0.63
3	Fishing	5.00	4.24	2.83	5.71	5.02	7.96	4.36
1 to 3	Agriculture, Animal Husbandry, Forestry and Fishing	2.60	1.40	1.59	2.95	3.25	4.95	2.56
4	Mining and Quarrying	4.05	4.54	3.79	5.74	4.41	17.40	5.11

Cont'd...

...Cont'd

1 to 4	Sub-Total Primary	2.64$	1.51$	1.67$	3.10$	3.33$	6.09$	2.69$
5	Mnf Regd.	6.01	4.74	5.13	8.86	6.19	12.94	6.23
6	Mnf Unregd.	5.20	3.48	4.50	5.77	6.15	9.01	4.62
5 and 6	Manufacturing	5.58	4.14	4.85	7.56	6.18	11.52	5.52
7	Electricity, Gas and Water-Supply	6.89	9.52	6.33	9.62	6.60	5.18	7.50
8	Construction	5.62	6.79	3.11	3.61	4.75	14.82	4.75
5 to 8	Sub-Total – Secondary	5.62$	5.20$	4.36$	6.53$	5.84$	12.06$	5.37$
9	Trade	5.04	4.46	4.92	5.85	8.33	14.11	5.62
10	Hotel and Restaurants	4.69	4.23	4.47	5.75	10.59	13.47	5.77
9 and 10	Trade, Hotel and Restaurants	5.02	4.44	4.90	5.85	8.46	14.07	5.63
11	Railways	6.42	3.68	4.50	5.41	3.02	7.82	4.40
12	Transport by Other means	5.75	4.70	7.77	5.72	7.93	15.46	6.46
13	Storage	1.99	3.12	12.92	3.19	1.46	8.00	4.40
14	Communication	6.51	6.68	6.03	5.17	15.86	16.23	7.63
11 to 14	Transport, Storage and Communications	5.97	4.68	6.72	5.51	8.71	14.75	6.24
9 to 14	Tertiary	5.25	4.50	5.37	5.75	8.53	14.26	5.79
15	Banking and Insurance	7.69	5.00	7.80	12.39	11.43	11.07	8.71
16	Real Estate, Ownership of Dwellings and Business Services	3.16	3.30	3.29	10.68	6.34	12.91	5.93
15 and 16	Financing, Insurance, Real Estate and Business Services	4.36	3.86	5.16	11.46	8.99	11.92	7.04
17	Public Administration	4.91	7.58	4.76	7.29	6.46	9.41	6.29
18	Other Services	2.83	3.88	2.81	5.54	7.13	10.00	4.61
17 and 18	Community, Social and Personal Services	3.39	5.11	3.60	6.33	6.84	9.75	5.21
9 to 18	Sub Total – Finance and Services	4.43$	4.61$	4.73$	6.99$	8.14$	12.45$	5.86$
1 to 18	NDP	3.53$	3.07$	3.23$	5.30$	6.03$	10.75$	4.40$
	Per Capita NDP	1.57$	0.84	0.93	3.08$	3.96$	—	1.81$^

^ 1950–51 to 1999–2000

Source and all other notes: Same as Table 3.5.

Appendix A3.8 *Trend in percentages distribution of workers in the rural and urban areas of India and Odisha from 1977–78 to 2004–05*

	All India			*Rural India*			*Urban India*		
NSS Round (year)	*Male*	*Female*	*Total*	*Male*	*Female*	*Total*	*Male*	*Female*	*Total*
32 (1977–78)	54.3	29.7	42.3	55.2	33.1	44.4	50.8	15.6	34.1
38 (1983)	53.8	29.51	42	54.7	34	44.5	51.2	15.1	34
43 (1987–88)	53.1	28.5	41.2	53.9	32.3	43.4	50.6	15.2	33.7
50 (1993–94)	54.5	28.6	42	55.3	32.8	44.4	52.1	15.5	34.7
55 (1999–00)	52.7	25.9	39.7	53.1	29.9	41.7	51.8	13.9	33.7
61 (2004–05)	54.7	28.7	42	54.6	32.7	43.9	54.9	16.6	36.5
	All Odisha			*Rural Odisha*			*Urban Odisha*		
NSS Round (year)	*Male*	*Female*	*Total*	*Male*	*Female*	*Total*	*Male*	*Female*	*Total*
32 (1977–78)	55.7	26.2	41.1	56.4	27.4	44.9	50.4	14.9	34.1
38 (1983)	56.5	27.3	42.1	57.5	29.4	43.4	50.1	11.1	32.1
43 (1987–88)	55.6	25.7	40.8	56.6	27.6	42.2	49.3	12.5	32.2
50 (1993–94)	55.7	29.6	42.8	56.6	31.7	44.2	51	15.7	34.1
55 (1999–00)	53.9	27.7	41.0	55.1	29.9	42.3	47.5	14.5	31.7
61 (2004–05)	57.2	29.5	43.6	58.6	32.2	42.2	50.4	14.8	33.4

Source: Sarvekshana Reports of the various quinquennial NSSO rounds.

Appendix A3.9 *Percentages distributions of workers by rural urban location in Odisha from 1977–78 to 2004–05*

		Agri	M and Q	Primary	Manf.	EGWC	Secondary	T and H	Trans	FIPAD	Service	Total
Rural	1977–78	97.88	82.04	97.75	76.29	65.56	74.29	67.59	37.87	58.64	59.98	91.22
	1983	97.97	55.78	97.47	78.91	76.01	78.47	75.54	45.82	63.21	66.00	90.78
	1987–88	98.08	79.24	97.67	80.32	82.71	81.08	68.68	47.03	60.69	62.40	89.87
	1993–94	97.63	80.19	97.35	75.31	66.91	73.16	68.29	42.97	53.97	57.33	88.91
	1999–00	97.89	69.63	97.66	76.27	69.84	74.30	60.37	55.42	57.18	58.39	88.51
	2004–05	97.32	84.66	97.16	85.21	78.16	82.68	67.57	62.88	57.02	62.76	88.01
Urban	1977–78	2.12	17.96	2.25	23.71	34.44	25.71	32.41	62.13	41.36	40.02	8.78
	1983	2.03	44.22	2.53	21.09	23.99	21.53	24.46	54.18	36.79	34.00	9.22
	1987–88	1.92	20.76	2.33	19.68	17.29	18.92	31.32	52.97	39.31	37.60	10.13
	1993–94	2.37	19.81	2.65	24.69	33.09	26.84	31.71	57.03	46.03	42.67	11.09
	1999–00	2.11	30.37	2.34	23.73	30.16	25.70	39.63	44.58	42.82	41.61	11.49
	2004–05	2.68	15.34	2.84	14.79	21.84	17.32	32.43	37.12	42.98	37.24	11.99

Notes: Agri. – Agriculture, M and Q – Mining and quarrying, Manf. – Manufacturing, EGWC – Electricity, Gas, Water supply and Construction, T and H – Trade and Hotel, Trans – Transport, FIPAD – Finance, Insurance and public administration, etc.

Source: Sarvekshana Reports of the various quinquennial NSSO rounds.

Appendix A3.10 *Trend in percentages distribution of workers across the sectors by rural urban location in Odisha and India from 1977–78 to 2004–05*

A		Agri	M and Q	Primary	Manf.	EGWC	Secondary	T and H	Trans	FIPAD	Service	Total
Rural	1977–78	85.12	0.61	85.73	4.88	0.96	5.84	3.20	0.43	4.80	8.43	100
	1983	79.17	0.54	79.71	8.38	1.45	9.83	4.29	0.58	5.70	10.47	100
	1987–88	75.91	1.37	77.28	7.83	3.71	11.54	4.95	0.82	5.49	11.18	100
	1993–94	80.97	1.06	82.03	6.34	1.94	8.28	4.38	0.95	4.60	9.69	100
	1999–00	78.08	0.46	78.54	8.14	3.29	11.43	4.63	1.17	4.23	10.03	100
	2004–05	68.91	0.76	69.67	11.02	5.67	16.69	6.81	1.98	4.85	13.64	100
Urban	1977–78	19.13	1.40	20.53	15.77	5.23	21.01	15.96	7.35	35.16	58.47	100
	1983	16.18	4.18	20.36	22.04	4.51	26.55	13.67	6.75	32.67	53.09	100
	1987–88	13.16	3.18	16.34	17.01	6.89	23.90	20.03	8.19	31.54	59.76	100
	1993–94	15.76	2.11	17.87	16.67	7.68	24.35	16.31	10.05	31.42	57.78	100
	1999–00	12.97	1.54	14.52	19.50	10.95	30.45	23.40	7.25	24.38	55.04	100
	2004–05	13.93	1.02	14.95	14.04	11.62	25.66	23.99	8.58	26.82	59.39	100
Total	1977–78	79.33	0.68	80.01	5.84	1.33	7.17	4.32	1.04	7.46	12.82	100
	1983	73.36	0.87	74.23	9.64	1.73	11.37	5.15	1.15	8.19	14.40	100
	1987–88	69.55	1.55	71.11	8.76	4.03	12.79	6.48	1.57	8.13	16.10	100
	1993–94	73.73	1.18	74.91	7.49	2.57	10.06	5.71	1.96	7.57	15.02	100
	1999–00	70.60	0.58	71.19	9.44	4.17	13.61	6.79	1.87	6.54	15.20	100
	2004–05	62.32	0.79	63.11	11.38	6.38	17.76	8.87	2.77	7.48	19.12	100
B							*All India*					
Rural	1977–78	83.51	0.41	83.92	6.25	1.45	7.70	3.31	0.82	4.48	8.60	100
	1983	81.26	0.50	81.76	6.74	1.84	8.58	3.45	1.08	4.86	9.39	100
	1987–88	78.26	0.59	78.85	7.23	3.54	10.77	4.03	1.32	5.16	10.50	100
	1993–94	78.40	0.60	79.00	7.00	2.60	9.60	4.30	1.40	5.70	11.40	100
	1999–00	76.30	0.50	76.80	7.40	3.50	10.90	5.10	2.10	5.20	12.40	100
	2004–05	72.70	0.50	73.20	8.10	5.10	13.20	6.10	2.50	5.00	13.60	100

Cont'd...

...Cont'd

		Agri	M and Q	Primary	Manf.	EGWC	Secondary	T and H	Trans	FIPAD	Service	Total
Urban	1977–78*											
	1983	14.7	1.0	15.7	26.7	5.5	32.3	18.0	8.1	25.0	51.1	100
	1987–88	13.5	1.2	14.7	26.0	6.3	32.4	19.0	7.8	26.3	53.2	100
	1993–94	12.3	1.2	13.5	23.6	7.3	30.9	19.4	7.9	28.2	55.5	100
	1999–00	8.8	0.8	9.6	22.7	8.7	31.4	26.9	8.7	23.6	59.2	100
	2004–05	8.8	0.8	9.6	24.6	8.7	33.3	24.6	8.6	24.0	57.2	100
Total	1977–78	65.76	0.49	66.25	12.82	2.05	14.87	6.42	1.93	10.53	18.88	100
	1983	68.64	0.60	69.24	10.61	2.51	13.12	6.31	2.44	8.89	17.64	100
	1987–88	64.89	0.72	65.61	11.07	4.11	15.17	7.09	2.65	9.47	19.22	100
	1993–94	63.96	0.71	64.67	10.63	3.63	14.26	7.58	2.86	10.62	21.07	100
	1999–00	60.29	0.57	60.86	10.99	4.66	15.65	10.26	3.67	9.56	23.49	100
	2004–05	56.39	0.55	56.94	12.17	5.92	18.10	11.08	4.04	9.84	24.96	100

Notes: Agri. – Agriculture, M and Q – Mining and quarrying, Manf. – Manufacturing, EGWC – Electricity, Gas, Water supply and Construction, T and H – Trade and Hotel, Trans – Transport, FIPAD – Finance, Insurance and public administration etc.* mismatch found in the figure, hence not reported

Source: Sarvekshana Reports of the various quinquennial NSSO rounds.

Appendix A3.11 *Percentages distributions of workers by gender (male and female) in Odisha from 1977–78 to 2004–05*

		Agri	*M and Q*	*Primary*	*Manf.*	*EGWC*	*Secondary*	*T and H*	*Trans*	*FIPAD*	*Service*	*Total*
Male	1977–78	66.96	64.32	66.94	68.10	84.99	71.24	69.85	98.40	76.53	76.05	68.41
	1983	65.52	81.96	65.71	64.11	78.38	66.28	74.83	97.63	81.56	80.30	67.88
	1987–88	66.47	76.68	66.69	57.09	77.74	63.61	82.55	97.78	80.13	82.80	68.89
	1993–94	62.60	65.98	62.65	62.55	80.61	67.17	80.76	97.88	78.16	81.50	65.94
	1999–00	64.21	86.93	64.39	50.71	79.51	59.54	86.91	97.50	77.15	84.01	66.71
	2004–05	61.86	70.55	61.97	55.41	79.81	64.17	88.10	98.70	72.84	83.67	66.51
Female	1977–78	33.04	35.68	33.06	31.90	15.01	28.76	30.15	1.60	23.47	23.95	31.59
	1983	34.48	18.04	34.29	35.89	21.62	33.72	25.17	2.37	18.44	19.70	32.12
	1987–88	33.53	23.32	33.31	42.91	22.26	36.39	17.45	2.22	19.87	17.20	31.11
	1993–94	37.40	34.02	37.35	37.45	19.39	32.83	19.24	2.12	21.84	18.50	34.06
	1999–00	35.79	13.07	35.61	49.29	20.49	40.46	13.09	2.50	22.85	15.99	33.29
	2004–05	38.14	29.45	38.03	44.59	20.19	35.83	11.90	1.30	27.16	16.33	33.49

Notes: Agri. – Agriculture, M and Q – Mining and quarrying, Manf. – Manufacturing, EGWC – Electricity, Gas, Water supply and Construction, T and H – Trade and Hotel, Trans – Transport, FIPAD – Finance, Insurance and public administration etc.

Source: Sarvekshana Reports of the various quinquennial NSSO rounds.

Appendix A3.12 *Trend in percentages distribution of workers across the sectors by gender in Odisha from 1977–78 to 2004–05*

		Agri	*M and Q*	*Primary*	*Manf.*	*EGWC*	*Secondary*	*T and H*	*Trans*	*FIPAD*	*Service*	*Total*
Male	1977–78	77.64	0.64	78.28	5.81	1.66	7.47	4.41	1.49	8.35	14.25	100
	1983	70.81	1.05	71.86	9.10	2.00	11.10	5.68	1.51	9.84	17.03	100
	1987–88	67.11	1.73	68.84	7.26	4.55	11.81	7.76	2.14	9.45	19.35	100
	1993–94	70.00	1.18	71.18	7.10	3.15	10.25	6.99	2.60	8.98	18.57	100
	1999–00	67.95	0.76	68.71	7.18	4.97	12.15	8.84	2.73	7.57	19.14	100
	2004–05	57.96	0.84	58.80	9.48	7.66	17.14	11.75	4.11	8.19	24.06	100
Female	1977–78	82.98	0.77	83.75	5.90	0.63	6.53	4.13	0.05	5.54	9.72	100
	1983	78.75	0.49	79.24	10.77	1.17	11.93	4.04	0.08	4.70	8.83	100
	1987–88	74.97	1.16	76.13	12.08	2.89	14.96	3.63	0.11	5.19	8.90	100

Cont'd...

...Cont'd

	1993–94	80.96	1.18	82.14	8.23	1.47	9.70	3.22	0.11	4.85	8.16	100
	1999–00	75.92	0.23	76.15	13.98	2.57	16.55	2.67	0.14	4.49	7.30	100
	2004–05	70.97	0.70	71.67	15.16	3.85	19.00	3.15	0.11	6.07	9.33	100
Total	1977–78	79.33	0.68	80.01	5.84	1.33	7.17	4.32	1.04	7.46	12.82	100
	1983	73.36	0.87	74.23	9.64	1.73	11.37	5.15	1.05	8.19	14.40	100
	1987–88	69.55	1.55	71.11	8.76	4.03	12.79	6.48	1.51	8.13	16.10	100
	1993–94	73.73	1.18	74.91	7.49	2.57	10.06	5.71	1.75	7.57	15.02	100
	1999–00	70.60	0.58	71.19	9.44	4.17	13.61	6.79	1.87	6.54	15.20	100
	2004–05	62.32	0.79	63.11	11.38	6.38	17.76	8.87	2.77	7.48	19.12	100

Notes: Agri. – Agriculture, M and Q – Mining and quarrying, Manf. – Manufacturing, EGWC – Electricity, Gas, Water supply and Construction, T and H – Trade and Hotel, Trans – Transport, FIPAD – Finance, Insurance and public administration, etc.

Source: Sarvekshana Reports of the various quinquennial NSSO rounds.

Appendix A3.13 *Trend in the Employment Growth in Odisha*

		Agri	*M and Q*	*Primary*	*Manf.*	*EGWC*	*Secondary*	*T and H*	*Trans*	*FIPAD*	*Service*	*Total*
Male	2nd to 3rd	0.50	13.06	0.63	11.97	6.31	10.82	7.66	2.65	5.79	6.08	2.37
	3rd to 4th	0.30	11.95	0.52	–3.10	19.49	2.65	7.92	8.63	0.57	4.01	1.39
	4th to 5th	2.97	–5.39	2.80	1.67	–5.16	–0.75	–0.02	6.22	1.06	1.27	2.11
	3rd to 5th	1.63	2.92	1.65	–0.74	6.45	0.94	3.88	7.42	0.82	2.63	1.75
	5th to 6th	0.58	–7.31	0.47	1.40	10.89	4.69	6.06	2.17	–2.21	1.80	1.19
	6th to 7th	–0.50	4.82	–0.44	8.60	11.97	10.03	8.73	11.46	4.36	7.52	2.71
	5th to 7th	0.04	–1.44	0.02	4.94	11.42	7.33	7.39	6.71	1.02	4.62	1.95
Female	2nd to 3rd	1.80	–6.03	1.74	16.04	16.24	16.06	2.42	11.21	–0.46	0.91	2.87
	3rd to 4th	–0.54	19.43	–0.36	2.78	20.38	5.09	–1.65	7.15	2.45	0.61	0.44
	4th to 5th	6.51	5.15	6.49	–2.85	–8.40	–3.82	2.40	5.26	3.49	3.08	4.89
	3rd to 5th	2.93	12.07	3.01	–0.07	5.01	0.54	0.36	6.20	2.97	1.84	2.64
	5th to 6th	–0.80	–27.56	–1.02	11.71	12.41	11.81	–3.24	5.65	–1.06	–1.72	0.49
	6th to 7th	1.53	28.56	1.66	4.59	11.55	5.79	6.39	–2.48	9.28	8.07	2.91
	5th to 7th	0.36	–3.50	0.31	8.09	11.98	8.76	1.46	1.50	3.98	3.06	1.69
Total	2nd to 3rd	0.94	7.70	1.00	13.33	8.05	12.42	6.19	2.81	4.45	4.93	2.53
	3rd to 4th	0.01	13.45	0.22	–0.83	19.68	3.50	5.82	8.60	0.93	3.37	1.09
	4th to 5th	4.22	–2.50	4.09	–0.17	–5.84	–1.82	0.42	6.20	1.56	1.59	3.01
	3rd to 5th	2.09	5.18	2.13	–0.50	6.15	0.80	3.09	7.39	1.25	2.48	2.04
	5th to 6th	0.08	–12.29	–0.08	5.74	11.19	7.24	4.52	2.25	–1.96	1.19	0.95
	6th to 7th	0.24	9.28	0.33	6.70	11.88	8.39	8.43	11.19	5.57	7.61	2.78
	5th to 7th	0.16	–2.09	0.13	6.22	11.53	7.82	6.46	6.63	1.74	4.35	1.86

Notes: Agri. – Agriculture, M and Q – Mining and quarrying, Manf. – Manufacturing, EGWC – Electricity, Gas, Water supply and Construction, T and H – Trade and Hotel, Trans – Transport, FIPAD – Finance, Insurance and public administration, etc.

Source: Sarvekshana Reports of the various quinquennial NSSO rounds.

Appendix A3.14 *Trend in percentages distribution of workers across different sectors by rural urban location in different regions of Odisha from 1983 to 2004–05*

		Agri	*M and Q*	*Primary*	*Manf.*	*EGWC*	*Secondary*	*T and H*	*Trans*	*FIPAD*	*Service*	*Total*
						Coastal						
Rural	1983	74.02	0.96	74.98	8.84	0.96	9.79	5.89	0.86	8.48	15.23	100
	1993–94	79.74	0.78	80.52	5.28	0.99	6.27	6.29	0.86	6.05	13.20	100
	1999–00	71.82	0.58	72.40	6.94	4.03	10.96	7.63	2.37	6.63	16.64	100
	2004–05	64.57	0.27	64.84	9.91	6.51	16.41	9.56	2.73	6.46	18.74	100

Cont'd...

...Cont'd

Urban	1983	13.77	0	13.77	15.45	5.44	20.88	18.15	4.89	42.31	65.35	100
	1993–94	15.48	0	15.48	9.55	6.89	16.45	18.02	11.36	38.69	68.07	100
	1999–00	7.39	0	7.39	22.03	6.31	28.34	24.77	8.61	30.89	64.27	100
	2004–05	15.86	0.00	15.86	11.47	10.13	21.59	29.69	7.63	25.23	62.54	100
Total	1983	67.44	0.85	68.3	9.56	1.45	11.0	7.23	1.3	12.17	20.70	100
	1993–94	71.87	0.69	72.56	5.80	1.71	7.52	7.72	2.14	10.05	19.91	100
	1999–00	61.41	0.49	61.90	9.38	4.40	13.77	10.40	3.38	10.55	24.33	100
	2004–05	58.97	0.24	59.21	10.09	6.92	17.01	11.88	3.29	8.62	23.78	100
							Northern					
Rural	1983	79.92	0.51	80.43	9.24	1.55	10.78	3.40	0.48	4.91	8.79	100
	1993–94	79.91	1.62	81.53	8.80	1.50	10.3	3.01	0.65	4.51	8.16	100
	1999–00	77.68	0.54	78.22	10.26	3.35	13.61	3.64	0.59	3.93	8.16	100
	2004–05	69.29	1.62	70.91	15.23	4.22	19.45	4.22	1.68	3.74	9.64	100
Urban	1983	17.32	8.65	25.97	29.66	4.44	34.10	9.45	8.24	22.25	39.94	100
	1993–94	5.99	5.62	11.62	24.14	8.43	32.58	16.73	12.10	26.98	55.81	100
	1999–00	15.36	4.01	19.37	19.56	14.26	33.82	21.63	7.19	17.99	46.81	100
	2004–05	11.97	2.32	14.30	15.45	11.94	27.40	18.51	10.11	29.68	58.31	100
Total	1983	71.89	1.55	73.44	11.86	1.92	13.77	4.18	1.48	7.13	12.78	100
	1993–94	72.83	2.00	74.83	10.27	2.16	12.44	4.32	1.74	6.66	12.73	100
	1999–00	69.93	0.98	70.91	11.42	4.71	16.13	5.88	1.41	5.68	12.97	100
	2004–05	62.61	1.70	64.31	15.26	5.12	20.38	5.89	2.66	6.76	15.31	100
							Southern					
Rural	1983	85.01	0	85.01	6.51	2.15	8.66	3.38	0	2.96	6.34	100
	1993–94	84.98	0.61	85.61	4.04	3.57	7.61	3.73	0.57	2.49	6.78	100
	1999–00	88.79	0	88.79	6.75	1.99	8.73	1.45	0.16	0.87	2.47	100
	2004–05	76.98	0.07	77.05	5.59	6.66	12.24	6.06	1.21	3.43	10.71	100
Urban	1983	19.97	0	19.97	10.88	2.23	13.11	15.71	6.11	45.09	66.92	100
	1993–94	39.64	0	39.64	20.31	8.87	29.18	11.12	1.81	18.25	31.18	100
	1999–00	26.14	0	26.14	10.76	17.63	28.39	23.22	2.65	19.60	45.47	100
	2004–05	13.95	0.00	13.95	18.55	16.17	34.73	23.28	6.22	21.83	51.32	100
Total	1983	81.38	0	81.38	6.75	2.15	8.90	4.04	0.34	5.31	9.72	100
	1993–94	81.64	0.56	82.23	5.24	3.96	9.20	4.27	0.66	3.65	8.58	100
	1999–00	83.89	0	83.89	7.06	3.21	10.27	3.15	0.35	2.33	5.84	100
	2004–05	73.47	0.07	73.53	6.31	7.19	13.50	7.02	1.49	4.45	12.97	100

Notes: Agri. – Agriculture, M and Q – Mining and quarrying, Manf. – Manufacturing, EGWC – Electricity, Gas, Water supply and Construction, T and H – Trade and Hotel, Trans – Transport, FIPAD – Finance, Insurance and public administration etc.* mismatch found in the figure, hence not reported.

Source: NSSO, Unit level data various quinquennial rounds.

Appendix A3.15 *Trend in percentages distribution of workers across different sectors by gender in different regions of Odisha from 1983 to 2004–05*

		Agri	*M and Q*	*Primary*	*Manf.*	*EGWC*	*Secondary*	*T and H*	*Trans*	*FIPAD*	*Service*	*Total*
							Coastal					
Male	1983	66.78	0.84	67.62	7.93	1.77	9.71	7.89	1.69	13.09	22.68	100
	1993–94	69.35	0.84	70.18	5.55	2.04	7.60	8.57	2.80	10.85	22.22	100
	1999–00	58.95	0.52	59.47	7.94	5.14	13.08	12.27	4.39	10.79	27.45	100
	2004–05	55.09	0.19	55.28	9.06	7.58	16.64	14.64	4.44	9.00	28.08	100

Cont'd...

...Cont'd

Female	1983	69.85	0.89	70.54	14.93	0.37	15.30	5.03	0	9.13	14.17	100
	1993–94	79.57	0.21	79.80	6.56	0.73	7.29	5.17	0.15	7.63	12.95	100
	1999–00	68.82	0.40	69.22	13.69	2.14	15.84	4.78	0.33	9.84	14.95	100
	2004–05	68.91	0.36	69.27	12.71	5.25	17.96	4.79	0.33	7.65	12.77	100
Total	1983	67.44	0.85	68.3	9.56	1.45	11.0	7.23	1.3	12.17	20.7	100
	1993–94	71.89	0.69	72.58	5.81	1.71	7.52	7.73	2.14	10.05	19.92	100
	1999–00	61.41	0.49	61.90	9.38	4.40	13.77	10.40	3.38	10.55	24.33	100
	2004–05	58.97	0.24	59.21	10.09	6.92	17.01	11.88	3.29	8.62	23.78	100
						Northern						
Male	1983	68.77	1.99	70.76	11.85	2.23	14.07	4.47	2.10	8.59	15.16	100
	1993–94	68.84	1.93	70.77	9.44	2.77	12.21	5.84	2.79	8.39	17.02	100
	1999–00	67.95	1.49	69.44	7.88	5.34	13.22	8.08	2.19	7.07	17.35	100
	2004–05	59.19	2.11	61.30	11.40	6.71	18.11	8.55	4.37	7.67	20.59	100
Female	1983	78.01	0.69	78.71	11.88	1.31	13.18	3.59	.26	4.26	8.11	100
	1993–94	79.20	2.11	81.32	11.60	1.20	12.80	1.91	.08	3.90	5.89	100
	1999–00	73.02	0.18	73.19	16.92	3.73	20.65	2.45	0.20	3.51	6.16	100
	2004–05	67.80	1.09	68.89	21.13	2.71	23.83	1.85	0.05	5.37	7.28	100
Total	1983	71.89	1.55	73.44	11.86	1.92	13.77	4.18	1.48	7.13	12.78	100
	1993–94	72.83	2.00	74.83	10.27	2.17	12.44	4.32	1.75	6.66	12.73	100
	1999–00	69.93	0.98	70.91	11.42	4.71	16.13	5.88	1.41	5.68	12.97	100
	2004–05	62.61	1.70	64.31	15.26	5.12	20.38	5.89	2.66	6.76	15.31	100
						Southern						
Male	1983	78.41	0	78.41	6.99	2.42	9.41	4.04	0.60	7.53	12.18	100
	1993–94	77.46	0.60	78.08	5.78	5.44	11.23	5.02	1.14	4.56	10.72	100
	1999–00	82.14	0	82.14	5.42	4.33	9.74	4.44	0.59	3.09	8.11	100
	2004–05	67.37	0.00	67.37	6.24	9.24	15.49	9.46	2.34	5.34	17.14	100
Female	1983	85.22	0	85.22	6.45	1.81	8.25	4.09	0	2.43	6.53	100
	1993–94	87.40	0.51	87.87	4.50	1.93	6.43	3.26	0	2.41	5.66	100
	1999–00	86.54	0	86.54	9.56	1.51	11.08	1.20	0.00	1.18	2.38	100
	2004–05	84.03	0.18	84.21	6.42	3.62	10.04	2.80	0.03	2.91	5.74	100
Total	1983	81.38	0	81.38	6.75	2.15	8.90	4.06	0.34	5.31	9.72	100
	1993–94	81.65	0.56	82.20	5.24	3.96	9.21	4.28	0.66	3.65	8.59	100
	1999–00	83.89	0	83.89	7.06	3.21	10.27	3.15	0.35	2.33	5.84	100
	2004–05	73.47	0.07	73.53	6.31	7.19	13.50	7.02	1.49	4.45	12.97	100

Notes: Agri. – Agriculture, M and Q – Mining and quarrying, Manf. – Manufacturing, EGWC – Electricity, Gas, Water supply and Construction, T and H – Trade and Hotel, Trans – Transport, FIPAD – Finance, Insurance and public administration, etc.

Source: NSSO, Unit level data various quinquennial rounds.

Appendix A3.16 *Trend in the Employment Growth among different regions of Odisha from 1983 to 2004–05*

	Year	*Primary*	*Secondary*	*Service*		*Primary*	*Secondary*	*Service*
				Coastal				
Rural	3rd to 5th	2.36	–2.79	0.19	Male	2.37	–0.48	1.78
	5th to 6th	–0.61	13.52	6.34		–1.42	13.59	6.30
	6th to 7th	–0.94	9.78	3.71		0.18	6.66	2.11
	5th to 7th	–0.77	11.64	5.01		–0.62	10.07	4.19
Urban	3rd to 5th	4.97	1.31	4.18	Female	3.11	–5.43	0.93
	5th to 6th	–10.40	15.82	2.69		–0.99	18.97	4.84
	6th to 7th	20.18	–2.31	2.59		1.53	4.10	–1.63
	5th to 7th	3.77	6.37	2.64		0.26	11.29	1.55

Cont'd...

...Cont'd

Total	3rd to 5th	2.54	–1.89	1.52	Total	2.54	–1.89	1.53
	5th to 6th	–1.30	14.99	6.05		–1.31	14.99	6.04
	6th to 7th	0.69	5.97	1.12		0.69	5.97	1.12
	5th to 7th	–0.31	10.39	3.56		–0.32	10.39	3.55
				Northern				
Rural	3rd to 5th	1.70	1.10	0.81	Male	1.75	0.32	2.93
	5th to 6th	0.69	7.34	1.52		1.27	3.28	2.05
	6th to 7th	–0.70	8.77	4.70		–1.12	7.97	4.91
	5th to 7th	–0.01	8.05	3.10		0.07	5.60	3.47
Urban	3rd to 5th	–5.21	2.26	6.22	Female	2.04	1.41	–1.50
	5th to 6th	13.62	3.35	–0.97		–0.41	11.92	2.63
	6th to 7th	–3.87	–2.06	6.74		0.21	4.38	4.88
	5th to 7th	4.51	0.61	2.81		–0.10	8.09	3.75
Total	3rd to 5th	1.92	0.70	1.69	Total	1.92	0.70	1.69
	5th to 6th	0.60	7.11	2.07		0.60	7.11	2.07
	6th to 7th	–0.55	6.26	4.83		–0.55	6.26	4.83
	5th to 7th	0.02	6.68	3.44		0.02	6.68	3.44
				Southern				
Rural	3rd to 5th	1.96	0.58	2.57	Male	1.87	3.73	0.62
	5th to 6th	2.64	4.73	–16.74		2.94	–0.95	–3.63
	6th to 7th	–1.26	8.68	36.21		–2.36	11.46	17.99
	5th to 7th	0.67	6.69	6.50		0.26	5.07	6.63
Urban	3rd to 5th	9.45	10.71	–5.32	Female	2.22	–0.60	0.46
	5th to 6th	–5.78	1.84	10.42		1.67	13.71	–14.24
	6th to 7th	–9.80	6.48	4.78		1.15	–0.28	21.28
	5th to 7th	–7.81	4.13	7.56		1.41	6.48	1.98
Total	3rd to 5th	2.02	2.25	0.65	Total	2.01	2.26	0.66
	5th to 6th	2.35	4.21	–5.61		2.36	4.19	–5.63
	6th to 7th	–1.00	7.36	19.23		–1.00	7.36	19.23
	5th to 7th	0.66	5.77	6.09		0.67	5.76	6.08

Source: NSSO, Unit level data various quinquennial rounds.

Appendix A5.1 *Percentages distributions of populations of Odisha by social groups, 2001*

	Rural			*Urban*			*Total*		
	SC	*ST*	*Others*	*SC*	*ST*	*Others*	*SC*	*ST*	*Others*
Anugul *	17.67	12.19	70.14	14.32	8.39	77.28	17.20	11.67	71.13
Balangir	17.15	22.62	60.24	15.20	5.39	79.42	16.92	20.63	62.45
Baleshwar	19.79	11.60	68.61	11.07	8.73	80.20	18.84	11.28	69.88
Bargarh *	19.52	20.43	60.05	17.53	6.52	75.95	19.37	19.36	61.27
Baudh *	21.86	13.00	65.14	22.36	2.10	75.55	21.88	12.47	65.65
Bhadrak *	22.78	1.90	75.32	10.65	1.77	87.58	21.50	1.88	76.62
Cuttack	21.18	4.47	74.35	13.52	1.20	85.28	19.08	3.57	77.34
Debagarh *	15.19	34.49	50.32	17.55	22.39	60.06	15.37	33.60	51.03
Dhenkanal	18.77	13.49	67.74	15.58	5.48	78.94	18.49	12.79	68.71
Gajapati *	6.52	56.21	37.28	16.20	2.97	80.83	7.50	50.78	41.72
Ganjam	19.89	3.35	76.76	12.36	0.65	86.99	18.57	2.88	78.56
Jagatsinghapur *	22.21	0.55	77.25	10.49	3.28	86.23	21.05	0.82	78.13
Jajapur *	23.32	7.81	68.87	16.07	6.53	77.39	22.99	7.76	69.25
Jharsuguda *	18.05	40.07	41.88	15.36	16.13	68.50	17.07	31.34	51.59
Kalahandi	17.72	30.42	51.86	17.08	6.76	76.15	17.67	28.65	53.68
Kandhamal	16.61	54.71	28.67	20.74	14.22	65.04	16.89	51.96	31.15

Cont'd...

...Cont'd

Kendrapara *	20.62	0.49	78.89	18.87	1.05	80.08	20.52	0.52	78.95
Kendujhar	11.25	47.81	40.93	13.93	23.56	62.51	11.62	44.50	43.88
Khordha *	16.71	6.06	77.23	9.33	4.00	86.67	13.54	5.18	81.28
Koraput	12.92	56.46	30.62	13.63	15.76	70.61	13.04	49.62	37.34
Malkangiri *	20.99	60.56	18.46	26.31	14.93	58.76	21.35	57.43	21.22
Mayurbhanj	7.47	59.51	33.02	10.53	17.90	71.57	7.68	56.60	35.72
Nabarangapur *	13.63	57.78	28.59	21.73	10.21	68.06	14.10	55.03	30.87
Nayagarh *	13.83	6.09	80.08	18.79	1.24	79.96	14.04	5.88	80.08
Nuapada *	13.29	36.34	50.37	19.25	7.54	73.21	13.62	34.71	51.66
Puri	19.65	0.31	80.04	9.21	0.21	90.58	18.23	0.30	81.47
Rayagada *	13.94	62.79	23.26	13.76	12.13	74.11	13.92	55.76	30.32
Sambalpur	17.60	43.05	39.35	15.56	11.52	72.92	17.04	34.50	48.46
Sonapur *	23.53	10.25	66.22	24.76	3.90	71.34	23.62	9.78	66.60
Sundargarh	8.56	66.41	25.04	8.73	19.24	72.03	8.62	50.19	41.19
Odisha	12.75	8.10	79.16	17.19	24.61	58.20	16.53	22.13	61.34

Source: Census of India, 2001.

Appendix A5.2 *Percentage distribution of work participation rates, main workers and agricultural labourers across the social groups by different districts of Odisha, 2001*

	Work Participation Rate				*Main Workers*				*Agricultural Labour*			
	Total	*GEN*	*SC*	*ST*	*Total*	*GEN*	*SC*	*ST*	*Total*	*GEN*	*SC*	*ST*
Anugul	39.8	38.3	40.4	47.7	65.9	69.6	61.1	54.2	28.00	20.40	42.70	46.80
Balangir	41.9	40.5	41	46.7	62.8	65.5	60.2	57.7	40.10	35.00	46.80	48.30
Baleshwar	31.9	30.4	31	42.2	74.9	78.1	72.9	63.2	33.00	24.30	43.90	58.40
Bargarh	44.1	41.9	45.9	49.2	64.8	68	61.9	58.9	41.50	30.80	55.00	57.90
Baudh	45.7	44.5	47.3	49.6	62	63.6	59.5	58.4	39.50	35.30	50.00	42.10
Bhadrak	28.9	28	30.8	40.7	78.3	79.7	74.7	71.7	28.50	21.30	46.50	75.10
Cuttack	33.9	32.1	38.6	47.7	77.3	80.5	69.1	66.4	22.90	15.20	44.70	42.20
Debagarh	46.1	44.1	46.7	48.8	55.3	59.9	54	49.7	43.70	31.20	52.40	57.20
Dhenkanal	33.4	30.1	37.7	45	73.5	78	66.9	65.2	34.30	21.90	50.70	58.90
Gajapati	53.1	46	51.8	59.1	65.3	67.7	64.2	64	45.20	37.50	48.50	49.70
Ganjam	41.3	39.7	46.2	53.8	62.8	64.2	58.8	55.2	38.60	36.00	46.20	48.60
Jagatsinghapur	31.2	29.8	35.8	48.6	75.1	77.7	66.7	79.5	25.30	17.90	48.40	24.50
Jajapur	27.5	25.9	29.4	36.4	77.8	81.7	73	64.5	30.20	18.90	53.50	46.80
Jharsuguda	37.2	34.1	37.1	42.4	70.2	79.2	66.2	60.3	24.30	13.60	31.00	35.30
Kalahandi	46.5	44.3	44.6	51.7	61.5	64.1	59.7	58.3	50.30	45.40	56.20	55.00
Kandhamal	47.2	42.6	45.9	50.4	57.5	63.1	56.4	55	36.10	33.40	43.20	35.30
Kendrapara	29.8	29.2	31.8	45.3	75.6	76.6	72.2	71.8	25.50	18.80	48.40	42.30
Kendujhar	39.8	36.5	35.3	44.2	63.6	71.6	69.7	55.8	36.40	22.90	38.50	46.90
Khordha *	30.6	29.3	34.5	41.5	83.5	86.8	75	65.1	16.40	11.80	29.00	40.10
Koraput	48.3	42.3	45.7	53.5	61.9	70.3	60.1	57.4	40.20	28.50	44.00	46.30
Malkangiri	49.1	43	43	53.6	62.3	69.2	62.8	60	25.90	21.40	18.80	29.40
Mayurbhanj	46.2	41.4	45.2	49.4	60.2	66.5	59.3	56.9	38.40	25.60	37.40	45.30
Nabarangapur	49.5	46.3	45	52.4	52.2	56.6	54.9	49.4	52.30	44.30	48.90	57.00
Nayagarh	33.3	31.5	37.3	48.5	71.4	74.7	64.3	54.9	32.70	30.20	41.10	39.80
Nuapada	46	43.8	44.2	50.1	53.8	58.2	52.1	48.8	46.30	41.40	51.80	50.60
Puri	30	28.6	36.1	38.6	82.9	85.7	73.2	75.1	25.00	17.00	52.90	49.50
Rayagada	48	39.7	44.9	53.3	62.6	73.4	61.9	58.4	46.10	26.60	48.50	53.40
Sambalpur	45	40.1	46.1	51.5	67	74.7	64.9	59.5	31.80	18.20	37.70	44.20
Sonapur	43.7	42.5	45.2	48.8	62.7	64.4	59.7	59.6	44.60	37.70	58.50	53.90
Sundargarh	40.4	35.2	41.2	44.4	65	77.8	62.8	57	29.80	15.70	35.10	38.00
Odisha	38.8	35	39.3	49	67.2	72.9	64.8	57.4	35.00	25.80	45.70	46.90

Source: Census of India, 2001.

Appendix A5.3 *Percentage distribution of literacy rate across the social groups*

	Total	*General*	*SC*	*ST*
Anugul *	68.8	75.20	57.00	45.30
Balangir	55.70	60.20	53.50	43.60
Baleshwar	70.60	79.30	59.40	31.90
Bargarh *	64.00	70.80	55.60	50.20
Baudh *	57.70	62.40	49.90	46.60
Bhadrak *	73.90	78.90	59.40	27.40
Cuttack	76.70	82.10	61.40	35.70
Debagarh *	60.40	72.10	52.80	45.30
Dhenkanal	69.40	78.20	56.20	39.40
Gajapati *	41.30	57.40	37.10	27.80
Ganjam	60.80	65.40	44.60	35.50
Jagatsinghapur *	79.10	82.80	66.30	48.60
Jajapur *	71.40	81.00	54.60	31.40
Jharsuguda *	70.70	80.80	64.00	57.20
Kalahandi	45.90	51.70	47.10	34.20
Kandhamal	52.70	65.60	53.40	44.50
Kendrapara *	76.80	81.10	60.70	40.10
Kendujhar	59.20	75.90	64.40	40.30
Khordha *	79.60	83.80	65.00	49.90
Koraput	35.70	57.50	35.40	18.70
Malkangiri *	30.50	53.20	49.90	14.70
Mayurbhanj	51.90	71.20	53.60	38.80
Nabarangapur *	33.90	45.90	45.40	24.00
Nayagarh *	70.50	74.50	57.30	47.10
Nuapada *	42.00	47.40	44.00	33.10
Puri	78.00	81.10	64.00	58.70
Rayagada *	36.10	64.00	35.20	20.20
Sambalpur	67.30	79.30	61.50	52.70
Sonapur *	62.80	66.50	56.80	52.20
Sundargarh	64.90	79.70	61.50	52.80
Odisha	63.10	73.90	55.50	37.40

Source: Census of India, 2001.

Appendix A5.4 *Housing conditions across the districts of Odisha by different social groups, 2001*

District	*All*			*SC*			*ST*		
	Good	*Livable*	*Dilapidated*	*Good*	*Livable*	*Dilapidated*	*Good*	*Livable*	*Dilapidated*
Odisha	25.64	64.41	9.95	19.06	64.82	16.12	17.11	74.63	8.25
Bargarh *	21.59	70.28	8.13	14.88	72.92	12.20	12.80	75.88	11.32
Jharsuguda *	28.97	66.02	5.01	21.79	71.25	6.97	18.91	74.77	6.32
Sambalpur	29.37	63.86	6.77	23.57	65.60	10.83	18.44	74.57	7.00
Debagarh *	15.04	75.51	9.45	12.11	75.47	12.42	10.15	78.69	11.15
Sundargarh	29.99	64.75	5.26	26.23	65.06	8.71	20.07	74.45	5.48
Kendujhar	23.12	68.05	8.84	20.52	66.80	12.68	16.24	73.09	10.67
Mayurbhanj	20.33	70.35	9.32	20.18	65.96	13.86	15.00	74.64	10.36
Baleshwar	23.47	61.40	15.13	18.12	63.02	18.86	8.94	66.60	24.46
Bhadrak *	20.80	65.93	13.27	13.74	67.80	18.46	9.61	56.20	34.18
Kendrapara *	21.61	63.70	14.69	14.96	64.06	20.98	12.40	63.10	24.50
Jagatsinghapur *	22.45	55.58	21.98	14.60	55.72	29.68	15.82	57.29	26.89
Cuttack	31.78	51.88	16.34	18.93	54.85	26.22	11.82	59.74	28.44

Cont'd...

...Cont'd

Jajapur *	23.42	58.25	18.33	13.56	58.33	28.11	13.57	62.78	23.65
Dhenkanal	24.12	63.65	12.23	15.00	64.09	20.91	9.24	72.20	18.56
Anugul *	29.82	60.32	9.86	19.82	62.50	17.68	15.60	69.24	15.16
Nayagarh *	22.01	67.89	10.10	14.92	66.34	18.73	12.53	76.82	10.65
Khordha *	38.17	51.54	10.29	24.21	57.80	17.99	15.06	68.50	16.44
Puri	24.13	59.78	16.09	14.37	59.03	26.60	19.97	59.42	20.61
Ganjam	33.74	57.45	8.81	21.88	63.06	15.06	15.08	73.20	11.72
Gajapati *	24.10	71.28	4.62	22.35	69.19	8.45	16.66	79.80	3.54
Kandhamal	25.70	70.17	4.13	24.77	69.87	5.36	22.33	74.02	3.65
Baudh *	20.04	74.71	5.25	16.02	74.57	9.40	18.07	77.00	4.94
Sonapur *	19.61	74.62	5.76	14.68	76.59	8.73	13.81	78.38	7.82
Balangir	23.12	70.91	5.98	20.83	71.73	7.44	17.56	75.88	6.57
Nuapada *	20.72	75.36	3.92	20.78	74.45	4.78	18.90	77.78	3.32
Kalahandi	21.95	71.58	6.47	21.59	71.42	6.99	17.45	77.09	5.46
Rayagada *	24.31	72.73	2.96	21.91	73.66	4.43	18.90	78.78	2.33
Nabarangapur *	20.88	73.66	5.46	21.56	70.10	8.34	17.88	77.28	4.83
Koraput	28.39	67.42	4.19	28.82	65.34	5.84	21.39	74.49	4.12
Malkangiri *	21.28	72.23	6.50	24.38	67.06	8.56	17.87	76.28	5.85

Source: Census of India, 2001.

Annexures

ANNEXURE A: TOPOLOGY OF ODISHA

Geographically, the state as a whole is a mixture of both hilly and plain areas. While the eastern region of the state comprises of fertile plain areas, the southern-western part is dense with hilly areas. Odisha can be divided into five different regions on the basis of morphological peculiarities and geographical, climatic and edaphic conditions. These are (i) the coastal plains, (ii) the river valleys and flood plains, (iii) the rolling uplands, (iv) the plateaus and (v) the mountainous ranges.

The coastal plains also termed as land of 'six deltas', because these areas are the gift of six major rivers, which bring silt from their catchment and reclaimed areas from the shore of the Bay of Bengal. The whole coastal plains have been divided into three parts: (i) The northern coastal plains or the Balasore plains, which comprises the flood plains and deltas of the Subarnarekha and the Budhabalanga and roughly covers the whole of Balasore districts. However, economically, this area remained underdeveloped compared with other coastal area, as the region not only lacks irrigation but is also affected by drought and floods. The region also comes in the direct path of cyclonic storms from the Bay of Bengal. (ii) Middle coastal plains which geographically fall between the north coastal plains and the south coastal plains of the Rusikulya river. This area comprises of the combined deltas of Salandi, Baitarini and Brahmani in the north and Mahanadi in the south. Presence of an extensive irrigated area and insurance against the annual inundations by well coordinated embankment system lead to the maximum pressure of population and high yield per unit of land of Odisha. More than one crop is undertaken in extensive areas in this region. Cuttack, Puri, Kendrapara and Jajpur are the oldest towns in this region. (iii) Southern coastal plains or the Berhampur coastal plains are completely cut off from the northern plains by Chilika lake and spurs from the eastern Ghats. The fluviatile deposits of the Rushikulya river build up the region. The region not only had limited modern irrigation facility due to undulating topography but the higher frequency in failures of rainfall and rainy days often leads to crop failures. Berhampur controls the economic activity of this region.

The river valleys and flood plains in Odisha lie in between the 76 metre and 153 metre contour lines. The soil of these regions is of lower fertile compared to the coastal plains. The whole river valley and flood plains are broadly divided into mainly six sub-categories, such as (i) Baripada plains, (ii) Anandapur plains, (iii) Talcher and Bhuban plains, (iv) Sonepur and Athagarh plains, (v) Saria and Kharia plain and (v) Aska plains.

'The rolling uplands are included in the sub-montane zones and have mostly a rolling undulating topography with elevations varying from 152 to 305 metres. Different uplands are: (i) Rairangpur rolling uplands, (ii) Rourkela uplands, (iii) Jharsuguda rolling uplands, (iv) Bargarh uplands, (v) Bolangir uplands, (vi) Rairakhole uplands, and (vii) Malkanagiri uplands.

There are two plateau regions: The Keonjhar plateau, which is drained by Baitarani and present features of high plateau. The flat topography in this region is limited to the mountain slopes only. Rice is extensively cultivated in this region but yield per acres has remained very low because of the low fertility of the soil. The Nabarangapur plateau has a range of 600 metres to 900 metres. This region receives maximum rainfall in Odisha and neither faces either drought or flood. In fact, this is a surplus rice-producing region, but lack of good transport system stands in the way of its development.

But absence of well-planned irrigation facility in the state, in general, and in the hilly region in particular, remain the major reason behind the low productivity. The coastal region has fertile plains and irrigated land, but it is beset with occurrence of natural calamities like floods and occasional cyclones, while the advantages of the non-coastal districts can be found in presence of natural resources like minerals and forests. The geographical specificity therefore, was not only the cause of low economic growth in the state, and thereby bringing about regional disparities. Given the broad geographical specificity, we have presented the level of land utilisation pattern and agro-climatic conditions in the state and across the districts as both of these factors can influence the hunger and deprivation of any agrarian economy to a large extent.

ANNEXURE B: LAND REVENUE SYSTEM ODISHA

Land Revenue System in North Odisha

After occupying Odisha (Cuttack, Puri and Balasore here), the main problem faced by the Britisher was the problem relating to land, i.e. (i) geographical division and different land revenue system, (ii) the issues of land ownership and the system of land revenue. After the downfall of Odisha (1568), and before the Odisha division axed to the British during 1803, the state was part of Maughals Empire initially and later on of Marathas. Therefore, the land revenue systems in this part of Odisha were also different and the Britisher finds it difficult to identify an ideal land revenue system. In such situations they adopted a land revenue system, which was combination of native, Mughul, Marathas and British, which had bitter impacts on Odisha's economy; in fact, the volume of man-made hunger has been started increasing since then.

The Process of Change of Land Revenue System in Odisha

During the pre-British period there was no established code of rules giving the right ownership to any one party. Even there was no evidence that – Hindu king being claiming exclusive ownership right over land. 'It was generally accepted that state had as much right on share of the produce as revenue as the cultivator had the right to cultivate (Jena, K.C., p. 30, 1968). During that time, different officers, who were originally elected by the people and not hereditary or nominated by the kings, collected the land revenue. These officers were in charge of revenue collections but never claimed the right of ownership over the land. Hence, those land revenue officers could not be termed as Zamindars in the same sense as the British or Mughuls understood. They were mere servants of the state with jageers granted, but jageers could be confiscated for failure in duty. G. Toynbee cleared the position and said that, there were no intermediary agencies like Zamindars claiming any exclusive ownership over land during the pre-Mughul period of Odisha. Jena (1968) pointed out till the period Akbar there was no significant changes has been brought as the ownership right of land is concerned. But, after Akbar, when the Mughul rulers found difficulties to pay the militaries troops, a practice of granting loan to the military chief for rent was developed. In course of time the doner claimed ownership right. Over a period of time as the Mughul rulers becoming weaker, the private ownership right became stronger. And finally, the collectors of land revenue formed themselves into a group as distinct from state and the cultivators. The collection of discriminatory rate of land revenue started from this time. Since the military chiefs were generally from Khandayat caste, that might be a reason that this caste in general might be revenue collectors.

After the Mughuls, the Maratha ruled Odisha. Since, there aim was to collect revenue; it was immaterial for them to know the conditions of the cultivators and quality of land once the revenue collection was over. Hence, due to collection of high revenue, the total production was concentrated between two groups – the rulers and the collectors, which also lead to the development of different distinct class in Odisha's agrarian economy. As it was pointed out, the Marathas created a period of acute suffering and extreme poverty for the cultivating class of Odisha (Jena, 1968, p. 35). In fact, that was the period in which the seed of man-made hunger has been put into the soil, which had a spurt when the British government brought the landlords from other provinces to make them the Zamindars of Odisha, who enjoyed the status of ownership right and gave birth a different kind of agrarian structure and different consequences in Odisha's economy. And subsequently, the landlords' class became the more powerful economically as well socio-politically.

The Zamindary System of (North) Odisha

In the 19th century itself, the British had two phases and two kinds of land revenue administrations – from 1803 to 1836–37 and from 1837 to1866–67, the latter was further extended to 1897. The first phase can be called as period of experiments or the short-term settlements; the other one is of the permanent settlements. 'In the permanently-settled estates, the sums to be paid by the estates holders remained fixed in perpetuity while in temporarily-settled estates land revenue was to be revised at the lapse of thirty years' (Samal, 2000). Though in both the cases Zamindars were the real masters and ryots held the land under tillage at the pleasure of the land lords, an important difference between these two categories of settlements is the fixations of rent either by the officials or by the Zamindars? As many as, there were nine ways the rent was fixed for different kinds classes.

In the permanent settlement the burden was always on ryots as the land has been permanently under the control of Zamindars. And since the land-revenue was fixed once, the British rulers seldom bothered about the improvement of the land or tenants and they left the matter to the Zamindars, who leased out the cultivated land to the tenants under arbitrary terms and conditions reserving absolute right to revise these conditions according to their own convenience. Hence, from the agrarian development point of view there was less effort taken either by the landlords or governments or by the tenants.

However, in the temporary settlement, the risks are like to be more on the landlords as in each period the amount of rent fixation was revised. However, the Zamindars can shift such incidence to the ryots. But from agrarian development point of view, the Settlement Officer and his staff used to do land-survey, classifications of land, fixation of rent on the basis of productivity and preparations of records in every thirty years. However, how much of such improvements go to the hand of ryots is not clear. Apart, in the mughalbandi area also, some agrarian reforms have been brought during 1859, based on which the occupancy right has been given to the non-resident cultivators, who had been in possession of their holdings for twelve years or so. Grant of such occupancy right has been spread to other parts of the division and the struggle of the peasants' class against the landlord's class has continued. It also gave birth to many organisations and movements. On 1 September, there was a big demonstration by the peasants class demanding, 'Rent to be refused, Zamindary system to be abolished' etc. In return the first Congress Ministry paid special attention to it and by the Odisha Tenancy Act of 1938 some of the immediate measures were taken where, 'the occupancy right were declare transferable and no consent of landlords was required to the transfer nor any amount was to be payable to him for recognising transfer; and the occupancy ryots got the right to the trees on his home, to plant trees, enjoy fruits and cut and utilise the timber of trees felled. In the process, the harassment by the Zamindars to the ryots had reduced. The particular consequences of all types of land revenue system have been discussed together later. Below we have discussed the land revenue system in Western part of Odisha.

Land Revenue System in Sambalpur (Central Provinces)

Prior to colonial rule Sambalpur was mostly under the suzerainty of Marahatta (Sahu, 1971). As the pre-colonial land revenue system of Sambalpur was concerned, at that time there were two types of villages, such as Khalsa and Zamindari. Khalsa was under the direct control of the king through the intermediary of Gaontia (village headman). This was the most closely cultivated and bulk of the land revenue was derived from this proportion. Whereas, in the zamindari area, the extent of cultivation was much less and this was controlled by tribal Zamindars (landlords) or cadets (a younger son or daughter) of the royal family, with village headman under them. But the land revenue system is almost similar for both the cases.

In the Zamindari system, the zamindar used to collect revenue through an intermediary called gaontia (village headman) and paid some tributary fee to King. The gaontia had to pay two types of revenue. One is nazrana (lease renewal fee), which was used to collect during the period of granting the pattas (deed) to the goanties. Failing or not agreeing to pay the amount of nazrana was generally disposed off from the gaontee. In theory though the headman was supposed to bear the nazrana alone, in practices the share was contributed by each cultivator (Babu, 1984). The other form of revenue was called Malguzar (land revenue), which the gaontia was collecting directly from the ryots (cultivators) on behalf of the King/Zamindars. In lieu of this service (including some other services) the gaontia was enjoying rent free bhogra (home-farm) land.

The land revenue system of Sambalpur called as Mahalwari system, as the Sambalpur district was part of Central Provinces, where the Mahalwari system was prevailing. The first and foremost activity in this system is to decide the level of revenue demand from a village by the agents (king, zamindar, gaontia, and ryot). Then decide about the relative value of each holding, which was decided upon depending on the supposed productivity of the land. And on the basis of the relative value, the rent was distributed among the ryots. But here it is not clear that whether the entire households in villages (except the king, Zamindars and gaontias) are ryots? The major feature of this system was that revenue was paid in kind. There were as many as six different classes of proprietors with different degree of proprietorships. Apart from this there were also village service tenants where the people rendering their service to the village have been given tenantships. Some of the main features can be seen as follows:

(1) Since mode of payment of revenue was in kind, level of market exchange was minimal.

(2) There was a sort of community property right over the land than the present system of individual property right.

(3) Reducing risk to uncertainties: Since the rent was fixed as a proportion of output, and since there was redistribution of land among the ryots, each of the agents (king, zamindar, gaontia, and ryots) was equally exposed to any uncertainties. However, if there would have been crop failures, the ryots might be facing the problem of hunger and deprivation.

The problems in this system were, (i) the short-term settlements, which was troublesome and expensive, and (ii) the tenants had to render free labour commonly known as bethi-begar and had to pay fee known as nazarana to secure the consent of the Zamindars to transfer of the land, which all bring resistance from the tenants and after the independent Odisha, attempts were taken to bring reforms in this part of Odisha as well.

Land Revenue System in South Odisha (Madras Presidency)

In the southern part of Odisha, both Zamindary and Rawyatwari land revenue system have existed. However, this Zamindary system that came from Madras presidency was different in its process from that which existed in the northern Odisha. The system ensured more legal arrangement, e.g. there has to be a written agreement between the Zamindar and tenants on the fixed land dues. The succession of a Zamindar was governed by the law of primogeniture of the Madras Presidency Settlement Regulation XXV of 1802. Land had been also given to the sacerdotal class of temple institutions, which in many cases became the property of Brahmins later on. One of the differences between the land system of coastal Odisha and South Odisha is in the Odisha Tenancy Act, an under-ryot was recognised but in Ganjam and Koraput, the tenant holding was governed by contract and had no recognised status. Where Rawyatwari existed later on in south Odisha was also under the Zamindary system but due to the rebellion of Zamindars of Ganjam against British, the government introduced Rawyatwari system in Chhatrapur, Berhampur and Ghumsar, where the revenue from the ryots was collected by the official and not by landlords.

However, here too the system was not less harmful for the tenants. The single most harmful features of the system was the heavy amount of collection of rent, which was often more than half the worth or produce from the tillage. Even at the time of drought and flood when the farmers received scanty production they were pressed for the stipulated amount. The other harmful feature of this system was the unlawful extractions of land in the form of begari and gothi (free labour) for which there was no legal precautionary. In the rayatwari system too there were many defects and great inequality in the incidence of rents on tenants as the rent was not correlated to the productivity of land. However, situation has improved in the Rawyatwari system as these ryots were having direct contact with the government. Despite the demerits that the Rayatwari system was having, as Samal (2000) pointed out – 'though the fixation of rent seems high and disproportionate and the conditions of cultivators not much encouraging, ryotwari system seems to be first important administrative reforms in revenue matter. In this system the government directly dealt with its subjects without inserting any injurious intermediary, and thus ensured the subjects of their sense of property right. Scientific principle of revenue assessment was for the first time introduced in contrast against arbitrary assessment prevailing in the Zamindary areas of Ganjam district. In several respects, it proved better than Zamindary system, a mindless, unscrupulous machine of exploitation'.

Irrespective of the land revenue system of different parts of Odisha, a common features in all is that it is the intermediaries or the landlord class who were held responsible for the poor conditions of peasants and labour class. Some of the economic consequences have been narrated in the text.

ANNEXURE C: HISTORY OF ODISHA AS AN ADMINISTRATIVE UNIT

Emerging from prehistoric times with references in the Vedic and Epic literature, the people of ancient Kalinga distinguished themselves as pioneers in maritime activities and overseas colonisation in early days and took to empire making under the Jain king Kharavela, before the Christian era (Census Report, 1951, pp. 4). After the Kharavela, Kalinga was under the subjugation of foreigners, viz. by the Kushans and Guptas, but mostly had dynasties of its own ruler over the land. At the beginning of the 9th century, Kalinga had already her native powerful ruler, who builds different monuments till 16th century. Those monuments though may not confirm people's standard of living does however, confirm the prosperity in the state. By the time of 15th century AD, the land of Oriya speaking people was well known in the Indian history as Udisa or Odisa. During that time the Gajapati king of Odisha ruled over a kingdom, extending from the Ganges in the north to Kaveri in the far south. But after the downfall of the kingdom in 1568, Odisha has been broken and annexed into different regions by different dynasties and ruled by them for around 230 years. Finally, in the year 1803, East India Company conquered Odisha and ruled for another 145 years. But the British's occupied Odisha was initially (in19th century) confined only to the three coastal districts of Puri, Cuttack and Baleshwar. Ganjam and other Oriya-speaking areas south of the Chilika Lake remained tagged to Madras; Midnapore to Bengal, Singhbhum, Seraikela and Kharsawan to Chota Nagpur division; and Sambalpur and Chhatisgarh feudatory states with the central provinces. This might be the reason of limited consolidated work (covering all the areas) about the 19th century of Odisha and probably might not be such work with regards to hunger and deprivation in the colonial Odisha (not Odisha division). It was in the third decade of 20th century, the reunification of Odisha happened.

The East India Company took more than half a century to conquer the whole of Odisha but the company had no intention to unify the Oriya speaking territories. On the contrary, after further reorganisation of the Oriya-speaking areas, which lay outside the Odisha division, they were administered by five separate political authorities. However, an awakening attempt for the integration of Oriya-speaking people (a language movement) was just after the 'Na Anka' famine of 1866. The intensity of the suffering of that famine might have an additional reason for such moment. In that moment, the main manifesto was language. The attempts of amalgamation were initially within Odisha division, but gradually it had spread into other parts as well, and many associations and organisations were formed for the mentioned purpose. The attempts initially remained at the ground and non-official level but later on, it was continuously appealed to the official authority. And finally, it was around 1905; the first amalgamation started when Sambalpur district was transferred to Odisha division. The process gradually moved forward with the continuous efforts from grassroot level. And it was 'in the lines of the recommendations of the committee, the Government of India Bill 1935 was passed by the parliament and new province of Odisha as an administrative unit came into being on the 1 April 1936'. But there were only six districts, viz, Cuttack, Puri, Baleshwar, Sambalpur, Ganjam and Koraput. The old district of Angul was split into two statutory districts (Angul and Khondmals) and remained under the administrative supervision of Cuttack and Ganjam respectively. The Rajas and Maharajas governed rest of the provinces. This continued till 1947, and it was after independence that Sardar Vallabhbhai Patel, the then Union Home Minister, merged all princely states with old British administered districts of Odisha. Finally, during 1950 the reformation of the present Odisha was completed with 13 districts. But for the purpose of better administrations, the large sized districts were again subdivided and at present there are 30 districts in Odisha. How the present Odisha has been reformulated after 1803 has been presented below.

Odisha After the Downfall of Native Ruler During 1568

Years	*Regions Includes*
1568	Last native king of Odisha (Ganga to Godavari).
1568–1751	Fragmented. Under the control of Pathan sultan of Bengal.
1751–1803	Under the control of Marhattas by Alivardi Khan and Nawab of Bengal.
1803	British conquered Odisha, which was attached to the province of Bengal.
1804–05	First, the territory was demarcated, and subsequently constituted into Cuttack.

Cont'd...

...*Cont'd*

1823	The demarcated area split up into three regulation, Cuttack, Balasore, Puri and 17 non-regulation tributary Mahals administered by native chiefs under the responsibility of a commissioner.
1835	Khondmals was ceded to the tributary chief of Baudh state.
1847	Angul was annexed to Odisha:
1905	(a) Sambalpur was transferred from the Central provinces to Odisha.
	(b) Five tributary states (Patna, Kalahandi, Sonepur, Bamra and Rairakhol) were also transferred from central provinces.
	(c) Two states of Gangpur and Bonai were also transferred from Chottanagpur division.
1912	Odisha was transferred to Bihar from a division of Bengal.
1936	Some portions of the Central provinces (Nawapara subdivision) and the Madras provinces (Ganjam and Koraput) districts were added to the bound of Odisha.
April, 1936	Constitution of Odisha as a separate province with six districts (Cuttack, Balasore, Puri, Sambalpur, Ganjam and Koraput). And 24 Oriya speaking feudatory states placed under the political Agent – in charge of the Eastern Zone stationed at Sambalpur.
November, 1947	Adivasi problem.
January, 1948	The merger of 23 states with Odisha.
January, 1949	Mayurbhanj merger with State of Odisha.
1950	Reformation of present Odisha.
At present (2003), Odisha covers almost same square kilometres, but for the smoothness of administration, many large sized districts have been divided. And the number of districts has been increased from 13 to 30.	

Source: Extracted from Census Report, 1951, 2001.

Changes in the Area of Odisha Since 1872 (in Square Metres)

	Area in square miles		
	Odisha Proper	*Odisha state*	*Total*
1872	7,717	16,184	23,901
1881	9,053	15,187	24,240
1891	9,853	14,387	24,240
1901	9,841	14,387	24,228
1911	13,743	28,046	41,789
1921	13,736	28,046	41,782
1931	13,706	28,046	41,752
1941	32,198	25,194	57,392
1951	60,136	—	60,136
2001	60,119		

Notes: During 1941, Odisha proper constituted with Cuttack, Balasore, Puri, Sambalpur, Ganjam and Koraput, whereas Odisha states constituted with 23 feudatory states.

Source: Census of India, 1951, Volume XI, Odisha, Part-I Report, pp. 7.

ANNEXURE D: NOTES ON STRUCTURAL TRANSFORMATION

Structural changes do not only characterise economic development; they also are necessary and important for sustaining economic growth and reducing poverty. Kuznets, one of the structuralist economists has empirically demonstrated that growth is brought about by changes in sectoral composition. Mainly, two sets of reasons have been stated for the growth and structural changes – the demand side and the supply side (Fisher, 1939, 1946; Clark, 1940, 1984).[1] The demand side reasons state – income elasticity of demand for agricultural product is lower than that for industrial, particularly manufacturing commodities, and for services, it is still higher. Therefore, when there is increase in income, the demand for agricultural products declines and that for industrial goods increases and after reaching a reasonably high level of income, demand for services increases sharply. Hence, this can be inferred, the share of different sectors in total output is determined by the pattern of demand, which again depends on incremental income of the people. On the supply side, the relative share of agriculture comes down because agriculture mainly depends on a fixed factor of production – land, which faces a limit on its growth because of the early operations of the law of diminishing returns. Industry, especially manufacturing, has a larger scope for using capital and technology, which further can be augmented with human efforts. Constraint of labour supply in industry may be substituted with labour-saving technology. Similarly in service sector too, an application of technology has a much larger scope. Additional reasons stated for the increase of the share of service sector with industrial development are the technological developments and economic and institutional arrangements compelled by them (Kuznets, 1973). Technological developments facilitate large-scale based production and hence geographical concentration, which together lead to the larger requirements of transport, storage, communications etc. This requires manpower, and manpower in turn requires facilities like housing, education, health, entertainment etc. All these together boost the service sector.

However, there have been debates among the economists, who placed different emphasis on the demand and supply side explanations of structural shifts.[2] One of these studies, which best present the view is by Kuznets, 1971, who stated, 'the driving force for changes in sectoral compositions of output is differences in income elasticity of demand for products of different sectors, but caused by differential growth of productivity in different sectors and increase in per capita income'.

Irrespective of these positions taken by the economist in the said debates and irrespective of the primary and secondary factors that cause structural transformations, there is hardly any disagreement among them that structural changes (both in terms of output and employment) inevitably accompany, and bring about economic growth. Therefore, when an economy would progress, not only the share of output of agriculture sector would decline, the share of employment in agriculture is also expected to decline, as there may be transfer of labour from agriculture to industry. Lewis, 1954 considered such transfer of labour as a source of capital accumulations, as these transfers would (a) increase the labour productivity in agriculture by withdrawing the surplus or disguised labour force, who were contributing to the production insignificantly but claiming their full shares in consumptions; (b) these labour get absorbed in the industrial sectors, where their contributions lead to a net addition to the output of the economy without significantly increasing the cost of labour.[3] But some of the assumptions of the Lewis model have been widely debated. However, this can be inferred that economic development of an underdeveloped country not only involves but requires shifts of labour force from agriculture to industry. But how much labour force can be shifted, depends on how much technology has been introduced in the economy and how much industrial base has been created in the economy, which can absorb theses surplus/disguised labour force. The next shift of labour force is to service sector, which comes after the first phase of shift to industry. When the second shift of labour force will occur is also not clear, but Clark has argued that demand for manufactured goods saturates at around 20 to 25 per cent and with continuing decline in the demand for agricultural products the demand for that of services rises. But other economists have stated that the increase in the employment in the service sector is due to productivity differentials in industry and service sectors.

ANNEXURE E: EVOLUTION OF PUBLIC PROVISIONING IN INDIA

In India, provision of social security in the post-independence period occurred with the constitutional provisions and legislative enactment. The Fundamental Rights and Directive Principles of State Policy state that – the State shall strive to promote the welfare of the people by securing and protecting them. 'The citizen (Men and Women) equally has the right to an adequate means of livelihood ...' (Planning Commission, 1950:1). For making such commitment fruitful, Planning Commission had been set up in March 1950, by the Government of India. Since then, the Planning Commission has been involved in various ways to ensuring the livelihoods of the Indian citizens. Though the inceptions may be traced to one of the most popular food security schemes known as Public Distribution Schemes, in the later period various schemes have been introduced, which covers food and nutrition, health and education, assets and income, housing and other social security schemes. Each scheme is discussed below.

Considering the government's implicit and explicit efforts to provide social security, the whole periods (from first Plan to 10th Plan) can be divided into two phases – in the first phase (first to third Plan period), there was no explicit or targeted schemes, rather most schemes were addressed under a holistic banner known as social services, which includes food, education, health housing and other schemes. During the fourth plan period, it was realised, that not much satisfactory results have been achieved in reducing hunger and deprivation. Increased unemployment, livelihood insecurity problem (56.4 per cent of the rural India were living below the poverty line), and high landlessness were features of great concern.[4]

The 'trickle down hypotheses[5] failed to trickle and ameliorate livelihood insecurity of the poor as expected by planners. Many researchers found inverse relationship between rural poverty and agricultural production. All these together led to introduction of many targeted schemes to ensure livelihood security for the poor. Hence, from fourth plan onwards, it could be a different phase, as explicit schemes have been introduced for different groups in both rural and urban areas. Table E-1 summarises introduction, objectives and functioning of these schemes, which covers the area of mainly (i) food and nutrition supporting schemes (ii) health and education supporting schemes (iii) assets and income generating schemes (iv) housing schemes and (v) other schemes.

Table E-1: *Introduction, objectives and functioning of different welfare schemes in rural and urban India and Odisha*

Programmes name and year of introduction	*Objectives*	*Functioning*
	Food and Nutrition	
Public Distribution System (PDS), since 1939 but in an organised and large scale since 1950 onwards. Reforms in the schemes produced Revamped PDS (1992), Targeted PDS (1997) and Antyodaya Arna Yojona (AAY, 2000).	1. Providing universal food security. 2. Procuring food and maintain a buffer stock. 3. Stabilise the food prices.	It has been ensuring the food security to a larger sections of the poor though there have been different issues need to correct in the system. Such issues includes regional biasness, limited quantity of cereals, quality of cereals, targeting error etc.
Annapurna Schemes (2000–2001)	Meant for the indigent senior citizens or 65 years of age or above who though eligible for old age pension under the National Old Age Pension Scheme (NOAPS) but are not getting the pension, are covered and 10 kgs of food grains per person per month are supplied free of cost under the scheme.	The total coverage under the scheme was 6,75,000 including the enhancement of 35,000 made during 2004–05. During 2006–07, ₹ 8,100.00 lakhs has been provided in the State's Non-Plan Budget for payment of pension to the above targeted 6,75,000 beneficiaries under the scheme in Odisha
Mid-Day Meal (MDM) Scheme 15 August 1995	For the benefit of students in primary schools under Employment Assurance Scheme (EAS)/earlier Revamped Public Distribution System (RPDS) blocks (2,368). The Scheme covers students of Class I-V in the Government Primary Schools/Primary Schools aided by Govt. and the Primary Schools run by local bodies.	The MDM in the state cover almost all the block but while in half of the total block (157) provides cooked meal in other half of the block they distribute dry food, which may affects the objective of achieving nutritional security.

Cont'd...

...Cont'd

Integrated Child Development Services, 1974–75	Aims to provide health care services covering supplementary nutrition, immunisation, pre-school education, health check up, referral services and health education to children within the age group of 0–6 years. In that process the scheme also tries to provide nutritional support to the pregnant and nursing mother and health support to adolescent girls.	In Odisha there is an increasing tendency in the number of beneficiaries from ICDS.
	Primary Education and Primary Health	
Universal Elementary Education (UEE, 1950), District Primary Education Programme (DPEP, 1996–97). Sarva Shiksha Abhiyan (SSA, 2003–04). National Programme of Education for Girls Child at Elementary level (NPEGEL, 2003), Education Guarantee Scheme (EGS, 2001–02), National Literacy Mission (NLM, 1991–92).	Aims to provide universal elementary education.	In both health and education the state remained backward.
Public Health System	Aims to minimum health facility to all.	
	Assets and Income (employment)	
Integrated Rural Development Programme 1979 (Self Employment Programme)	An integrated approach for rural development and targeting the poor. This programmes also imbibed the objectives of Small Farmer Development Agencies. It aims to provide subsidy and credit to the rural poor households to take up self-employment and income augmenting activities. The cost of the programme was to be shared equally between centre and state. National target of assisting 600 families per block annually was laid out.	Initiated as a pilot project in selected blocks, it was expended to cover the whole country in 1980. During the Sixth Plan the programme covered 15.13 million families with ₹ 1,500 crores and a similar number in the Seventh Plan was covered. ₹ 1,186 crores for Central assistance was allocated in the seventh Plan. By the end of March 97, over 50 million families with 50,000 crores have been covered since inception.
Development of Women and Children in Rural Areas (DWCRA) – 1983	To provide income generating skills and activities to poor women in rural areas. Social and economic empowerment leading to improved areas to health, education etc.	Started as a pilot scheme in 50 districts in 1983, it is now implemented all over the country. As on January, 1997, 1.88 lakh groups with 29.54 members have been assisted to the tune of ₹ 220.43 crores. This is one of the fairly successful schemes.
Training of Rural Youth in Self Employment (TRYSEM) 1979.	Providing technical and entrepreneurial skills to youth from the poor families to enable them to take up self-employment.	Started as a centrally-sponsored programme on 15 August 1979. As on December 96, 37.09 lakh youth have been trained but a very large number of them could not settle in the trades in which they were trained. Lack of credit facilities and demand for trade are some of the reasons.

Cont'd...

...Cont'd

National Rural Employment Guarantee Programme (NREGP) 2005	An Act to provide for the enhancement of livelihood security of the households in rural areas of the country by providing at least one hundred days of guaranteed wage employment in every financial year to every household whose adult members volunteer to do unskilled annual work and for matters connected therewith or incidental thereto.	In Phase I with effect from February 2005, the NREGP is implemented in 19 districts of Odisha. These districts are Balangir, Boudh, Deogarh, Dhenkanal, Gajapati, Ganjam, Jharsuguda, Kalahandi, Kandhamal, Kendujhar, Koraput, Malkanagiri, Mayurbhanj, Nabarangapur, Nuapada, Rayagada, Sambalpur, Sonepur and Sundargarh.
	Housing Schemes	
Indira Awas Yojana (1985–86)	Aim is to assist the rural BPL persons to construct or upgrade their dwelling (should include sanitary latrine and smokeless chullah) by proving about ₹ 20,000.	In Odisha, during the Ninth Five-Year Plan Period (1997–2002), 45 lakh houses were constructed under IAY, which mostly benefited the poor though among the poor there was undue preference to certain beneficiaries (Planning Commission, 2005)
	Old Age Pension	
State Old Age Pension (SOAP) 1.4. 1975 National Old Age Pension (NOAP), 15 August 1995	The SOAP aims to provide some financial assistance (₹ 100 per month revised to ₹ 200) to the old and destitute persons of 60 years of age (65 for national schemes) and above, to leprosy patients with visible signs of deformity and to destitute widows irrespective of age. The NOAP scheme also aims to provide some financial assistance to the old persons (65 and above) living below the poverty line. The Government of India provides ₹ 75 per month and the State Government adds ₹ 25 per month to each beneficiary under the above scheme.	In the SOAP, till 2004–05, against the target of 6,75,000 beneficiaries; 6,74,722 beneficiaries including 3,43,414 women have availed the facility. In the NOAP, till 2004–05, as against the target of 4,93,400 beneficiaries, 4,93,273 beneficiaries including 2,51,344 women, have been covered.
During April 1999, most of the rural self-employed programmes have been imbedded into SGSY and the urban self-employment has been embedded into Swarn Jayanti Shahari Rojgar Yojana. Hence, at the current period understanding of these two schemes may help us to understand the public provisioning to a greater extent.		

Notes: SGSY includes IRDP, TRYSEM, DWCRA, SITRA, GKY and MWS.

Sources: Complied from Planning Commissions documents and <http://www.Odisha.gov.in/p&c/eco2006/annexture.htm>.

ANNEXURE F: DIFFERENT PUBLIC PROVISIONING SCHEMES

Food and Nutritional Support Schemes

Public Distribution System, Revamped Public Distribution System, Targeted Public Distribution System and Antyodaya Anna Yojana: The Public Distribution System (PDS) took its inception during the II World War as a food rationing system to manage the scarcity; and distributed food grains at affordable prices especially in urban scarcity areas. Over the time periods, the PDS has become an integral part of a welfare policy of the Government of India for management of food and food prices in the country and by 1960s the schemes not only spread over the country but it had substantially contributed to the food security of the food deficit region like Kerala. In the mid-1960s, with the increase in the agricultural production mostly due to green revolution, the outreach of PDS was extended both in terms of coverage and quantity allotted (Swaminathan, 2000).

During June 1992, the first reforms in the PDS had been brought when the Revamped Public Distribution System (RPDS) was launched in 1,775 blocks throughout the country, where the aim was to strengthen and streamline the PDS, as well as to improve its reach in the far-flung, hilly, remote and inaccessible areas where a substantial section of the poor live. The RPDS took an area approach to ensure effective reach of the PDS commodities to the poor. In these 1,775 blocks, area-specific programmes such as the Drought Prone Area Programme (DPAP), Integrated Tribal Development Projects (ITDP), Desert Development Programme (DDP) and certain Designated Hill Areas (DHA) were identified in consultation with State Governments for special focus, with respect to improvement of the PDS infrastructure. In the RPDS, the food grains for distribution were issued to the States at 50 paise below the Central Issue Price and the amount of issue per card was up to 20 kg. However, it was the responsibility of the State Governments to deliver the commodities at the doorstep of fair price shops (FPSs) in the identified areas, additional ration cards to the left out families, infrastructure requirements like additional Fair Price Shops, storage capacity etc. and additional commodities such as tea, salt, pulses, soap etc. for distribution through PDS outlets.

During 1997, another reform has been found in PDS, which mainly come because of the wide criticisms to PDS for its failure to serve the population Below the Poverty Line (BPL), its urban bias, limited coverage in the States with high concentration of the rural poor and lack of transparent and accountable arrangements for delivery. During, June 1997, the Government of India launched the Targeted Public Distribution System (TPDS) with special focus on the poor.

The PDS has been operated under the combined responsibility of the Central and the State Governments. The Central government, through Food Corporation of India (FCI), has assumed the responsibility for procurement, storage, transportation and bulk allocation of food grains to the State Governments. It is the responsibility of the State Governments to operationalise the policy including allocation within State, identification of families below the poverty line, issue of Ration Cards and supervision of the functioning of FPS, etc. Under the PDS, the commodities presently supply distributed are wheat, rice, sugar and kerosene. Some States/UTs distribute additional items of mass consumption through the PDS outlets such as cloth, exercise books, pulses, salt and tea, etc. In TPDS, the state has to play a bigger role. States are required to formulate and implement foolproof arrangements for identification of the poor for delivery of food grains and for its distribution in a transparent and accountable manner at the FPS level.

The TPDS was intended to benefit about 6 crore poor families by supplying 72 lakh tonnes of food grains annually, which is around 10 kilograms a month per family (or annually 120 kg). The identification of the poor in the TPDS was worked out with a general consensus at the Food Minister's conference held in August 1996, for adopting the methodology used by the expert groups set up by the Planning Commission under the Chairmanship of Late Prof. Lakdawala. The population projections of the Registrar General of India for 1995 and the statewise poverty estimates of the Planning Commission for 1993–94, which stated there was 596.23 lakh households under the below poverty line. For the identification of the BPL HHs at the ground level, the State Governments were also provided guidelines and advised to involve the Gram Panchayats, Gram-Sabhas and Nagar Palikas. In TPDS, the allocation of food grains to the respective States/UTs was made based on average consumption in the past i.e. average annual off-take of food grains under the PDS during the past ten years at the time of introduction of TPDS.

While identifying the BPL HHs, the thrust should be to include the really poor and vulnerable sections[6] of the society. By 1.12.2000, the number of BPL families increased to 652.03 lakh as against 596.23 lakh families that were estimated when TPDS was introduced in June 1997. Efforts were made to increase the quantity of food grains. For better target of food grains to the BPL families and better target of food subsidy, the

Government of India increased the allocation to BPL families from 10 kg. to 20 kg of food grains per family per month at 50% of the economic cost and allocation to APL families at economic cost w.e.f. 1.4.2000. But the allocation to the APL families was retained at the same level as at the time of introduction of TPDS but the Central Issue Prices (CIP) for APL were fixed at 100% of economic cost from that date so that the entire consumer subsidy could be directed to the benefit of the BPL population. However, the final retail price to be charged to the households is fixed by the States/UTs after taking into account margins for wholesalers/retailers, transportations charges, levies, local taxes etc. Further, under the TPDS the States were requested to issue food-grains at a difference of not more than 50 paise per kg over and above the CIP for BPL families. However, flexibility in TPDS by the States/UTs. has allowed in the matter of fixing the retail issue prices by removing the restriction of 50 paisa per kg over and above the CIP for distribution of food grains under TPDS. The exception is Antyodaya Anna Yojana, where the end retail price is to be retained at ₹ 2a kg. for wheat and ₹ 3 per kg. for rice.

Reforms brought into the PDS are in the form of Antyodaya Anna Yojana (AAY). As the National Sample Survey points out that, about 5 per cent of the total population in the country does not get two square meals a day. Hence, in order to make TPDS more targeted towards this category of population, the 'Antyodaya Anna Yojana' (AAY) was launched in December 2000 for one crore poorest of the poor families initially. The States/UTs are required to bear the distribution cost, including margin to dealers and retailers as well as the transportation cost. Thus, the entire food subsidy is being passed on to the consumers under the scheme. Over a period, although different reforms have been brought in AAY, e.g. by 1 April 2002, the scale of issue that was initially 25 kg per family per month has increased to 35 kg per family per month. The first expansion in the AAY during 2003–2004 added another 50 lakh BPL households headed by widows or terminally ill persons or disabled persons or persons aged 60 years or more with no assured means of subsistence or societal support. With this increase, the total number of families in AAY became 1.5 crore (i.e. 23% of BPL). The second phase of expansion of AAY comes in line with the National Common Minimum Programme (NCMP) of the UPA Government. The Hon'ble Finance Minister in the Union Budget 2004–05, announced the Antyodaya Anna Yojana (AAY)[7] *and has been expanded with effect from 1 August 2004 by another 50 lakh BPL families by including, inter-alia, all households at the risk of hunger. With this increase, the total number of AAY families has increased to 2 crore (i.e. 30.66% of BPL) families. The third phase of expansion in the AYY comes with the announcement of the Union Budget 2005–06, where the AAY has further been expanded to cover another 50 lakh BPL households thus increasing its overage to 2.5 crore households.(i.e. 38% of BPL) and at present around the monthly allocation of food grains under AAY is around 7.27 lakh tones.*

As the quantity issued of food grains under TPDS is concerned, the scale of issue increased to 20 kg per family per month in 1997, with effect from 1 April 2000. The allocation for APL families has retained at the same level as at the time of introduction of TPDS (i.e. 10 kg. per family per month). The allocation of food grains for the BPL families has further increased from 20 kg to 25 kg. per family per month with effect from July, 2001. Initially, the Antyodaya families were provided 25 kg. of food grains per family per month at the time of launching of the scheme. The scale of issue under APL, BPL and AAY has been revised to 35 kg per family per month with effect from 1 April 2002 with a view to enhancing the food security at the household level.

The Central Issue Price (CIP) of food grains being supplied under TPDS: (Figure in ₹ per kg)

Commodity	*APL*	*BPL*	*AAY*
Rice	8.30	5.65	3.00
Wheat	6.10	4.15	2.00

With the reforms in PDS, the programme has become pro-poor. Amount of food grains has increased, food insecurity. The CIP of food grains for BPL families has not been revised since 25.07.2000. To create a hunger free India, the numbers of AAY households have been increased to 2.5 crores and the CIP for AAY category was kept lowest i.e. ₹ 2 per kg for wheat and ₹ 3 per kg for rice. Recently, certain measures have been taken to strength the PDS.

(1) The Citizens' Charter was issued in November 1997, for adoption by the State Governments to provide services in a transparent and accountable manner under PDS.

(2) The PDS (Control), Order of 2001, which is inter alia, covers a range of areas relating to correct identification of BPL families, issue of Ration Cards, proper distribution, and monitoring of PDS related operations. Contraventions of the provisions of the Order are punishable under the Essential Commodities Act, 1955.

During 2000–2001, the Ministry of Rural Development launched another scheme, which also provides food security named Annapurna scheme. The Annapurna schemes meant for the indigent senior citizens or 65 years of age or above who though eligible for old age pension under the National Old Age Pension Scheme (NOAPS), but are not getting the pension, are covered and 10 kgs of food grains per person per month are supplied free of cost under the scheme. The applicants must be 'destitute' in the sense of having little or no regular means of subsistence from his/her own source of income, or financial support from family members or other sources. From 2002–2003, the scheme has been transferred to State Plan along with the National Social Assistance Programme comprising the National Old Age Pension Scheme and the National Family Benefit Scheme. The Ministry of Finance as Additional Central Assistance (ACA) to the State Plan is releasing the funds for the transferred scheme and the States have the requisite flexibility in the choice of beneficiaries and implementation of the Scheme. Therefore, the implementation of the Scheme at the ground rests with the States/UTs.

The PDS also functions in Odisha for maintaining three major objectives of ensuring price stability, adequate availability of food grains, sugar and kerosene oil to consumers, and implementing a special programme for drought-prone and tribal areas. The system runs through a network of fair shops. By the end of December 2004, there was a network of 26,217 fair price shops, out of which 22,322 were in rural areas and 3,895 in urban Odisha. In Odisha, the total blocks of 314 have been divided into two groups one group, where the Integrated Tribal Development Programme (ITDP) and Drought Prone Area Programme have been implemented to give more focus to the poor. There are 143 Integrated Tribal Development Programme and Drought Prone Area Programme (ITDP) blocks in the state with effect from 1 June 1997. The other group is the remaining 171 non-ITDP/non-DPAP blocks.

In the TPDS schemes, all the rural below poverty line families enlisted in the BPL list of all the 143 ITDP and DPAP blocks are supplied with 16 kg of common rice per family per month at the specially subsidised price of ₹ 4.75 per kg and additional 9 kg of rice at ₹ 6.30 during 2004–05. On the other hand, the remaining 171 non-ITDP/non-DPAP blocks, all the rural below poverty line families who are enlisted in the BPL list were supplied with 25 kg of common rice per family per month at the rate of ₹ 6.30 per kg during 2004–05. However, in Kalahandi, Balangir and Koraput (KBK) region, the APL families were also supplied with rice 25 kg per month at BPL rate of ₹ 6.30 per kg during 2004–05. Earlier, the monthly quota of BPL rice per family was 16 kg, which also has been enhanced from 16 kg to 25 kg since September, 2002. There was also eliminations of private retailers for selling of rice under the Targeted PDS in the ITDP and DPAP blocks and Government have engaged: (i) Gram Panchayats (ii) Large-sized Adivasi Multi Purpose Cooperative Societies (LAMPs). (iii) Rural Consumers' Co-operative Stores, (iv) Service Co-operative Societies (v) Model Fair Price Shops run by Odisha Civil Supply

Corporation Ltd. and (vi) Registered Mahila Mandals. The central assistance of food grains to be distributed through the PDS is presented in Table F-1. The Table depicts a marginal decline of food grains in all the commodities.

Table F-1 *Central assistance of food grains distributed through PDS in Odisha*

Sl. No.	*Commodities*	*Unit*	*Allotment during*				
			2000–01	*2001–02*	*2002–03*	*2003–04*	*2004–05 (p)*
1	Sugar	Lakh MT	1.75	1.08	1.04	0.27	1.08
2	Wheat	Lakh MT	0.90*	—	3	3.6	3.6
3	Rice (APL BPL)	Lakh MT	9.95	9.97	24.99**	22.80**	22.80**
4	Rice (AP)	Lakh MT	—	—	0.08	0.08	0.08
5	Rice (AAY)	Lakh MT	—	—	—	2.12	2.61
6	Kerosine Oil	Lakh KL	4.91	4.24	4.08	4	4.05

Notes: P – Provisional, * for BPL beneficiaries, ** For APL + BPL beneficiaries.
Source: Food Supplies and Consumer Welfare Department, Odisha.

Mid-Day Meal Scheme

The Mid-Day Meal (MDM) Scheme was another food and nutritional security providing scheme launched by the Ministry of Human Resource Development (Department of Education) with effect from 15 August 1995 to benefit the students of primary schools under Employment Assurance Scheme (EAS)/earlier Revamped Public Distribution System (RPDS) blocks (2,368). The Scheme covers students of Class I-V in the Government Primary Schools/Primary Schools aided by Govt. and the Primary Schools run by local bodies. The amount of food grains (wheat and rice) are supplied free of cost at 100 gram per child per school day, where cooked/processed hot meal is being served with a minimum content of 300 calories and 8–12 gms of protein each day of school for a minimum of 200 days. About 3 kgs per student per month for 9–11 months in a year, where food grains are distributed in raw form. In drought-affected areas, the mid-day meal is distributed in summer vacations also. To cut down delays in implementation of the scheme, Department of Elementary Education and Literacy has been authorised to make State/UT-wise allocation of food grains under intimation to this Department. Food Corporation of India (FCI) releases food grains to States/UTs at BPL rates as per allocation made by Department of Elementary Education and Literacy.

In Odisha, MDM Programme was introduced in 1995 to provide a cooked noon meal to primary school children of all government and government-aided schools studying in Class I to V for 210 working days in a year. The objectives of MDM are to increase the enrolment and reduce the number of school dropouts and improving the nutritional status of the children. However, the MDM in the state functions in two ways. The KBK districts and the ITDA Blocks under the MDM Programme, provide cooked food. Since the July 2001, cooked meal to the primary school children were provided only in the rural areas of the 8 KBK districts (80 Blocks including 44 ITDA Blocks) and in 74 ITDA Blocks of the Non-KBK Districts. In the other districts of the State, dry ration, at 3 kgs of rice per beneficiary per month was being supplied. The cooked meal system under the MDM Programme was extended to the primary school children in the backward district of Boudh with effect from 1 April 2002. Therefore, out of 314 blocks, 157 Blocks in the State were covered under the 'cooked meal system' while the remaining 157 Blocks were covered under the 'dry ration system till 31 August 2004. With effect from 1.9.04, cooked meal at noon time is being provided to all students of all the Govt./Govt. aided Primary Schools as per direction of the Hon'ble Supreme Court. Till 2004–05, the MDM schemes, the total coverage was 51,51,346 including 11,69,832 students from the KBK districts. This has gone up to 51,56,154 in 69,700 schools including 11,56,426 students in 18,486 schools the KBK districts by financial year 2005–06.

For the management of the MDM, the Government of India, Ministry of Human Resource Development, Department of Elementary Education and Literacy, New Delhi are providing free rice at 100 grams per beneficiary per day for 210 days every academic year. Transportation charges for transportation of the rice from the FCI Depot to the school point at ₹ 75 per quintal; cooking cost at ₹ 1.00 per beneficiary per day (w.e.f. 1.9.04) But the state government add another ₹ 0.64 per beneficiaries, so accordingly the cooking cost becomes ₹ 1.64. The Government of Odisha also provides funds towards purchase of the pulses, vegetables, oil, condiments and transportation charges under the scheme at the rate of 0.64 paise per beneficiary per day. The revised rations cost is presented in Table F-2.

Table F-2 *Revised Ration Cost (in ₹) in Odisha, 2006*

1.	Supervision charges per child	0.30
2.	Pulses	0.50
3.	Vegetable, Condiments and Salt etc	0.16
4.	Oil (1gm)	0.05
5.	Fuel and Stationery	0.18
6.	T.C. etc	0.10
7.	Egg	0.35
	Total	1.64

NB: GoI provides rice @ 100 gm per child per school day free of cost.

Source: http://www.Odisha.gov.in/wcd/index.htm and http://fcamin.nic.in

However, the functioning of MDM schemes at the micro-level is not known well.

Integrated Child Development Programme

ICDS

The Integrated Child Development Services (ICDS) was introduced in India in 1972. It was found that the incidence of mortality, morbidity and malnutrition among children continues to be high, despite the significant achievements in India in the first four Plans in all spheres of development. Though the infant mortality rate varies in different parts of the country, yet it remained at more than 100 per thousand for many states, due to various reasons such as physical, geographical, economic and social factors and the level of social and economic development. Various surveys' results reported that there was a high incidence of malnutrition among pre-school children. Vitamin A deficiency, which lead to blindness, occurrence of diarrhea, dysentery, parasitic infection, skin diseases, respiratory infections, whooping cough, measles, etc. were also fairly common among the children. Hence, the physical growth (height and weight according to age) and the development of mental capacity of children were affected consequently.

Therefore, the then Minister of Planning had, in 1972, suggested to implement the Integrated Child Care Services in all States. Eight inter-ministerial study teams were constituted by the Planning Commission and based on their findings a proposal for integrated child care services was made for pre-school children, which should cover supplementary nutrition feeding, immunisation, health care including referral services, nutrition education of mother, pre-school education and recreation, family planning and provision of safe drinking water. As per the advice of the steering Group set up by the Planning Commission, it was suggested to adopt an integrated approach to early childhood services in the Fifth Plan. The planners accepted those recommendations and the scheme of Integrated Child Development Services was included in the Plan in the Social Welfare Sector to give a definite focus to the development of services for early childhood.

The ICDS was started only after a series of inter-ministerial discussions on the details of the ICDS scheme and initially 33 experimental ICDS projects were started in 1975–76. Further evaluation of these projects showed improvement in the health and nutritional status of children, in general, of children who belonged to the scheduled castes, scheduled tribes and other poorer sections of society in particular, which has encouraged the Government to expand the programme and the schemes has been functioning yet. Some of the objectives of the ICDS are: (i) to improve the nutritional and health status of children in the age-group 0–6 years; (ii) to lay the foundation for proper psychological, physical and social development of the child; (iii) to reduce the incidence of mortality, morbidity, malnutrition and school dropout; (iv) to achieve effective coordination of policy and implementation amongst the various departments to promote child development; and (v) to enhance the capability of the mother to look after the normal health and nutritional needs of the child through proper nutrition and health education.

All those objectives were attempted to achieve in a package system by the ICDS, which includes (i) Supplementary nutrition; (ii) Immunisation; (iii) Health Check-up; (iv) Referral Services; (v) Nutrition and health Education; (vi) Non-formal education. Though the ICDS has evolved for child's status, it also provides some additional services. The scheme provides integrated services to pre-school children, i.e. below the age of six years. The pre-school age can be considered a definite phase in the development of the child. This phase can be further divided into different age groups. Since the mother has a key role in the physical, psychological and social development of the child, nursing and expectant mothers have been brought into the scheme, and attention has been given to mothers. Therefore, women of 15–45 years have also been brought into the domain of Integrated Child Development Services.

Different services provided to the beneficiaries of ICDS

Beneficiary	*Service*
Expectant and nursing mothers	Health check-up; Immunisation of expectant mothers against tetanus; Supplementary nutrition; Nutrition and Health education.
Other women 15–45 years	Nutrition and health education.
Children less than 1 year	Supplementary nutrition; Immunisation; Health check-up; Referral Services.
Children 1–2 years	Supplementary nutrition; Immunisation; Health check-up; Referral Services; Non-formal pre-school education.

As per the supplementary nutrition, efforts should be made to provide daily nutritional supplements to the extent of 300 calories and 10 grams of proteins per child, 500 calories and 15–20 grams of proteins per pregnant woman/nursing mother, and 600 calories and 20 grams of proteins per severely malnourished child. Nutrition and health education are supposed to be given to all women in the age-group 15–45 years but priority will be given to nursing and expectant mothers. The message of health and nutrition education are can be spread through various means of diffusion and dissemination.[8] In the project, immunisation against Diphtheria, Whooping Cough, Tetanus, Poliomyelitis and Tuberculosis of all infants (by first birth-day) has been also proposed. The health check-up and referral services of ICDS includes ante-natal care of expectant mothers, post-natal care of nursing mothers and care of new born infants and care of children under six years of age. Non-formal education through the institution of Anganwadi in each village and urban centre, for children of 3–6 years is another aspect of ICDS. The Anganwadi will develop desirable attitudes, values and behaviour patterns and aim at providing environmental stimulation in the child. It would also strive to satisfy the curiosity of the child and channel it in a creative direction.

The Anganwadi workers are to be responsible for: (a) non-formal pre-school education, i.e. organising pre-school activities in an anganwadi for about 40 children in the age-group 3–5 years of age; the toys, play equipment, etc. to be used should be rural in character and origin, in designing and making which the Anganwadi worker will play a leading roles; (b) organising supplementary nutrition feeding for children (6 months to 5+ years) and expectant and nursing mothers; in planning the menu, the priority will be given to locally available food and local recipes; (c) giving health and nutrition education to mothers; (d) making home visits for educating parents, particularly in the case of children attending the Anganwadi so that the mother of the child can be enabled to play an effective role in child's growth and development; (e) eliciting community support and participation in running the programme; (f) assisting the Primary Health Centre Staff in the implementation of the health component of Integrated Child Development Services Projects, viz., immunisation, health check-up, referral services, and health education; (g) maintaining routine files and records to enable measurement of the impact of the services; (h) reporting to the Child Development Project Officer (to be appointed for each block) the developments in the village which require further attention, particularly in regard to the working of the coordinating arrangements of different departments in the village; (i) maintaining liaison with other institutions in the village which have relevance to her functions (mahila mandal etc.); and (j) maintaining liaison with the lady school teacher for assistance in organizing pre-school activities and for participation of the primary/middle school girls (where they exist) in the programmes of the Anganwadi, thus enabling the girls to acquire work experience.

In Odisha, the ICDS has been implemented, throughout the state, i.e. 326 ICDS projects in entire 314 blocks and in 12 urban areas. The packages of services provided through this schemes in the state includes the health care services covering supplementary nutrition, immunisation, pre-school education, health check up, referral services and health education to children within the age group of 0–6 years. There are 34,201 Anganwadi centres under these ICDS projects. Table F-3 below describes the coverage of beneficiaries under different schemes of ICDS during 2000–01 to 2004–05, which depicts in almost all cases there is an increasing tendency in the number of beneficiaries.

Table F-3 *Beneficiaries covered under ICDS during 2000–01 to 2004–05 (in lakhs)*

Name of the Schemes	*2000–01*	*2001–02*	*2002–03*	*2003–04*	*2004–05*
Supplementary Nutrition	19.89	28.79	28.79	29.65	30.86
Immunisation	27.69	32.26	36.66	34.13	35.53
Health check-up	5.4	6.09	7.62	9.22	9.59
Nutrition and health check-up	3.87	4.63	5.34	5.37	5.91
Referral services	1.45	1.97	2.46	4.06	6.49
Pre-school education	8.92	9.81	9.97	10.61	10.96

Source: Women and Children Department, Odisha.

Primary Education and Health

Extension of Universal Elementary Education (UEE) facilities is one of the primary responsibilities of a democratic welfare state as education can create a common will to live better and can provide awareness about civic rights and obligations. Article 45 of the Indian Constitution has made education as the subject of Directive Principle of state policy, which states, 'The State shall endeavour to provide, within a period of ten years from the commencement of this Constitution, for free and compulsory education for children until they complete the age of fourteen years'. During 86th Constitutional Amendment the elementary education has been made Fundamental Right and the state governments are committed to the UEE in the state by 2010 with central government assistance. Since independence, there has been increasing spread of elementary education. However, 37 per cent of the people were illiterate in Odisha during 2001, which was just 15 per cent during 1951. Odisha has implemented a number of schemes to remove such educational deprivation.

It may be useful to examine those schemes here.

(1) Early Childhood Care Education (ECCE)[9]: ECCE was implemented in Odisha since 1982 to inducting children in the age group of 3–5, which would help for the Universalisation of the Elementary Education. The ECCE has been assisted by the UNICEF. The ICDS centres have also been strengthened in this regard.

(2) Universal Elementary Education: As mentioned above, the UEE is a constitutional obligation and national commitment. For achieving this, the UEE three aspects have been given importance, i.e. (i) universal access and enrolment (ii) universal retention of children up to 14 years of age and (iii) universal improvement in achieving the level/quality of education. For providing universal access to primary schools, the national guideline is to provide a school within one kilometre and for upper primary school within three kilometres from habitations having a population of 300 or more and 500, respectively. To achieve UEE and to improve quality of education, Government of Odisha also engaged 35, 824 para-teachers called Swechhasevi Sikhya Sahayak (SSS) against the vacant post of primary and upper primary school teachers and as additional teachers. Despite such attempts, the teacher to enrolment ratio has increased from 40 during 2001–02 to 51 during 2004–05. To attract students and restrict the dropout rate Mid-Day Meal schemes has also been implemented and as the table shows there is a sharp decline in the dropout shares across all social groups.

Table F-4 *Number of primary schools, enrolment, teachers and gross/net enrolment ratio in Odisha*

Year	*Number of School*	*Enrolment (000)*	*No. of Teachers*	*teacher/ enrolment*	*Gross Enrolment Ratio*	*Net Enrolment Ratio*
2001–02	42,824	4769	1,16,231	41.03036	98.27	87.25
2002–03	42,824	4606	83,652	55.06145	101.78	90.98
2003–04	44,416	5214	97,175	53.65578	103.48	91.51
2004–05	45,700	5215	99,079	52.63477	104.26	93.13

Source: Economic Survey of Odisha, 2006.

Table F-5 *Dropout in primary schools of Odisha (%)*

	All Categories			*SC*			*ST*		
	Boys	*Girls*	*Total*	*Boys*	*Girls*	*Total*	*Boys*	*Girls*	*Total*
2000–01	42.3	41.4	41.8	50.5	54.3	52.4	61.7	66.5	64.1
2001–02	42	40	41	50	52	51	61	65	63
2002–03	32.3	36.5	34.4	35.8	38.7	37.2	49.3	57.4	53.3
2003–04	31.9	35.4	33.6	34.6	36.6	35.6	48.2	56.6	52.4
2004–05	31.4	32.7	32	34	35.6	34.8	48	56	52

Source: Economic Survey of Odisha, 2006.

(3) District Primary Education Programme (DPEP): DPEP is another assisting scheme to achieve UEE and was launched in Odisha during 1996–97. However, the scheme has taken at a partial coverage. The phase I of DPED was assisted by the World Bank and the project was operated in eight districts such as Bargarh, Balangir, Dhenkanal, Gajapati, Kalahandi, Keonjhar, Rayagada and Sambalpur, which were selected on the basis of low female literacy rate and educational

backwardness. The project ended in June 2003 and the cost of the project was ₹ 229.75 Crore. The phase II of DPEP was assisted by DFID with a cost of ₹ 313.80 Crore and this project covers eight districts, such as Boudh, Kandhamal, Koraput, Malkanagiri, Mayurbhanj, Nabarangapur, Nuapada and Sonepur. At the end, DPEP has created a positive environment for primary education and by the end of 2004–05, it has created 734 new primary schools, 5,641 EGS centre and 200 AIE centres under DPEP I programme and 844 new primary schools, 5,631 EGS centres and 128 AIE centre in DPEP II programme. But how these schools would sustain further may be another issue to be looked into.

(4) Sarva Shiksha Abhiyan (SSA): SSA is another comprehensive scheme that was launched in all the districts of Odisha in 2003–04 to achieve the goal of UEE. One of the important features of the SSA is it 'aims to improve the performance of the school system through a community participation approach and to impart quality elementary education to all the children in the age group of 6–14 by 2010. The SSA is funded 75 per cent by Centre and 25 per cent by the State.

Apart, there are many other schemes (such as NPEGEL, Kasturba Gandhi Balika Vidyalaya, Education Guarantee Schemes), have been introduced in the state to achieve the goal of UEE. It may be useful to examine the functioning of these schemes at the ground level, which may bring some inputs to help the system and may help in achieving the UEE goal.

Ensuring of public health facilities is another primary responsibility of a democratic welfare state. In the pre-independence period, medical relief was no one's birthright but a kind gift from Government. But in the independent India, the responsibility of the state's responsibility in improving basic medical care through public health has been solemnly affirmed in Article 47 of Directive Principle of state policy of the Indian Constitution, which states, 'The State shall regard the raising of the level of nutrition and standard of living of its people and the improvement of public health as among its primary duties, and in particular, the State shall endeavour to bring about prohibition of the consumption, except for medicinal purposes of intoxicating drinks and drugs which are injurious to health'.

In Odisha, though there is an expansion of health care services, yet communicable and nutrition related diseases continue to be the major problem in the state especially in the tribal and backward regions.

The objectives of the health programmes undertaken by Government of Odisha include: (a) to provide adequate and qualitative preventive and curative health care, (b) to reduce maternal and infant mortality rate and to improve maternal and child health, (c) to ensure health care services to all, particularly to disadvantaged groups like Scheduled Tribes, Scheduled Castes, backward classes and women, (d) to provide affordable quality health care through allopathic, homeopathic and ayurvedic systems of medicine etc. (e) to impart training to doctors, nurses and other paramedical staff to upgrade their skill an knowledge, (f) to ensure greater access to primary health care by providing medical institutions as close to the people as possible or through mobile health units, particularly in the under-served and less developed districts, (g) to improve hospital services at primary, secondary and tertiary levels in terms of infrastructure, drugs and personnel, (h) to eliminate diseases like polio and leprosy from the State and prevent and control other communicable diseases and (i) to improve health care facilities in the KBK districts of the State.

Table F-6 and F-7 below shows, though there is improvement in crude birth rate, and infant mortality rate, the physical infrastructure for providing medical care is concerned, there is stagnation both in terms of number of institutions and number of beds. The total number of Ayurvedic and Homeopathic hospitals in Odisha, during the time period, has remained stagnant at 5 and 4 respectively and number of beds in these two hospitals are 203 and 125 respectively. It appears that there is a supply constraint of public health care system in the state.

Table F-6 *Birth rate, death rate and infant mortality rate in Odisha and all-India*

Years	*Odisha*			*All India*		
	CBR	*CDR*	*IMR*	*CBR*	*CDR*	*IMR*
1981	33.1	13.1	135	33.9	12.5	110
1991	28.8	12.8	124	29.5	9.8	80
1995	27.8	10.8	103	28.3	9	74
2000	24.3	10.5	96	25.8	8.5	68
2001	23.4	10.2	90	25.4	8.4	66
2002	23.2	9.8	87	25	8.1	63
2003	23	9.7	83	24.8	8	60

CBR Crude Birth rate, CDR – Crude Death Rate, IMR – Infant Mortality rate.
Source: SRS Bulletin, April 2005.

Table F-7 *Allopathic medical institutions run by the State Health Department*

Years	*Hospital*	*CHC*	*PHCs*	*PHCs (New*)*	*Mobile health unit*	*Total medical institutions*	*Total no. of bed*
2000	180	158	183	1,161	14	1,701	13,786
2001	180	158	183	1,161	14	1,701	13,786
2002	180	158	184	1,161	13	1,701	13,786
2003	174	231	120	1,162	14	1,701	13,786
2004	174	231	120	1,162	14	1,701	13,786
2005	174	231	120	1,162	14	1,701	not recorded

* Dispensaries, Additional PHCs, Subsidiary health centres and Medical aid centres were amalgam and registered as PHC (New) since 1996–97.
Source: Directorate of Health Service, Bhubaneswar.

Therefore, to improve the public health, various schemes have been implemented such as the Revised National Tuberculosis Control Programme (RNTCP), which is implemented in the State since 1997 to achieve 70% case detection, 90% sputum conversion and 85% cure rate by the end of 2005. As per the 2006 Economic Survey of Odisha, 102 sub-district T.B. units and 530 Microscopy Centres are engaged in providing diagnosis and treatment services under the revised strategy. It has detected and treated 1,56,044 patients. The National Filaria Control Programme (NFCP), is a Centrally Sponsored Plan on a sharing pattern of 50:50 between the Centre and the State, is implemented through a Filariasis Bureau comprising of a research unit at Puri, 15 Filaria Control Units and 15 Filaria clinics in urban areas to provide treatment for microfilaria. By 2004–05, 36,658 blood slides were examined and 774 persons were clinically found positive and 3,721 persons were treated. The National Programme for Control of Blindness (NPCB) is a World Bank assisted project introduced in Odisha during 1976, with an aim to reduce the prevalence of blindness from 1.4% to 0.3%. By 2004–05, 91,509 cataract operations were performed including 87,525 IOL cases against 82,652 cataract operations including 78,590 IOL cases during 2003–04. The National Iodine Deficiency Disorders Control Programme (NIDDCP) is implemented in the State since December 1989 as a Centrally Sponsored Scheme to prevent the Iodine Deficiency Disorders (IDD) like incidence of goitre, physical and mental disorders and to bring the incidence of IDD below 10% in the State. Use of common salt has been banned in the State from 18 October 2001. The IDD survey conducted in Baragarh district in 2002, shows, the IDD prevalence rate has come down to be 7.66% as against 10.8% recorded in the previous survey conducted in

1995–99. The National Leprosy Elimination Programme (NLEP) was implemented in the State from 1982–83. The programme is supported by WHO, World Bank and international agencies like Lepra India, HOINDA and the German Leprosy Relief Association to eliminate leprosy by detecting all the cases and bringing them under Multi Drug Therapy (MDT). Similarly, the National Anti-Malaria Programme (NAMP) renamed as National Vector Borne Disease Control Programme (NVBDCP) is being implemented in the State since 1953 to reduce its impact. Some other schemes are the Odisha Health System Development Project (OHSDP), a World Bank assisted project, is being implemented in the State since September, 1998.

The Pancha Byadhi Chikitsa is being implemented in the State since 2001, which guarantees free treatment including free medicine for five common communicable diseases, viz., malaria, leprosy, diarrhoea, acute respiratory infections and scabies. The National AIDS control programme (NACO) Phase-I has been undertaken in the State during 1992–1999 and Phase-II programme is in operation since April, 1999 with the support of NACO and DFID to reduce the spread of HIV infection in the State and to strengthen the State's capacity to respond to the HIV/AIDS on long-term basis.

The Infant Mortality reduction mission, launched in the State on 15 August 2001 to ameliorate (a) Immunisation in outreach inaccessible areas, (b) Chemoprophylaxix against malaria during pregnancy. (c) Institutional delivery to the pregnant women free of cost. (d) Free treatment service for slum dwellers and (f) Transport cost for free institutional delivery.

The Reproductive and Child Health (RCH) programme is a 100% centrally assisted programme implemented in the State for improving the reproductive health of men and women and the health of children and to ensure safe motherhood and child survival to create confidence for accepting family planning methods.

Assets and Income Generating Schemes

There are many integrated schemes to increase the assets and income of the poor people. One such assembled scheme is Integrated rural Development Programme (IRDP) in rural area, which (along with a few more schemes) has been recently replaced by Swarnjayanti Gram Swarozgar Yojana (SGSY) for rural self-employed. Similarly, the Nehru Rozgar Yojana is one of the targeted schemes in the urban area, which has been replaced by Sarna Jayanti Shahari Rojgar Yojana (SJSRY) in recent time (from April 1999). Brief reviews of these schemes have been provided below.

Integrated Rural Development Programmes (IRDP)

By the end of the 1960s, it became clear that the benefits of a purely growth oriented strategy do not percolate to the poorest of the poor (Khanna, 1990). Hence, since the start of the fourth plan in 1969, Government of India introduced many programmes to make the benefit of development reach to the poor directly. Such programmes include Small Farmers Development Agency (SFDA), Marginal Farmers and Agricultural Schemes (MFAL), Tribal Sub Plan, Sub Component Plan for Scheduled Caste, Antyodaya Scheme and Area Development Programmes like the Drought Prone Area Programme (DAPA), Desert Development Programme (DDP) Hill Area Development Programme (HADP).

Throughout the 1970s, these programmes grew in its size and coverages and by the beginning of the 6th Plan it has blossomed into a full-fledged strategy of direct attack on poverty. The attack was having three-fold rural development strategies including a perspective to reduce poverty ratio to 10 per cent by 1995.

The first strategy among them was: 'individual household and poverty group-oriented programmes of income generation through assets and skill endowment and direct supplemental wage employment through public works' (ibid.). This strategy has two components. First, it comprises of a strategy of providing self-employment opportunities to certain families and includes the integrated Rural Development Programme (IRDP, 1978–79) and its sub-components – Training of Rural Youth for Self-Employment (TRYSEM) and Development of Women and Children in Rural Area (DWCRA).

The second component of the strategy was wage employment programmes, which seeks to provide wage employment opportunities to the rural people and, in the process, build up community assets of a durable nature. This includes schemes like National Rural Employment Programme (NREP) and Rural Landless Employment Guarantee Programme (RLEGP). (Khanna, 1990)

The Programme Evaluation Organisations of the Planning Commission (1985) established IRDP as a sound programme for poverty reduction, which has been supported by many other studies and till 1990 it was a successful one (ibid.). However, they also found it is inadequate in providing assistance and in functioning. Some of the barriers observed were less involvement of the state governments, district and block office, which may be due to paucity of resources and infrastructure with the state governments. Hence, the failure was at the ground level. They also suggested for the annual adjustment of prices both for providing assistance and identifying the beneficiaries. Khanna (ibid.) also suggested to increase the amount of assistant to lift the household to the critical minimum.

Saxena (2007) stated, at the end of 1994, the total cumulative outlay had grown to ₹ 132 billion in credit and ₹ 82 billion in subsidy. However, the government evaluation showed that only 15 per cent beneficiaries could cross the poverty line; 24 per cent of the assets provided to beneficiaries did not generate any income and 50 per cent of the assets did not contribute to net disposable income. Some of the reasons stated by Saxena as a reason of the failure of IRDP are as follow:

(a) *Lack of constant interactions between beneficiaries and authority:* IRDP loan has remained as a one-time affair rather than a constant affair of the beneficiaries, bankers and the development officials involved in the process. One would expect, there should be constant interaction among these agents, till the beneficiaries are well above the poverty line and generate a good worth of assets/capital.
(b) *Supply-led activities:* IRDP loan has been mostly supply led. Banks were restricting the loan to 'approved' activities. As Dreze found, in Uttar Pradesh a bank gave 20 loans for setting up shops in a village of 143 households. Hence, before advancing the loan, 'final demand' for an activity has to be estimated so that the activity generated can be profitable and sustainable. In other words, there should not be over supply of any output or supply of such products for which demand is low.
(c) *Neglecting the existing micro-enterprises:* In IRDP schemes, loans are being advanced as a new loan for different activities. It did not look into the existing problems of the already established micro-enterprises.
(d) There was total absence of 'social intermediation' in IRDP.

Sainath (1996) also found the failure of IRDP in Odisha and the major reason he stated was – that one of the resources provided in this schemes is a hybrid variety of cows, which were unable to survive in the prevailing climate in Odisha. Seabright (1989) found, 'the benefits to households of investing in livestock through the IRDP have been significantly below those to livestock purchased outside the scheme, which was principally due to higher prices paid by IRDP participants in the livestock markets, prices which were not compensated by higher quality of the animals purchased'. Further, the author also found that households lacking land and underemployed household labour were not able to manage livestock investments economically at this small scale of production. For such demerits of the IRDP, the authors have suggested that there should have been fewer loans and loans could have been directed to those who possessed at least some land. In such cases, if collective facilities would have been offered for livestock fodder or cattleshed to them, probably both the problems could have been solved. Furthermore, if there would have been a commonplace of work for the beneficiaries, they would have been better organised, which could have shaped their thinking process.

Rath (1985) look IRDP from 'Garibi Hatao' perspective, which shows that 'the effort to give cattle and other assets too has yielded little'. One of the major demerits found by Rath is the targeting errors. He found on an average, there might be 15 per cent Type II errors (non-eligible included as beneficiaries). However, such errors may occur in any social security schemes if not implemented properly. Rath suggested eradication of poverty needs a multi-pronged strategy.

Dreze (1990) in his article, 'poverty in India and the IRDP delusion' stated that IRDP debates reveal a puzzling contrast between two groups of studied. The micro-level studies have tended to project a very negative view of the programme. Whereas, the large-scale household surveys carried out by the government or by banking institutions offers a heart-warming picture of the achievements of IRDP. However, given this impression that IRDP is not completely useless, Dreze suggested for employment generation schemes for poverty reduction. Building long-term assets in a household is a strategy for ensuring that households are better off for a sustainable period of time. Hence, the evaluation of IRDP has to be done from sustainable livelihood perspective. Furthermore, in the subsequent period, though the IRDP has been replaced by SGSY, the demerits pointed out in it, have to be taken care. Similarly, there are several studies carried on the functioning of IRDP but none of them have rejected the IRDP as a helping scheme to the poor, though some of them have suggested alternatives. However by 1999 April, the IRDP has been replaced by the SGSY.[10] But from livelihood point of view, what may be more desirable is to what extent SGSY overcomes the already existing problem in IRDP.

Swarnjayanti Gram Swarozgar Yojana (SGSY)

To bring proper social intermediation linkages among the earlier programmes and to generate income sustainability, the existing self-employed programmes (IRDP, TRYSEM, DWCRA, SITRA, GKY and MWS)[11] have been restructured into the SGSY from April 1999. SGSY is a holistic programme and covers all aspects of self-employment such as organisation of the poor into Self-Help Groups (SHGs), training, credit, technology, infrastructure and marketing. Further, the SGSY is to be funded by the Centre and States in the ratio of 75:25.

The main objective of the SGSY is to provide livelihood security, i.e. the assisted poor families have to be brought above the poverty line by ensuring an appreciable increase in incomes over a period. As the Economic Survey (2004–2005), shows, the total self-employed (*Swarozagaries)* in India has increased from 83 million in 2002–2003 to 90 million in 2003–2004 However, such aggregate figure may not speak out many logical flaws of the schemes. The fundamental demerits of the SGSY through SHGs is putting the condition to generate some decent amount of saving first (by forming a SHG group), then equal amount will be supplemented by a bank at a subsidies rate. They are called Below Poverty Line households, because by the Planning Commission's evaluation their monthly per capita expenditure (proxy for income) is inadequate to get even enough food for the household. However, the SHG strategy of development expects, within that inadequacy, a household has to generate some saving to become a member of a SHG and then reap the benefit.

From where do the households get additional amount to generate a revolving fund? One way may be to reduce their food expenditure further, which means making their condition further worse. Even if they can succeed to generate some saving it may be not enough to lift them to above the poverty line. As Mahajan (2005) pointed out, 'micro-credit is barely adequate even as an instrument for poverty alleviation, leave alone economic growth. Hence, in case of the BPL households some alternative mechanism may be developed.

In case of Odisha, there is not many evaluative study, how as the Economic Survey of the state provides, the schemes function in the states since 1999–2000, but to what extent it has an impact at ground level may need micro level study.

Nehru Rozgar Yojana (NRY)

In order to alleviate the conditions of urban poor, a Centrally Sponsored programme – Nehru Rozgar Yojana – was launched at the end of the Seventh Five-Year Plan (October 1989) with the objective of providing of employment to the urban unemployed and under-employed poor. The Central Government indicated its overall contribution, while the State Governments undertook essential task of identifying, earmarking and coordinating the relevant sectoral inputs. The NRY consisted of three schemes namely (i) the Scheme of Urban Micro Enterprises (SUME); (ii) the Scheme of Urban Wage Employment (SUWE); and (iii) the Scheme of Housing and Shelter Upgradation (SHASU).

Swarna Jayanti Shahari Rojgar Yojana (SJSRY)

Like the SGSY in the rural area, the Swarna Jayanti Shahari Rojgar Yojana (SJSRY) seeks to provide gainful employment to the unemployed or underemployed poor through encouraging the setting up of self-employment ventures or provision of wage employment. The self-employment ventures of the SJSRY runs through the Urban Self Employment Programmes. The target groups are:

(a) Individual urban poor: providing assistance to them for setting up gainful employment venture.
(b) Urban poor women and children: providing assistance to groups of urban poor women for setting up gainful self-employment venture. The scheme is called 'The Scheme for Development of Women and Children in the Urban Areas (DWCUA).

Tables F-3 and F-4 provide the physical performance of special employment in urban areas.

Recent data from Economic Survey, 2005–06, the table shows:

- the beneficiary under Urban Self-Employment Programme (USEP) has increased from 0.64 lakhs in 2003–04 to 0.23 lakhs in 2004–05.
- The beneficiaries assisted under DWCUA have also shown a marginal increase from .36 lakh in 2003–04 to .37 lakhs in 2004–05.
- And person trained in urban are increases from 1.22 lakhs in 2003–2004 to 1.60 lakhs in 2004–05.

Since, Odisha followed the central policies and hardly any welfare schemes of her own, we have directly explored the functioning and performances of those schemes in the state at the macro-level first.

Housing Assistance Schemes

Housing

Housing is one of the basic human needs and an essential component of the quality of life. As mentioned in Chapter 1, the housing conditions in rural Odisha, during 1973–74 or in 2002 has remained worse compared to other Indian states except Assam during 2002. As per NSSO data, during 2002 the per cent of pucca houses in rural Odisha is 18 per cent and 53 per cent of the houses were katcha.

During the same time in rural India the numbers of pucca houses are 36 per cent which is twice than that of Odisha and number of kutcha houses in rural India is 21 per cent. However, in case of urban area, Odisha is not only better from her rural counterpart but also better than many other Indian states. Further, as the 2001 census data shows, the lowest percentages of good houses are found in rural Odisha (22.6%), whereas the similar share for all-India is 45 per cent. On the other hand, as the percentages of dilapidated houses are concerns, it is 10.29 per cent in rural Odisha, which is also one of the highest among the other major Indian states. Therefore, considering the poor conditions of houses, the Central Government announced a National Housing and Habitat Policy for the below poverty population, which aims at providing 'Housing for All' by the end of the Tenth Plan period. In this scheme, the Rural Housing implemented are by Panchayat Raj Department and the Urban Housing are implemented by Housing and Urban Development Department. Broadly there are four types of rural housing schemes, such as Indira Awas Yojana (normal and additional), Gramin Awas, Odisha Rural Housing Development Corporation (ORHDC), Odisha Co-operative Housing Corporation (OCHC). Similarly, there are two schemes for urban housing in the state, such as Valmiki Ambedkar Awas Yojana (VAMBAY) and Economically Weaker Section (EWS) and Low Income Group (LIG) Housing Schemes.

The Indira Awas Yojana (IAY) scheme was implemented by Government of India during 1985–86 to assist the rural BPL persons to construct or upgrade their dwelling. The scheme provides assistance to the BPL rural households belonging primarily to SC/ST (60%) and others (40%). The funding pattern in this scheme is in the ratio 75:25 between Centre and State. In Odisha the houses are allotted in rural areas and includes the people living below poverty line, 60% is earmarked for SCs, STs, widows, people affected by natural calamities, freed bonded labourers, fire victim families. Grama Sabha/Palli Sabha selects beneficiaries under this scheme. The scheme provided unit cost of each house was ₹ 20,000 in plain areas and ₹ 22,000 in hilly/difficult areas, which has been enhanced to ₹ 25,000 and ₹ 27,500 respectively. The house also includes a sanitary latrine and smokeless chullah. The beneficiaries have to arrange his/her land and has to construct their houses and also have to make their own arrangement for the construction, engage skilled workmen on their own and also contribute family labour. In Odisha, during the Ninth Five-Year Plan Period (1997–2002), 45 lakh houses were constructed under IAY, which mostly benefited the poor, though among the poor there was undue preference to certain beneficiaries (Planning Commission, 2005). During 2004–05, 67,892 houses have been constructed by utilising ₹ 181.94 crore against the target of 74,735 houses and available fund of ₹ 192.54 crores. Further, Government India has sanctioned 4,923 numbers of additional Indira Awas Yojana for the flood and natural calamities victims. Another 6 lakh additional Indira Awas have been allotted by the Government after the super cyclone of 1999.

Other housing scheme also aims to provide housing facility in the state. Yet, the there is large number of houses with depletion conditions, which may be hinted for targeting error in the field. Therefore, a primary survey may bring out the details.

Other Promotional and Protectional Schemes

The Old Age Pension Schemes

In Odisha, the State Old Age Pension (SOAP) scheme was introduced with effect from 1 April 1975. In this scheme, the State Government was providing pension of ₹ 100 per month per beneficiary consisting of old and destitute persons of 60 years of age and above, to leprosy patients with visible signs of deformity and to destitute widows irrespective of age. A criteria that has been put was for all the above categories, the annual income of the beneficiaries must not exceed ₹ 3,200. However, from April, 2006 the State Government has enhanced the rate of pension under this scheme to ₹ 200 per month per beneficiary along with the pensions under National Old Age Pension and Odisha Disability Pension. The total coverage under the scheme was 6,75,000 including the enhancement of 35,000 made during 2004–05. During 2006–07, ₹ 8,100.00 lakhs has been provided in the State's Non-Plan Budget for payment of pension to the above targeted 6,75,000 beneficiaries under the scheme. There is also National Old Age Pension schemes run in the state, which aims to provide some financial assistance to the old persons (65 and above) living below the poverty line. The Government of India provides ₹ 75 per month and the State Government adds ₹ 25 per month to each beneficiary under the above scheme. Till 2004–05, as against the target of 4,93,400 beneficiaries, 4,93,273 beneficiaries including 2,51,344 women, have been covered under the scheme.

Woman Development

For female education, educational institutions including Kanyashrams have been established exclusively, particularly in low literacy tribal areas. Education for girls is free from primary to postgraduate level. Many ITIs and polytechnics were also opened exclusively for women, which provide them professional training. Various reservation rules have also been implemented strictly, such as the Odisha Civil Services (Reservation of Vacancies for Women in Public Services) Rules 1993 have been framed and are now in force. To increase female employment, it has been decided by government that all posts of Anganwadi workers, Supervisors and CDPOs will be filled by women, which has opened new job opportunities for women at village level.

Balika Samridhi Yojana (BSY)

The Balika Samriddhi Yojana is 100% Centrally Sponsored Scheme. The objectives of the scheme are: (1) to extend assistance to change negative family and community attitudes towards the girl child at birth and towards her mother; (2) to improve enrolment and retention of girl children in schools; (3) to raise the age at marriage of girls; (4) to assist the girl to undertake income generating activities, both in the rural and urban areas in all districts in India. But it aim to cover the girl children in families below the poverty line (BPL) who are born on or after 15 August 1997 as defined by the Government of India. The girl children who are eligible under BSY will be entitled to get the benefits of (i) a post-birth grant amount of ₹ 500 (ii) when the girl child is born on or after 15/8/1997 and covered under BSY starts attending the school, she will become entitled to annual scholarships for each successfully completed year of schooling.

The amount to be awarded are as follows:

Class	Amount Of Annual Scholarship
I–III	₹ 300 per annum for each class
IV	₹ 500 per annum
V	₹ 600 per annum
VI–VII	₹ 700 per annum for each class
VIII	₹ 800 per annum
IX–X	₹ 1,000 per annum for each class

However, the benefits under BSY will be restricted to two girl children in each household born on or after 15 August 1997 irrespective of the total number of children in the household.

ANNEXURE G: THE PANCHAYAT DESCRIPTION

The Panchayat Sason was established in 1951 with an area of roughly 26.64 square kilometres. At that time, along with the present Sason Gram Panchayat, there were also three more gram panchayats (Kilasama, Bishal Khenda and Kan Khenda) which were part of the old Sason gram panchayat. In 1961, the Sason gram panchayat was reorganised when two more Panchayats were formed from the old Sason. These two gram panchayats are Kilasama and Bishal Khenda. Again in 2002, one more Panchayat was formed from the original Sason gram panchayat and that is Kan Khenda. The basis of these reorganisations is population. It is found that a panchayat may have minimum of 2,000 population and maximum of 10,000 population. Hence, when it is expected that, some panchayat is going to exceed their population beyond 10,000, a new panchayat can be formed. Following such procedure, during 2005, Sason gram panchayat had revenue villages with 6,033 population and an area of roughly 17 square kilometres. Table G-1 presents some detailed information about the panchayat.

Table G-1 *General details about the Panchayat*

Name of the Panchayat	*Sason*
State	Odisha
District	Sambalpur
Block	Dhankauda
Area of the Panchayat	4190.71 Acres or 16.95 square kilometers
Number of Revenue villages	8
Number of wards	15
Total population	6033
Population density	355.92
Scheduled Caste population	1691 (28%)
Scheduled Tribe population	1872 (31%)
Sex ratio	947
Number of households	1324*
Average household size	4.68*
Number of members in the Panchayat	15
Literacy rate	Male – 77.0, Female – 51.6*
Work Participation rate	Total –51.6, Male – 56.6, Female – 46.4*
Cultivable area under irrigation	59%
Major economic activities	Agriculture
Percentages of land less labour	35

Source: Institutional survey at Sason Notes: star marks are calculated from 2001 census data.

As the Table G-1 depicts the area of the Panchayat is 16.95 square kilometres. There are 8 revenue villages, which has been divided into 15 wards. Around 59 per cent of the cultivable area is irrigated in the Panchayat. The work participation rate for male is 56.6 per cent while for female it is 46.1 per cent. The major economic activity of the people is agriculture. Before going to the other details, in the following we have discussed the topology and sociology of the panchayat.

Topology of the Panchayat

The panchayat is located in the southwest direction of the Sambalpur district in the western part of Odisha. The climatic condition in this region is during winter, it is extreme cold, in summer it is extreme hot. Variation in rainfall is high. However, since there is 59 per cent of the cultivable area are irrigated, falling down of rainfall does not create many problems in agricultural production, unless the canal is not working as it was for the two years in the recent past.

As can be seen from Table G-2, there are 67 per cent of the total area of the gram panchayat is cultivable land and of which 58.2 per cent are irrigated and the main crop for cultivation is paddy.

Rest of the area of the gram panchayat is of railway area, unsettled area and government area. The government areas are covered with thin forest. There is also a little rocky area in the panchayat. Apart from the canal, there also flows a river in the northern part of Sason gram panchayat which also helps both for cultivation and for bathing purposes.

Table G-2 *Area under Sason Panchayat across the villages (Area in Acres)*

Village name	*Area*	*Rayati*		*Irrigated land/rayati land*
		Total	*Irrigated*	
Chattargada	277.1	153.21	0	0.0
Sadhasingha	489.28	353.77	251.73	71.2
Ajitpur Sason	237.96	159.95	82.42	51.5
Nuan Khurigaon	1064.29	571.51	98.37	17.2
Rani Khenda	1255.4	858.92	595.65	69.3
Paab Pali	247.75	192.71	167.97	87.2
Kanta Pali	411.4	379.31	350.34	92.4
Gand Pali	207.53	154.94	96.64	62.4
Total	4190.71	2824.32	1643.12	58.2

Source: Institutional survey at Sason (Data collected from Revenue Office, Sason).

Overall, to our observations, the panchayat is a suitable place for cultivation and the economy of gram panchayat could be strong enough in agriculture. As it was learned from the revenue officers, the productivity of paddy varies from 12 to 15 quintal in irrigated area to 3 to 4 quintal in non-irrigated area. It was also learned that most of the farmers are middle class farmers and based on the rule of the Odisha government, nobody was legally allowed to own more than 10 acres of land in his/her name. It was also observed that there are 35 per cent of landless labour in the gram panchayat. This high magnitude (35%) of the supply of labour could be one reason why the agricultural labourer's wage rate in the panchayat is very less. However, these agricultural labour historically belong to a specific background and they are not randomly distributed across the castes. If the pattern is examined if one happen to belong to a particular community then, there is higher possibility that he/she would be agricultural labour. This has been discussed in the following paragraph. However, exceptions are there, i.e. one can find non-SC/ST population also in agricultural labour class, but the general pattern is that most of the agricultural labour is from SC/ST.

The Sociology and Demography of the Panchayat

There are multiple castes of people who have been living in Sason GP. The cultural classifications of these castes are – Brahmin, Karana (Mohanty), Gauda, Dhuba, Kandha, Luhura, Ganda, Ghansi, Doomal, Kulta, Teli, Munda, etc. Among these, the social as well economically backward castes are the Scheduled Caste (Ghansi, Ganda (traditionally culturally excluded and were known as untouchables), Dhuba, Luhura, etc. and Scheduled Tribes (Munda and Kandha).

Traditionally, it has been the other caste, i.e. Brahmins who posses most of the land and most fertile land. Scheduled Tribes populations are not land poor, like the Scheduled Caste populations but they are far behind than other superior caste in the social ladder. Scheduled caste people are usually agricultural labours but exceptions are there. These exceptions are due to the history. In the old age some of the Scheduled Caste populations have been brought to a village from another village to serve the village and in return, they have been given some land. That is how some of these people own land and are counted as marginal or small farmers or cultivators. But they are confined only to a few households. Further, over a period of time as the family size of their community increases, gradually they become either tiny landholders or landless labourers. Apart from the land possession, we have also tried to observe the distinct housing conditions of the people of different castes in Sason gram panchayat, which have been narrated below.

Sason gram panchayat is 17 km from the headquarter of Sambalpur district, towards the west south direction on National Highway No. 6. Sason village is 100 to 150 metres distance from the gram panchayat. Before entering into the village, we had to cross the train line (near by the railway station of Sason) to enter into the village. The village is combination of many types of houses. We found a stratum in the village, where most of the houses are pucca with two to three stairs. In the top of the building, there would be a water tank whose colour would be matching to the colour of the building. There would be also a big TV antenna. Most of the gate would be iron gate with concrete boundary around the house. The cowshed would be at a distant place in a well finished roof and wall of a house. While we were visiting these houses for different purposes, they were calling us inside the house offering a high valued chair under a fan (in some cases near the cooler) and also offered tea (prepared with milk) and snacks. While it was the treatment in 2–3 random houses from this stratum, we thought that the standard of living of this stratum may be better. Later on, we discovered, mainly Brahmins are staying in this stratum. In the local language, they identified the stratum as *Brahman para*.

Then we moved towards the left of the *Brahman para*, we observed, a number of houses with tile in the roof and some are cement wall and some are earthen wall but with a good finishing. Here, too we received good hospitality. They were offering us wooden chair and the discussion went on. Later on, we discovered that they are *Gaud para* and *Kandha para*.

Then again, we proceeded towards the left, where we discovered the worse types of the houses. There is no boundary, no gate, half of the wall of the house would be broken and covered with torn cement bag, other half would be protected with so many patches in the wall. There is no place to throw the used water of the house. The waterlogging would be just in front of the house. The house put the ashes that come from their choolha (cooking place) near by. Some of the roofs are covered with thatch and some are with plastic. Here too, we received good hospitality. A young man of around 22 years brought a cot (made up with branches of some trees and with the local rope, which was mostly torn). He put the cot under the shade of a tree nearby the house and send a boy to the tube well to collect some drinking water for us. To us, it was clear that, the standard of living of the people was to an extent readable. Instead of a gorgeous chair, there was a cot with torn rope, instead of fan or cooler, there was open air under a tree, and instead of tea there was water from the tube well. The discussion went on. Several times the young man was asking whether we are there to give some public assistance. We could read the desperate need for public assistance by the gentlemen, in particular and of the stratum, in general. However, in spite of all these we found two pucca houses. Our further investigation found that – one of these was out of the public assistance received from Indira Awas Yojana (IAY) and the other was a house of a retired peon. Later on, we discovered that this is *Ghansi para* of the Sason village.

In the extreme left, we found another stratum, where the houses are definitely better than *Ghansi para* but not to the extent than the house of *Kandha para* or *Gaud para*. Here too, we found a few pucca households, which are either due to IAY help or of some the government employee. This is *Ganda para*. To our observation if the interaction of sociological classification can put across the housing conditions, then in general the best houses are of Brahmins, then *gauda/kandha,* than *Gandapara* and the last *Ghansipara*.

As this can bee seen from Table G-3, there are 28 per cent of the population in Sason gram panchayat who belonged to SC and 31 per cent belonged to ST. As the Table G-4 shows, the overall sex ratio is marginally in favour of male, although, it is in favour female in Ajitpur Sason, Sadha Singha and Nua Khurigaon.

Table G-3 *Distribution of population across the villages by their castes and sex in Sason Gram Panchayat*

Village Name	*SC*			*ST*			*OC*			*Total*		
	Male	*Female*	*Total*	*Male*	*Female*	*Total*	*Male*	*Female*	*Total*	*Male*	*Female*	*Total*
Rani Khenda	179	193	372	311	320	631	391	359	750	881	872	1,753
Ajitpur Sason	134	134	268	56	48	104	161	176	337	351	358	709
Kanta Pali	137	124	261	49	37	86	257	255	512	443	416	859
Sadhasingha	154	164	318	199	212	411	102	87	189	455	463	918
Paab Pali	76	64	140	65	60	125	180	182	362	321	306	627
NuanKhurigaon	166	166	332	256	259	515	160	160	320	582	585	1,167
Total	846	845	1,691	936	936	1,872	1,251	1,219	2,470	3,033	3,000	6,033

Source: Institutional survey at Sason, (Recent data collected from Anganawadi workers).

Note: It was told that while Nua Khurigaon include Ganda Pali population, Sadha Singha includes Chatar gada population.

Table G-4 *Sex Ratio (Male/Female) across the villages by their castes in Sason Gram Panchayat*

Village Name	*SC*	*ST*	*OC*	*Total*
Rani Khenda	0.93	0.97	1.09	1.01
Ajitpur Sason	1.00	1.17	0.91	0.98
Kanta Pali	1.10	1.32	1.01	1.06
Sadhasingha	0.94	0.94	1.17	0.98
Paab Pali	1.19	1.08	0.99	1.05
NuanKhurigaon	1.00	0.99	1.00	0.99
Total	1.00	1.00	1.03	1.01

Source: Same as Table 3.

With this sociological description, the following paragraph describes what are the communications facilities available in the Panchayat.

Access to Sason Gram Panchayat

As the Table G-5 shows, there are three modes of infrastructure to access the panchayat: (i) the physical infrastructure, (ii) the soft infrastructure, and (iii) the access to the panchayat through correspondence.

Table G-5 *Access facility to the Sason GP*

Physical infrastructure	*Soft infrastructure*	*Correspondence*
Road: National highway is within 100 metres from the GP. Every half an hour there is bus service.	Telephone is the only soft infrastructure.	There is a main post office in the GP since 1961 and many other sub post office.
Pucca road to the GP and to its many villages as well.	No E-mail, No telegraph.	No other private correspondence services are available.
There is railway station within the GP. Every four hour there is train service.	There is wireless in the police station.	
There is not enough vehicle (autorickshaw or car) available in the gram panchayat head quarter to access other villages of the gram panchayat.		

Source: Field survey.

Both railway and roadways are close by the Sason gram panchayat and there is frequent bus/train services. So it is easy to access the panchayat by any of these roots. However, since there is not many autorickshaw or car services to the different villages of the panchayat, it is not easy to access them. Telephone is there in all the villages. There also exists a post office in the panchayat.

The above discussion on the sociology, topology and access to the gram panchayat enable us to say, there is possibility of social disparity and would be difficult to access other villages in the panchayat. Given the above description, the following paragraph tries to understand the panchayats economy.

Economy of the Sason Gram Panchayat

As discussed earlier, 67 per cent of the total area of the gram panchayat is cultivable land and of which 59 per cent is irrigated. Further, most of the irrigated areas are double cropping area, where the productivity is 3 to 4 times more than the dry area. Further given other infrastructural and transport facilities to sell these agricultural output, to our observation, the primary sector is the base of the economy in the panchayat.

But how far this primary sector is a help to the common people (i.e. to the 35% landless agricultural labour), require further study because even though the labour has been is demand twice in a year, the wage rates has remained low (₹ 35 per male and ₹ 30 per female per day). However, although some people (mostly women) are involved in Beedi making activities, it just ameliorates the conditions and is not enough to increase the agricultural wage rate. These people's conditions are very bad.

When they try to take some land on leased or think to start some informal activities, they may not get adequate assistance from formal sector such as commercial bank or co-operative banks. Because being resourceless, these people do not qualify to get a loan. Hence if they need some credit, they have to depend on the local moneylenders, where the rate of interest varies from 60 per cent to more than 100 per cent per annum. And the bank is there mostly for solving the credit requirements of the middle-class farmers, landlords or the businessmen. Apart the contributions of the primary sector to the GP's economy as can be seen from Table G-6, perhaps the secondary sector is also creeping up in the economy.

Table G-6 *Number of rice, ata and oil mill and poultry firm in sason Gram Panchayat*

Rice Mill	7
Ata (flour) Mill	2
Oil Mill	2
Poultry firm	3

Source: Field survey.

As this can be seen from Table G-6, the Panchayat is not only having more paddy production as discussed earlier, but value addition has also happened in that product. There are 7 rice mills, which convert the paddy into rice (a lower value product to a higher value product). Two of them are at least 25 years old. However, from the investigation we, found that after the value addition the product have to be sold at a far distance place like Kolkata. Hence, although there are number of rice mills and it yield revenue for the owners, it may not directly help the common men. Similar things happen in the case of flour and oil mills. It is also found three poultry firms, which are 10 years, 7 years and 5 years old respectively. It is observed that although the poultry firms are located at Sason GP, they also produce for other places as well. In a sense, there is a drain of different output from the GP.

Overall, it seems the secondary sector in the panchayat is creeping up. However, it is difficult to write much on service sector of the gram panchayat although we have come across many small hotels, shop, public institutions like, educational institutions, Anganwadi, post office, banks, etc. Hence, our general impression is that the panchayat is supposed to be a good and better panchayat than many other Panchayats in Sambalpur district (at least better among the four GPs that we have visited earlier).

Given this background picture of the Sason gram panchayat, the objective here is to understand the functioning of different institutions that provide social security, first, we have tried to understand the functioning of the gram panchayat itself and followed by the functioning of different institutions in the gram panchayat.

Functioning of the Sason Gram Panchayat

The local governments have a three-tier structure – the district Panchayat (Zilla Parishad) at the district level, the block (Panchayat Samiti) at the block level and the Gram Panchayat at the Panchayat or at the grassroot level. Along with this, the urban areas of the states have separate local governments, namely corporations and municipalities.

Sason GP of Dhankauda block in Sambalpur district of Odisha also follows the same procedure as described above. The Panchayat is located in the Ajitpur Sason village of the Sason GP. The Panchayat office has three rooms. One is the office room (roughly 8*8 feet), one hall (14*22 feet), where all the ward members and Sarapanch sit together and discuss various issues, and a storehouse (20*30 feet), where all the assistance (mainly food assistance) received from the block office are kept and delivered to different institutions later on. Besides the ward members and Sarpanch, the only active and regular members in the gram panchayat are the panchayat secretary and the peon at the office. The panchayat secretary has hardly any power, but a lot of work to do. He is the only person to record all the official minutes of various meetings, maintain the accounts and the store and so on. With an unhappy mood, the panchayat secretary quoted an Oriya poem by late Gopabandhu Das from the poem 'Bandira Atma Katha' – 'Dekhaee Netra Jaha Chu Digara, Se Sabura Atai mu Adhiswara'. An understandable translation may be 'I am the all in all of all these things till the horizon that I can see'.

Apart from the executive member, there are other legislative members in the panchayat, they are 15 ward members and the Sarpanch. They get elected in the democratic process. Apart from the executive functions, the main function that a Panchayat needs to do is the financial management. In Sason panchayat, it received funds from the centre, state and collections of her own resources and it gets spent on various sectors with proper meetings and decisions at various levels. Different meetings at different levels have been discussed below.

First, there would be Palli sabha meetings, where the ward member will preside the meeting and the Panchayat secretary will take note of the meetings. In that meeting with the participation of people of that ward, there would be discussion on what are the things to be done in the ward with the allotted funds for the ward. The amounts of allotment of funds depend on the population, area, and SC/ST population and with few other criteria. There would be attempts to follow a complete democratic procedure but as learnt from the secretary, a few middle men including contractors interfere in the meeting indirectly and the outcome of the meeting is deviated from the right decision. For example, sometimes the local contractor influence the people at the Palli sabha in such a way that the youth of the ward will take the contract to do the work and later on the same would be handed over to the same contractor to get some thousand of money for the youth club. And even if the contractor completes the work with low quality, nobody can raise voice against him. In some other cases, these contractor would have a golden hand shake with the panchayat secretary, but as it was mentioned by the panchayat secretary in Sason gram panchayat, such matters have not been encouraged. However, it was learnt although powers have been given in the hands of the common people, it actually gets exercised by the middlemen and contractor.

Similar meeting would be there in all the wards and the Panchayat secretary would keep the record of the entire meeting. At the second level, there would be the gram sabha meeting where the sarpanch will preside and all the outcome of the meeting would be discussed with the presence of the ward members. After the meeting, the outcome would be sent to the third level, i.e. the Panchayat Samiti meeting. All the proposals that were submitted by different gram panchayats would be discussed here, and after the decision of the panchayat samiti, the plan gets finalised and fund would be released for the work from the block office.

On the whole, the procedure is fine. In the sense, power has been given to all the people equally, but unfortunately the power only gets exercised by a few influential people. Hence, the whole development plans gets slow. However, in this process the panchayat functions.

Notes

1. For a lucid understanding in this regards, please see Papola, 2005 and Hari, 2003.
2. For debates in this regards, please see Fishers, 1939, 1946; Kuznets, 1971, 1964; Kaldor, 1966, 1967; Bamoul, 1967, 2001 and Fuchs, 1968.
3. Fei and Ranis, 1964 also discussed similar kind of dual model and economic development.
4. The annual average growth rates of Net National Product at factor cost and per capita NNP at constant prices were 2.5 and .2 in the third plan, which is lower than the same in the first and second Plan period; there was huge inequality – the extent of landless households based on operational holding is as high as 27 per cent; the magnitude of agricultural labour increases to 26 per cent in 1971 from 16 per cent in 1961.
5. Trickle down hypothesis assumed that the benefit of growth will percolate down.
6. Landless agricultural labourers, marginal farmers, rural artisans/craftsmen, potters, toddy-tappers, weavers, blacksmith, carpenters etc. in the rural areas and slum dwellers and persons earning their livelihood on daily basis in the informal sector like potters, rickshaw-pullers, cart-pullers, fruit and flower sellers on the pavement etc. in urban areas.
7. The guidelines considered here were Landless agriculture labourers, marginal farmers, rural artisans/craftsmen, such as potters, tanners, weavers, blacksmiths, carpenters, slum dwellers, and persons earning their livelihood on daily basis in the informal sector like porters, coolies, rickshaw-pullers, hand cart pullers, fruit and flower sellers, snake charmers, rag pickers, cobblers, destitute and other similar categories irrespective of rural or urban areas. Households headed by widows or terminally ill persons or disabled persons or persons aged 60 years or more with no assured means of subsistence or societal support. Widows or terminally ill persons or disabled persons or persons aged 60 years or more or single women or single men with no family or societal support or assured means of subsistence. All primitive tribal households.
8. Use of mass media and other forms of publicity; special campaigns at suitable intervals aimed at saturating the project area; home visits by the Anganwadi workers; specially organized short courses in the village for about 30 women at a time; demonstration of cooking and feeding; and utilisation of the health and nutrition education Programmes of the Ministry of Health and Family Welfare, Ministry of Agriculture (especially Department of Food and Directorate of Agricultural Extension), etc. (at the Centre and in the States/Union Territories).
9. During 2004–05 the ECCE has made the following achievements. Arunima training has been imparted to all 34,401 Anganwadi workers of ICDS; Teaching Learning Materials (TLM) have been provided to all the 34,201 Anganwadi centres; 1,728 new ECCE centres have been opened under SSA and 1,432 centres under National Programme of Education for Girls at Elementary level (NPEGEL); 3,160 new ECCE instructors have been engaged; 2,026 newly engaged

instructors have been imparted Arunima Induction Training; 3,160 Mother-Teacher Associations have been engaged in different activities of ECCE.

10. The reasons of replacement as have been specified are the multiplicity of programmes of the IRDP, TRYSEM, DWCRA, SITRA, GKY, are 'being viewed as separate programmes in themselves, resulting in a lack of proper social intermediation, absence of desired linkages among these programmes *per se* and the implementations being more concerned with achieving individual programme target rather than focusing on the substantive issue of sustainable income generation' (GOI, 1999).
11. The former schemes will be no longer in operation (the unspent balances as on 1.4.1999, under these erstwhile programmes, will be pooled under the head of SGSY).

References

Abel-Smith, B. and P. Townsend (1965) *The Poor and the Poorest: A New Analysis of the Ministry of Labour's Family Expenditure Surveys of 1953–54 and 1960*, G. Bell and Sons Ltd., London.

Agarwal, B. (2001) 'Economics and Other Social Sciences: An Inevitable Divide?', *Contributions to Indian Sociology*, Vol. 35, No. 3, pp. 388–399.

Agarwal, B. and Pradeep Panda (2007) 'Toward Freedom from Domestic Violence: The Neglected Obvious', *Journal of Human Development*, Vol. 8, No. 3, November, pp. 359–388.

Ahluwalia, M.S. (1978) 'Rural Poverty and Agricultural Performance in India', *Journal of Development Studies*, Vol. 14, No. 2, April, pp. 298–323.

Appasamy, Paul, S. Guhan, R. Hema, Manabi Majumdar and A. Vaidyanathan (1995) 'Social Exclusion in Respect of Basic Needs in India', in Gerry Rodgers et al. (eds), *Social Exclusion: Rhetoric, Reality, Responses*, International Institute for Labour Studies, Geneva.

Babu, S.M. (2005) 'India's Recent Economic Growth: Some Limits and Limitations', *Economic and Political Weekly*, 23 July, 40(30), pp. 3249–3252.

Babu, A.K. (1984) Paddy marketing in Sambalpur district: Historical antecedents and contemporary structure (1949–81), Unpublished M.Phil. thesis, CDS, Thiruvananthapuram.

Bagchi, A.K. (2004) 'Agrarian Transformation and HD: Instrumental and Constitutive Links', in V.K. Ramachandran and Madura Swaminathan (eds), *Agrarian Studies: Essays on Agrarian Relations in Less Developed Countries*, Tulika Print Communication, Services, Kolkata.

Bailey, F.G. (1963) *Politics and Social Change: Orissa in 1959*, University of California Press, Berkeley.

Banerjee, K. (1984) *Regional Political Parties in India*, B.R. Publishing Corporation, Delhi.

Bardhan, P. (1984) *The Political Economy of Development in India*, Oxford University Press, New Delhi.

——. (1989) 'The Third Dominant Class', *Economic and Political Weekly*, Vol. 24, No. 3, January, pp. 155–156.

Bardhan, P.K. (1970) 'On the Minimum Level of Living and the Rural Poor', *Indian Economic Review*, Vol. 6, No. 1.

Bardhan, P.K. et al. (1974) *Poverty and Income Distribution in India*, Statistical Publishing Society, Calcutta.

Barik, R.K. (1987) 'Subaltern Politics and the Unrecognized Intelligentsia in Orissa', *Social Science Probings*, September.

Basu, K. (ed.) (2003) *Readings in Political Economy*, Blackwell Publishing, Malden, USA.

Baumol, William J. (1967) Macroeconomics of Unbalanced Growth: The Anatomy of Urban Crisis, *The American Economic Review*, Vol. 57, No. 3 (June 1967), pp. 415–426.

Behuria, N.C. (ed.) (1990) *Orissa State Gazetteers*, Vol. I, Cuttack, p. 387.

Behuria, N.C. (ed.) (1991) *Orissa State Gazetteer*, Vols. I, II and III, Gazetteers Unit, Department of Revenue, Government of Orissa, Cuttack.

Bergmann, T. (1984) *Agrarian Reform in India*, Agricole Publishing Academy, New Delhi.

Besley, T. and R. Burgess (2000) 'Land Reform, Poverty Reduction and Growth: Evidence from India', *The Quarterly Journal of Economics*, Vol. 117, No. 4, May, pp. 389–430.

Bhatia, B.M. (1967) *Famines in India, 1850–1945*, 2nd edition, Asia Publishing House, Bombay.

Bhatta, B.B. (1997) *The Natural Calamities in Orissa in the 19th Century*, Commonwealth Publishers, New Delhi.

Boserup, E. (1983) 'The Impact of Scarcity and Plenty on Development', in R.I. Rotberg and T.K. Rabb (eds), *Hunger and History: The Impact of Changing Food Production and Consumption on Society*, Cambridge University Press, Cambridge, pp. 185–210.

Booth, Charles (1889–1902) *Life and Labour of the People in London*, 17 Vols., Macmillan and Co. Ltd., London.

Byres, T.J. (1974) 'Land Reforms, Industrialisation and the Marketed Surplus in India: An Essay on the Power of Rural Bias', in D. Lehmann (ed.), *Agrarian Reform and Agrarian Reformism*, Faber and Faber, London.

Census of India, (1951, 1961, 1971, 1981, 1991, 2001 and 2011), The Registrar General and Census Commissioner, India, New Delhi, Ministry of Home Affarirs, Government of India.

Central Statistical Organisation (2006) *National Accounts Statistics*, Department of Statistics, Ministry of Planning, Government of India, New Delhi.

——. (various years) *Annual Survey of Industries: Summary Results*, Planning and Program Implementation, Department of Statistics, New Delhi.

——. (various years) *Statistical Abstract of India*, Department of Statistics, Ministry of Planning, Government of India, New Delhi.

Central Water Commission (1990) *National Register of Large Dams*, Government of India, New Delhi.

Cernea, M. (1995) 'Understanding and Preventing Impoverishment from Displacement: Reflections on the State of Knowledge', *Refugee Studies*, Vol. 8, No. 3, pp. 245–264.

Chadha, G.K. and A.N. Sharma (1997) *Growth, Employment and Poverty: Change and Continuity in Rural India*, Indian Society of Labour Economics, Vikas Publishing House Pvt. Ltd., New Delhi.

Chadha, G.K. and P.P. Sahu (2002) 'Post-Reform Setbacks in Rural Employment: Issues That Needs Further Scrutiny', *Economic and Political Weekly*, 37(21).

Chakravarty, S. (1993) *Development Planning: The Indian Experience*, OUP, New Delhi.

Chatterjee, P. (2004) *Politics of the Governed*, Columbia University Press, New York.

Chaudhury, P. (1981) 'Impact of Forced Commerce on the Pattern of Emigration from Orissa, 1901–1921', Mimeo, Paper presented in the Seminar on Commercialisation in Indian Agriculture held at the Centre for Development Studies, Trivandrum, 23–25 November.

Chenery, H. and Syrquin, M. (1989), 'Patterns of Development, 1950 to 1983', The World Bank Discussion Paper.

Clark, C. (1932) *The National Income, 1924–31*, Macmillan, London.

Clark, C. (1940) *The Condition of Economic Progress*, Macmillan, London, p. 504.

Currie, B. (2000) *The Politics of Hunger in India: A Study of Democracy, Governance and Kalahandi's Poverty*, Macmillan India Ltd., Chennai.

Dalua, A.K. (1991) 'Displacement and Rehabilitation of the Tribal People in the Event of Implementation of Irrigation Projects in Orissa', *Adivasi*, Vol. XXXI, Tribal and Harijan Research-cum-Training Institute, Bhubaneswar, June.

——. (1993) *Environmental Impact of Large Reservoir Projects on Human Settlement*, Ashish Publishing House, New Delhi.

Dandekar, V.M. and N. Rath (1971) *Poverty in India*, Indian School of Political Economy, Pune.

Dando, W.A. (1980) *The Geography of Famine*, V.H. Winston and Sons, New York.

Dantwala, M.L. (1973) *Poverty in India: Then and Now (1870–1970)*, MacMillan, Delhi.

Das, K. (2002) 'Social Mobilisations for Rehabilitations: Relief Work in Cyclone Affected Orissa', *Economic and Political Weekly*, Vol. 37, No. 48, pp. 4784–4788.

Dasgupta, P. (1993) *An Inquiry into Well-being and Destitution*, Clarendon, Oxford.

Dash, B. (2002) 'Lessons from Orissa Super Cyclone: Need for Integrated Warning System', *Economic and Political Weekly*, 19 October.

Datta, K.L. (2006) 'The Debate on the Poverty Estimates of 1999–2000', Working Paper No. 188, ICRIER, New Delhi.

De Haan, Arjan and Amresh Dubey (2005) 'Poverty, Disparities or the Development of Underdevelopment in Orissa', *Economic and Political Weekly*, Vol. 40, Nos. 22–23, 28 May–4 June.

De Haan, Arjan and Pulin Nayak (1995) 'Social Exclusion and South Asia', Discussion Paper No. 77, International Institute of Labour Studies, Geneva.

Deaton, A. and J. Dreze (2004) 'Poverty and Inequality in India: A Re-examination', *Economic and Political Weekly*, Vol. 37, No. 36, 7 September, pp. 3729–3748.

Deolalikar, A. (1988) 'Nutrition and Labour Productivity in Agriculture: Estimates for Rural South India', *Review of Economics and Statistics*, 70, pp. 406–413.

Deshpande, L.K. (1983) *Segmentation of Labour Market: A Case Study of Bombay*, Gokhale Institute of Economics and Politics, Pune.

Dev, S.M., P. Antony, V. Gayathri and R.P. Mamgain (2001) *Social and Economic Security in India*, Institute for Human Development, New Delhi.

Devereux, S. (1993) *Theories of Famine*, Harvester Wheatsheaf, London.

Downing, T.E. (2002) *Avoiding New Poverty: Mining-induced Displacement and Resettlement*, Vol. 52, International Institute for Environment and Development, London.

Dreze, J. and A. Sen (1989) *Hunger and Public Action*, Clarendon Press, Oxford.

——. (1990) Poverty in India and the IRDP Delusion, *Economic and Political Weekly*, Vol. 25, No. 39, (29 September 1990), pp. A95–A104.

——. (1991) *The Political Economy of Hunger*, Clarendon Press, Oxford.

——. (1999) *India: Economic Development and Social Opportunity*, Oxford University Press, Delhi.

El-Ghonemy, M.R. (1990) *The Political Economy of Rural Poverty*, Routledge, New York.

Engels, F. (ed.) (1886) *Karl Marx: Capital: A Critical Analysis of Capitalist Production*, Vol. I, Foreign Languages Publishing House, Moscow.

Fei, John C.H. and Gustav Rains (1964) *Development of the Labour Surplus Economy: Theory and Policy*, Richard A. Irwin, Inc, Homewood, IL.

Fernandes, W. (1994) 'An Activist Process Around the Draft National Rehabilitation Policy', *Social Action*, Vol. 45, July–September, pp. 277–298.

——. (1994a) *Development-Induced Displacement in the Tribal Areas of Eastern India*, Indian Social Institute, New Delhi.

Fernandes, W. and Mohammed Asif (1997) *Development-induced Displacement and Rehabilitation in Orissa 1951–995: A Database on its Extent and Nature*, Indian Social Institute, New Delhi, mimeo.

Fisher, A.G.B. (1946) Tertiary Production and Post-war International Economic Problem, *Review of Economics and Statistics*, Vol. 28, pp. 146–151.

——. (1935) *The Clash of Progress and Security*, Macmillan, London, p. 234.

——. (1939) 'Production: Primary, Secondary and Tertiary', *The Economic Journal*, Vol. XV.

Fogel, Robert W. (1994) 'Economic Growth, Population Theory and Physiology: The Bearing of Long-term Processes on the Making of Economic Policy', *American Economic Review*, Vol. 84, No. 3, June, pp. 369–395.

Foster, P. and H.D. Leathers (1999) *The World Food Problem: Tackling the Causes of Undernutrition in the Third World*, 2nd edition, Lynne Rienner Publishers, Boulder, CO.

Franke, R.W. (1996) *Life is a Little Better: Redistribution as a Development Strategy in Nadur Village, Kerala*, Promila and Co. Publishers, New Delhi.

Frankel, F.R. (2008) *India's Political Economy, 1947–2004: The Gradual Evolution*, Oxford University Press, New Delhi.

Fuchs, V.R. (1968) *The Service Economy*, Columbia University Press, New York.

Ghurye, G.S. (1969) *Caste and Race in India*, Popular Prakashan, Bombay.

Government of India (1993) Report of the Expert Group on Estimation of Proportion and Number of Poor, Perspective Planning Division, Planning Commission, New Delhi.

——. (1951, 1961, 1971, 1981, 1991, 2001) Census of India, various reports.

——. (1999), Economic Survey 1998–99, Ministry of Finance, New Delhi, GOI.

Government of Orissa (2006) Orissa Resettlement and Rehabilitation Policy, Revenue Department, No. 18040/R and REH.-1/2006R, Bhubaneswar.

——. (2007) Economic Survey of Orissa, Department of Economic and Statistics, Bhubaneswar.

——. (various years) Annual Report on Food Supplies and Consumer Welfare Department, Bhubaneswar.

Government of West Bengal (2004) West Bengal Human Development Report, Development and Planning Department, Kolkata.

Griffin, K. (1976) *Land Concentration and Rural Poverty*, The Macmillan Press Limited, London.

Gunewardena, D. (2004) *Poverty Measurement: Meaning, Methods and Requirements*, National Library of Sri Lanka-Cataloguing-in-Publication Data, Centre for Poverty Analysis, Colombo.

Hanstad, T. and S.B. Lokesh (2002) 'Allocating Homestead Plots as Land Reform: Analysis from West Bengal', RDI Reports on Aid and Development No. 115.

Hari, K.S. (2003) 'Structural Transformation in an Agrarian Economy: A Study of India 1970–2000', *Indian Journal of Labour Economics*, 46, pp. 945–59.

Harriss, J., K.P. Kannan and Gerry Rodgers (1990) *Urban Labour Market Structure and Job Access in India: A Study of Coimbatore*, International Institute for Labour Studies, Geneva.

http://fcamin.nic.in

http://www.ohpcltd.com

http://www.orissa.gov.in/p&c/eco2006/annexture.htm

http://www.orissa.gov.in/wcd/index.htm

Hunter, W.W. (1877) *History of Orissa*, Vol. 2, Trubner and Company, London.

Jena, K.C. (1968) *Land Revenue Administration in Orissa During the Nineteenth Century*, S. and Chand Co., New Delhi.

Jena, M. (2006) 'Orissa: Draft Resettlement and Rehabilitation Policy, 2006', *Economic and Political Weekly*, Vol. XLI, No. 5, 4 February.

Jit, N.K. (1984) *The Agrarian Life and Economy of Orissa: A Survey 1833–1897*, Punthi Pustak, Calcutta.

Kaldor, N. (1966) *Causes of the Slow Growth in the United Kingdom*, Cambridge University Press, Cambridge.

——. (1967) *Strategic Factors in Economic Development*, Cornell University Press, New York.

Kannan, K.P. (1988) *Of Rural Proletarian Struggles: Mobilisation and Organisation of Rural Workers in South-West India*, Oxford University Press, Delhi.

——. (1992) 'Labour Institutions and the Development Process in Kerala', in T.S. Papola and Gerry Rodgers (eds), *Labour Institutions and Economic Development in India*, International Institute for Labour Studies, Geneva.

——. (1994) 'Levelling Up or Levelling Down?', *Economic and Political Weekly*, Vol. XXIX, No. 30, 23 July.

——. (2002) 'The Welfare Fund Model of Social Security for Informal Sector Workers: The Kerala Experience', *The Indian Journal of Labour Economics*, Vol. 45, No. 2, pp. 243–272.

——. (2004) 'Shining Socio-spatial Disparities', *Seminar*, No. 537, May.

Kannan, K.P. and K. Pushpangadan (1988) 'Agricultural Stagnation in Kerala: An Exploratory Analysis', *Economic and Political Weekly*, Vol. 23, No. 39, 24 September.

——. (1990) 'Towards Understanding the Dynamics of Rural Labour Market: An Approach Based on Indian Evidence', *The Indian Journal of Labour Economics*, Vol. 33, No. 3.

——. (1995) 'Public Intervention and Poverty Alleviation: A Study of the Declining Incidence of Rural Poverty in Kerala, India', *Development and Change*, Vol. 26, No. 4, pp. 701–727.

Kannan, K.P. et al. (1991) Health Status in Rural Kerala: A Study of the Linkages between Socio-economic Status and Health Status, Integrated Rural Training Centre of Kerala Sastra Sahitya Parishad.

——. (2003) *Towards a Food Secure India: Issues and Policies*, Institute of Human Development, New Delhi.

Kedir, A.M. (1997) 'Modelling Poverty and its Determinants in Addis Ababa: A Focus on Multinomial Logit Selection Model', *Ethiopian Journal of Economics*, Vol. VI, No. 2, October.

Kishore, A. (1995) *Land, State and Poverty*, Aalekh Publishers, Jaipur.

Krishna, A., M. Kapita, M. Porwal and V. Singh (2005) 'Why Growth is Not Enough: Household Poverty Dynamics in Northeast Gujarat, India', *Journal of Development Studies*, 41 (7).

Kumbhar, R.K. (2001) 'Food Security in a Regional Perspectives: A Case of PDS in Orissa, India', Unpublished M.Phil thesis, Thiruvananthapuram.

——. (2006) 'The Relationship between Land Possession/Landlessness and Poverty: Experiences from Orissa, Kerala and West Bengal', in D.M. Melis et al. (eds), *Putting Land First: Exploring the Links between Land and Poverty*, Centre for Poverty Analysis, Sri Lanka.

Kuznets, S. (1964) 'Quantitative Aspects of the Economic Growth of Nations: 9, Level and Structure of Foreign Trade: Comparisons for Recent Years.' *Economic Development and Cultural Change*, Vol. 13, No. 1, pt. 2 (October).

——. (1971) *Economic Growth of Nations: Total Output and Production Structure*, Harvard University Press, Cambridge, MA.

——. (1973) 'Modern Economic Growth: Findings and Reflections', *The American Economic Review*, Vol. 63, No. 3, pp. 247–258.

Lal, Deepak and H. Myint (1996) *The Political Economy of Poverty, Equity, and Growth: A Comparative Study*, Clarendon Press, Oxford.

Leibenstein, H. (1957) *Economic Backwardness and Economic Growth*, Wiley, New York.

Lewis, W.A. (1954) 'Economic Development with Unlimited Supplies of Labour', *The Manchester School of Economic and Social Studies*, Vol. 22, No. 2, pp. 139–191.

Lieten, G.K. (1996) 'Land Reforms at Centre Stage: The Evidence on West Bengal', *Development and Change*, Vol. 27, No. 1, pp. 111–130.

Lipton M. and M. Ravallion. (1995) 'Poverty and Policy' in *Handbook of Development Economics*. Vol. IIIB. Elsevier, Amsterdam.

Lipton, M. (1985) 'Land Assets and Rural Poverty', World Bank Staff Working Paper No. 744, The World Bank, Washington, D.C.

Mahajan, V. (2005) From Microcredit to Livelihood Finance, *Economic and Political Weekly*, Vol. 40, No. 41 (8–14 October), pp. 4416–4419.

Mahapatra, L.K. (1989) 'Rehabilitation of Tribals Affected by Major Dams and Other Projects in Orissa', in A.P. Fernandez (ed.), *A Report on the Workshop on Rehabilitation of Persons Displaced by Development Projects*, Institute for Social and Economic Change, Bangalore.

——. (1995) 'The Orissa Resettlement and Rehabilitation of Projects Affected Persons Policy, 1994: A Critique', *Social Action*, Vol. 45, No. 3, July–September, pp. 318–326.

Mahendra Dev, S. (1988) 'Poverty of Agricultural Labour Households in India: A State Level Analysis', *Indian Journal of Agricultural Economics*, Vol. 43, No. 1, January–March.

Mahtab, H.K. (1960) *History of Orissa*, 2 Vols., Cuttack.

Malthus, T.R. (1798) *An Essay on the Principle of Population, as it Affects the Future Improvement of Society*, J. Johnson, London.

Marx, K. (1956), *The Poverty of Philosophy*. Lawrence & Wishart, London.

Meher, S. (2003) 'Dynamics of Economic Development in Orissa', *Review of Development and Change*, Vol. 13, No. 2, July–December, pp. 221–248.

Mellor, John W. and G.M. Desai (1985) *Agricultural Change and Rural Poverty: Variations on a Theme by Dharm Narain*, International Food Policy Research Institute, The Johns Hopkins University Press, Baltimore.

Minhas, B.S. (1970) 'Rural Poverty, Land Distribution and Development Strategy: Facts and Policy', *Indian Economic Review*, Vol. V, No. 16, pp. 97–128.

Mining and Industrialisation Update, Orissa (2006) *Environmental Information on Industries and Mining in Orissa*, Vol. 2, No. 3, Bhubaneswar, November.

Mishra, A. (2005) 'Local Perceptions of Famine: Study of a Village in Orissa', *Economic and Political Weekly*, Vol. 40, No. 6, 5 February, pp. 572–578.

Mitra, A. (1990) 'Structural Aspects of Growth: A State-Level Analysis', *The Indian Journal of Social Science*, Vol. 2, No. 3.

Moene, K.O. (1992) 'Poverty and Land Ownership', *American Economic Review*, Vol. 82, No. 1.

Mohanty, B. (1993) 'Orissa Famine of 1866: Demographic and Economic Consequences', *Economic and Political Weekly*, Vol. 28, Nos. 1–2, January, pp. 55–66.

——. (2005) 'Displacement and Rehabilitations of Tribes', *Economic and Political Weekly*, Vol. 40, No. 13, 26 March.

Mohanty, M. (1984) 'Social Roots of Backwardness in Orissa: A Study of Class, Caste and Power', *Social Science Probing*, Vol. 1, No. 2, June.

——. (2002) *Trade Union Movement in Orissa*, Antiquarian Booksellers and Publishers, Kolkata.

Myrdal, G. (1968) *Asian Drama: An Inquiry into the Poverty of Nations*, Penguin Books, Massachusetts.

——. (1970) *The Challenge of World Poverty: A World Anti-Poverty Program in Outline*, The Johns Hopkins University School of Advanced International Studies, Washington, D.C.

Nagraj, R. (2006) 'Public Sector Performance since 1950: a Fresh Look', *Economic and Political Weekly*, 24 June: 2551–7.

National Commission for Scheduled Castes and Scheduled Tribes (1998) Government of India.

National Family Health Survey of India (1992–93, 1998–99, 2005–06) International Institute of Population Sciences (IIPS), Bombay.

National Sample Survey Organisation (NSSO) Various Rounds on Employment and Unemployment, Consumption Expenditure Survey, and Health Care and Morbidity, Department of Statistics, Government of India, New Delhi.

Nayak, P. and Sarangi (1993) *Net State Domestic Product (NSDP) of Orissa 1950–51 to 1988–89: Trends and Patterns*, Nabakrushna Choudhury Centre for Development Studies, Odisha.

Nayak, P. (1995) Social Exclusion in India, International Institute of Labour Studies, Geneva.

Newman, L.F. (1990) *Hunger in History*, Basil Blackwell, Oxford, UK.

Nolan, B. and C.T. Whelan (1996) *Resources, Deprivation and Poverty*, Clarendon Press, Oxford.

Nozick, R. (1973) 'Distributive Justice', *Philosophy and Public Affairs*, Vol. 3, No. 1, pp. 45–126.

Oommen, M.A. (1975) *A Study on Land Reforms in Kerala*, Oxford and IBH Publishing Company, New Delhi.

Ota, A.B. (1996) 'Countering the Impoverishment Risk: The Case of Rengali Dam Project', in A.B. Ota and A. Agnihotri (eds), *Involuntary Displacement in Dam Projects*, Prachi Prakashan, New Delhi.

——. (2001) Orissa Drought 2001: Its Magnitude and Consequences, http://www.anthrobase.com/txt/o/ota_A_01.htm

Padhi, S. (1999) *Land Relations and Agrarian Development in India: A Comparative Historical Study of Regional Variations*, Centre for Development Studies, Thiruvananthapuram, Kerala.

——. (2000) Orissa Economy: A Database, Mimeo, Nabakrushna Choudhury Centre for Development Studies, Bhubaneswar, December.

Panda, P.K. (2003) 'Rights-Based Strategies in the Prevention of Domestic Violence', CDS Working Paper No. 344, March.

Panda, P.K. and Bina Agarwal (2005) 'Marital Violence, Human Development and Women's Property Status in India', *World Development*, Vol. 33, No. 5, pp. 823–850.

Panda, P.K. (1997) 'Female Headship, Poverty and Child Welfare: A Study of Rural Orissa, India', CDS Working Paper No. 280.

Pandey, B. (1996) 'Impoverishment Risk: A Case Study of Five Villages in Coal Mining Areas of Talcher, Orissa', Paper presented at the Workshop on Involuntary Resettlement and Impoverishment Risks, New Delhi, March.

——. (1998) *Depriving the Underprivileged for Development*, Institute for Socio-Economic Development, Bhubaneswar.

Papola, T.S. (2005) 'Emerging Structure of Indian Economy: Implications of Growing Inter-Sectoral Imbalances', Presidential Address of 88th Annual Conference of Indian Economic Association, Vishakhapatanam, 27–29 December.

——. (2005) 'Structural Change in the Indian Economy: Some Implications of the Emerging Patterns', *Artha Beekshan*, Vol. 13, 4.

Papola, T.S. and Gerry Rodgers (eds) (1992) *Labour Institutions and Economic Development in India*, International Institute for Labour Studies, Geneva.

Parasuraman, S. (1999) *The Development Dilemma: Displacement in India*, St. Martin's Press, New York.

Parida, J. (2006) 'Politics of Development, Problems of Displacement of Tribals: A Case Study of a Project in Orissa', JNU (thesis), New Delhi.

Parthasarthy, G. and C.S. Murthy (1997) 'Land Reform, New Economic Policy and Poverty' in G.K. Chadha and Alok Sharma (eds), *Growth, Employment and Poverty: Change and Continuity in Rural India*, Vikas Publishing House, New Delhi.

Pati, B. (2006) 'Tatas and the Orissa Model of Capitalist Development', *Social Scientist*, Vol. 34, Nos. 3–4, March–April, pp. 37–42.

Patnaik, J.K. (2004) 'Human Rights: The Concept and Perspectives: A Third World View', *The Indian Journal of Political Science*, Vol. LXV, No. 4, October–December.

Patnaik, N.R. (ed.) (1997) *Economic History of Orissa*, Indus Publishing Company, New Delhi.

Planning Commission (1951) First Five Year Plan, Government of India.

Planning Commission. (1979) Report of the Task Force on Projections of Minimum Needs and Effective Consumption Demand, Perspective Planning Division, Government of India, New Delhi.

——. (1993) Report of the Expert Group on Estimation of Proportion and Number of Poor, Perspective Planning Division, New Delhi.

——. (2004) 'Poverty and Inequality in India', *Economic and Political Weekly*, Vol. 37, No. 36.

——. (2007) 'Poverty Estimates for 2004–05', Press Information Bureau, Press Release, Government of India, New Delhi, 21 March.

Praxis (2001) *The Politics of Poverty: A Tale of the Living Dead in Bolangir*, Books for Change, Bangalore.

Press Information Bureau (1997) 'Estimates of Poverty', Government of India, 11 March.

——. (2001) 'Poverty Estimates for 1999–2000', Government of India, 22 February.

Raj, K.N. (1974) 'Agricultural Development and Distribution of Land Holdings', *Indian Journal of Agricultural Economics*, Vol. 30, No. 1, pp. 1–13.

——. (1990) *Organisational Issues in Indian Agriculture*, Oxford University Press, Delhi.

——. (1992) 'Land Reform and Contemporary Measures in Kerala', *IASSI Quarterly*, Vol. 10, No. 3, January–June, pp. 20–24.

Rammohan, K.T. (2005) 'Economic History as Human Science', *Economic and Political Weekly*, Vol. 40, No. 26, 25 June.

Rangarajan, L.N. (1987) *Kautilya: The Arthashastra*, Penguin Books, New Delhi.

Rao, R.S. (1995) *Towards Understanding Semi-Feudal Semi-Colonial Society (Studies in Political Economy)*, Perspective, Hyderabad.

Rao, V.K.R.V. (1987) 'Growth and Structural Changes in India Economy', in Brahmananda, P.R. and Panchamukhi, V.R. (eds) *The Development Process of the Indian Economy*. Himalaya Pub. House, Bombay.

Rath, S.N. (1977) The Development of the Welfare State in Orissa, 1950–60, S. Chand and Company Ltd., Ram Nagar, New Delhi.

Rath, N. (1985) Garibi Hatao: Can IRDP Do It? *Economic and Political Weekly*, Vol. 20, No. 6, (9 February 1985), pp. 238–246.

Ravallion, M. and G. Datt (1996) 'How Important to India's Poor is the Sectoral Composition of Economic Growth?', *World Bank Economic Review*, Vol. 10, No. 1, pp. 1–25.

Rawls, J. (1971) *A Theory of Justice*, Harvard University Press, Cambridge, Massachusetts.

Raychoudhuri, A. (2004) *Lessons from the Land Reform Movement in West Bengal, India*, Department of Economics, Jadavpur University, Kolkata.

Reddy, I.U.B. (1993) *Displacement and Rehabilitation*, Mittal Publications, New Delhi.

Reddy, U.K.P. (1997) 'Involuntary Resettlement and Marginalisation of Project-Affected Persons: A Comparative Analysis of Singrauli and Rihand Power Project', in H.M. Mathur (ed.), *Impoverishment Risks in Resettlement*, Sage Publications, New Delhi.

Registrar General of India (1999) Compendium of India's Fertility and Mortality Indicators, 1971–1997, New Delhi.

——. (2006) SRS-Based Abridged Life Tables, 1999–2003, New Delhi.

Rima, I.H. (2001) *Development of Economic Analysis*, Routledge, Taylor and Francis Group, London.

Rodgers, G.C. and J.B. Figueiredo (eds) (1995) *Social Exclusion: Rhetoric, Reality, Responses*, International Institute for Labour Studies, ILO, Geneva.

Rowntree, B. Seebohm (1902) *Poverty: A Study of Town Life*, Macmillan and Co. Ltd., London.

Roy, B.C. and S. Pal (2002) 'Investment, Agricultural Productivity and Rural Poverty in India: A State-Level Analysis', *Indian Journal of Agricultural Economics*, Vol. 57, No. 4, October–December.

Runciman, W.G. (1966) *Relative Deprivation and Social Justice*, Routledge and Kegan Paul, London.

Saha, G. (1991) 'Social Backwardness and Politics of Reservation', *Economic and Political Weekly*, Annual Number.

Sahoo, A. and B. Sahoo (1997) 'Dimension of Development: A Study on Inter District Disparity in Orissa', *Vision*, Vol. XVIII, Nos. 3–4.

Sahu, J.K. (1971) 'Chauhan Rule in Western Orissa', in N.K. Sahu (ed.), *New Aspects of History of Orissa*, Sambalpur University, Sambalpur.

Sahu, N.R. (ed.) (1997) *Economic History of Orissa*, Indus Publishing Company, New Delhi.

Sainath, P. (1996) *Everybody Loves a Good Drought: Story from India's Poorest Districts,* Penguin Books, New Delhi.

Sainath, P. (1996) *Everybody Loves a Good Drought: Story from India's Poorest Districts,* Penguin Books India (P) Ltd., New Delhi.

Samal, J.K. (2000) *Economy of Colonial Orissa, 1866–1947*, Munshiram Manoharlal Publishers Pvt. Ltd., New Delhi.

Samal, K.C. (1998) 'Alternative to World Bank Prescription for Poverty Alleviation in Post-Liberalisation Era: Study of a Tribal Block in Orissa', NCDS, Working Paper.

——. (1998a) 'Ensuring Food Security to Alleviate Poverty and Hunger: A Case Study', *Man and Development*, September, pp. 62–70.

Sarangi, P. and Govinda Chandra Penthoi (2005) 'Economic Implications of Natural Disaster in Orissa: A Retrospective View', *Orissa Review*, Bhubaneswar, June.

Sarap, K. (1990) 'Interest Rates in Backward Agriculture: The Role of Economic and Extra Economic Control', *Cambridge Journal of Economics*, Vol. 14, No. 1, March.

——. (1991) *Interlinked Agrarian Markets in Rural India*, Sage Publications, New Delhi.

——. (2000) 'Food Security System and its Operation in Orissa: Evidence from Some Villages in Semiarid Area', Unpublished.

Satapathy, B. (1977) *Land Reforms Administration in Orissa*, Institute of Behavioural Sciences, Bodhgaya.

Satapathy, S. (1993) 'Floods and Flood Control Policies: An Analysis with Reference to the Mahanadi Delta in Orissa', Centre for Development Studies, Occasional Paper Series, Trivandrum.

Saxena, N.C. (2007) Rural Poverty Reduction through Centrally Sponsored Schemes, *Indian Journal of Medical Research*, October 2007, pp. 318–389.

Seabright, P. (1989) Failure of Livestock Investments under IRDP: Evidence from Two Villages in Tamil Nadu, *Economic and Political Weekly*, Vol. 24, No. 39 (30 September 1989), pp. 2203–2208.

Sebastian, T. (2002) 'Factors Contributing to Social Sector Expenditure: A Case Study of Indian States', Unpublished M.Phil thesis, CDS.

Seeta Prabhu, K. and R. Sundarshan (2002) *Reforming India's Social Sector: Poverty, Nutrition, Health and Gender*, Social Science Press, New Delhi.

Sen, A. (1983) 'Poor, Relatively Speaking', *Oxford Economic Papers* 35.

——. (1987) *On Ethics and Economics*, Basil Blackwell, Oxford and New York. Italian translation: Editori Laterza, 1988; Spanish translation: Alianza Editorial, 1987; Indian edition: Oxford University Press, New Delhi, 1990; French translation (with other selected essays), Presses Universitaires de France, 1993.

——. (2000) 'Social Exclusion: Concept, Application, Scrutiny', *Critical Quest*, New Delhi.

——. (1981) 'Public Action and the Quality of Life in Developing Countries', *Oxford Bulletin of Economics and Statistics*, Vol. 43, (4): 287–319.

——. (1982) *Choice, Welfare and Measurement*, Blackwell, Oxford and MIT Press, Cambridge, MA.

——. (1983) 'A Sociological Approach to the Measurement of Poverty', *Oxford Economic Papers* 37, pp. 669–676.

——. (1984) *Resources, Values and Development*, Blackwell, Oxford and Harvard University Press, Cambridge, MA.

——. (1985) 'A Sociological Approach to the Measurement of Poverty: A Reply to Professor Peter Townsend', *Oxford Economic Papers* 37, pp. 669–676.

——. (1987) *Commodities and Capabilities*, Oxford University Press, New Delhi.

——. (1999) *Development as Freedom*, Oxford University Press, New Delhi.

——. (1981) *Poverty and Famines: An Essay on Entitlement and Deprivation*, Clarendon Press, Oxford.

Sen, A. and Himanshu (2004) 'Poverty and Inequality in India', *Economic and Political Weekly*, Vol. 39, No. 38, 18 September, pp. 4247–4263.

Sen, S. (2001) 'Economic Growth and Social Opportunities Under Radical Political Regimes: A Comparative Study of Kerala and West Bengal', Unpublished M.Phil thesis, CDS.

Shariff, A. (2001) *India: Human Development Report*, Oxford University Press, New Delhi.

Sharma, H.R. (1994) 'Distribution of Land Holdings in Rural India, 1953–54 to 1981–82: Implications for Land Reforms', *Economic and Political Weekly*, Vol. 29, No. 13, 26 March.

Silver, Hilary (1995) 'Reconceptualising Social Disadvantage: Three Paradigms of Social Exclusion', in Gerry Rodgers et al. (eds), *Social Exclusion: Rhetoric, Reality, Responses*, International Institute for Labour Studies, Geneva.

Smith, Adam (1976) *An Inquiry into the Nature and Causes of the Wealth of Nations*, Book IV, Chapter One, republished by Home University Library, London.

Srinivas, M.N. (ed.) (1996) *Caste: Its Twentieth Century Avatar*, Viking, Delhi.

Stanley, W. (1996) 'Machkund, Upper Kolab and NALCO Projects in Koraput District, Orissa', *Economic and Political Weekly*, Vol. 31, No. 24, June, pp. 1533–1538.

Stiglitz, J.E. (2002) 'Employment, Social Justice and Societal Well-Being', *International Labour Review*, Vol. 141, No. 1–2, pp. 9–29.

Sundaram, K. and Suresh D. Tendulkar (2003) 'Poverty in India in the 1900s: Revised Results for All-India and 15 Major States for 1993–94', *Economic and Political Weekly*, 15–22 November, pp. 4865–4872.

Suri, S. (2000) *Odisha Disaster: Agony of the Living*, New Delhi: Authors Press.

Swaminathan, M. (2000) *Weakening Welfare: The Public Distribution of Food in India*, Leftword Books, New Delhi.

The United Nations (1948) Universal Declaration of Human Rights, Chicago.

The United Nations and Human Rights (1948) Universal Declaration of Human Rights, G.A. res. 217A (III), U.N. Doc A/810 at 71, Department of Public Information, United Nations, New York.

Thiesenhusen, W.C. and J. Melmed-Sanjak (1990) 'Brazil's Agrarian Structure: Changes from 1970 Through 1980', *World Development*, Vol. 18, No. 3.

Thorat, S. (2004) 'The Hindu Social System and Human Rights of Dalits', *Critical Quest*, New Delhi.

Thorner, D. (1980) *The Shaping of Modern India*, Allied Publishers Pvt. Ltd., New Delhi.

Timberlake, Lloyd (1985) *Africa in Crisis: The Causes, the Cures of Environmental Bankruptcy*. Earthscan, London.

Tiwari, A.K. (1992) *State Politics in India: A Case Study of Orissa, 1961–1971*, National Book Organisation, New Delhi.

Townsend, P. (1970) *The Concept of Poverty*, Heinemann, London.

——. (1979) *Poverty in the United Kingdom*, Penguin Books, Harmondsworth, Middlesex.

——. (1984) 'A Sociological Approach to the Measurement of Poverty', *Oxford Economic Papers* 37, pp. 669–676.

——. (1985) 'A Sociological Approach to the Measurement of Poverty: A Rejoinder to Professor Amartya Sen', *Oxford Economic Papers* 37, pp. 659–668.

——. (1987) 'Deprivation', *Journal of Social Policy*, Vol. 16, No. 2, pp. 125–146.

Toynbee, G.A. (1960–61) 'A Sketch of the History of Orissa, 1803–1828', *Orissa Historical Research Journal*, Vol. IX, Nos. 1 and 2 (April and June 1960) and Nos. 3 and 4 (October and January 1961).

Tripathy, S.N. (2003) *Economic Development and Problem of Displacement*, Anmol Publications Pvt. Ltd., New Delhi.

UNDP (1990 and 2000), *Human Development Report*, Oxford University Press, New York.

United Nations (UN/CDS) (1975) Poverty, Unemployment and Development Policy: A Case of Selected Issues with Reference to Kerala, Department of Economic and Social Affairs, New York.

Vaidyanathan, A. (2001) 'Poverty and Development Policy', *Economic and Political Weekly*, 26 May.

Vasundhara (2005) *Development Policies and Rural Poverty in Orissa: Macro Analysis and Case Studies*, Supported by Planning Commission, Government of India, Bhubaneswar, Orissa.

Visaria, P. (1978) 'Size of Land Holding, Living Standards and Employment in Rural Western India, 1972–73', Working Paper No. 3, World Bank.

World Development Report (2001) 'Attacking Poverty', World Bank.

www.iipsindia.org

www.indiastat.com

www.orissagov.nic.in

Index